不只做一个技术者，
更要做一个思考者！

谭勇德/Tom

咕泡学院 gupaoedu.com Java架构师成长丛书

Netty 4核心原理

与手写RPC框架实战

谭勇德（Tom）◎著

電子工業出版社
Publishing House of Electronics Industry
北京•BEIJING

内容简介

本书首先从硬件层面深入分析网络通信原理，结合Java对网络I/O的API实现，将理论与实践串联起来，帮助大家透彻理解网络通信的起源，然后介绍了Netty产生的背景并基于Netty手写Tomcat和RPC框架，帮助大家初步了解Netty的作用，接着分析Netty的核心原理和核心组件，基于Netty手写一个消息推送系统并进行性能调优，最后介绍了设计模式在Netty中的应用和经典的面试题分析。

如果你想深入了解网络通信原理，如果你还不知道Netty能做什么，如果你想深入了解Netty的实现原理，如果你看源码找不到入口，无从下手，如果你想了解设计模式在Netty中的应用，本书都能帮到你。

图书在版编目（CIP）数据

Netty 4 核心原理与手写 RPC 框架实战 / 谭勇德著. —北京：电子工业出版社，2020.4
（咕泡学院 Java 架构师成长丛书）
ISBN 978-7-121-38506-3

Ⅰ. ①N… Ⅱ. ①谭… Ⅲ. ①JAVA 语言—程序设计Ⅳ. ①TP312.8

中国版本图书馆 CIP 数据核字(2020)第 029046 号

责任编辑：董　英
印　　刷：三河市良远印务有限公司
装　　订：三河市良远印务有限公司
出版发行：电子工业出版社
　　　　　北京市海淀区万寿路 173 信箱　　邮编：100036
开　　本：787×980　1/16　印张：28.25　字数：628 千字
版　　次：2020 年 4 月第 1 版
印　　次：2020 年 4 月第 1 次印刷
印　　数：6100 册　　定价：108.00 元

凡所购买电子工业出版社图书有缺损问题，请向购买书店调换。若书店售缺，请与本社发行部联系，联系及邮购电话：（010）88254888，88258888。

质量投诉请发邮件至 zlts@phei.com.cn，盗版侵权举报请发邮件至 dbqq@phei.com.cn。

本书咨询联系方式：010-51260888-819，faq@phei.com.cn。

序　　言

在互联网分布式系统的推动下，Netty 作为一个能够支撑高性能、高并发的底层网络通信框架而存在。Netty 底层是基于 Java NIO 实现的，对 NIO 进行了非常多的优化，因此深受广大开发者尤其是一线大厂开发者的青睐。

作为一个 Java 开发者，如果没有研究过 Netty，那么你对 Java 语言的使用和理解可能仅仅停留在表面，会点 SSH，写几个 MVC，访问数据库和缓存，这些只是初级 Java 开发者做的事。如果你要进阶，想了解 Java 服务器的高阶知识，Netty 是一个必须要跨越的门槛。学会了 Netty，你可以实现自己的 HTTP 服务器、FTP 服务器、UDP 服务器、RPC 服务器、WebSocket 服务器、Redis 的 Proxy 服务器、MySQL 的 Proxy 服务器等。

如果你想知道 Nginx 是怎么写出来的，

如果你想知道 Tomcat 和 Jetty 是如何实现的，

如果你也想实现一个简单的 Redis 服务器，

……

那么你应该好好研究一下 Netty，它们高性能的原理都是类似的。

因为 Netty 5.x 已被官方弃用，本书内容基于 Netty 4 分析其核心原理，培养高级开发者自己“造轮子”的能力。本书不仅讲述理论知识，还围绕能够落地的实战场景，开创手写源码的学习方式，使读者学习源码更加高效。本书的主要特色是首次提供了基于 Netty 手写 RPC 框架、基于 Netty 手写消息推送系统等实战案例。

关于本书

适用对象	• 不知道Netty能做什么的人群 • 想深入了解Netty源码实现原理的人群 • 看源码找不到入口、无从下手的人群 • 想了解设计模式在Netty源码中如何应用的人群
CentOS版本	7.0
源码版本	Netty 4.1.6
IDE版本	IntelliJ IDEA 2017.1.4
JDK版本	JDK 1.8及以上
Gradle版本	Gradle 4.0及以上
Maven版本	3.5.0及以上

随书源码会在 https://github.com/gupaoedu-tom/netty4-samples 中持续更新。

读者服务

微信扫码回复：38506

- 获取博文视点学院 20 元优惠券。
- 获取免费增值资源。
- 加入读者交流群，与更多读者互动。
- 获取精选书单推荐。

关于我

为什么都叫我“文艺汤”？

我自幼爱好书法和美术，长了一双能书会画的手，而且手指又长又白，因此以前的艺名叫“玉手藝人”。中学期间，曾获市级书法竞赛一等奖，校园美术竞赛一等奖，校园征文比赛二等奖。担任过学生会宣传部长，负责校园黑板报、校园刊物的编辑、排版、设计。

2008 年参加工作后，我做过家具建模、平面设计等工作，亲自设计了咕泡学院的 Logo。做讲师之后，我给自己起了一个跟姓氏谐音的英文名字“Tom”，江湖人称“编程界写字写得最好的、书法界编程最牛的文艺汤”。

我的技术生涯

我的 IT 技术生涯应该算是从 2009 年开始的，在此之前做过 UI 设计，做过前端网页，到 2009 年才真正开始参与 Java 后台开发。在这里要感谢所有帮助我入门编程的同事和老师。2010 年至 2014 年担任过项目组长、项目经理、架构师、技术总监，对很多的开源框架建立了自己的独特见解。我会习惯性地用形象思维来理解抽象世界。譬如：看到二进制数 0 和 1，我会想到《周易》中的两仪——阴和阳；看到颜色值用 RGB 表示，我会想到美术理论中的太阳光折射三原色；下班回家看到炒菜流程，我会想到模板方法模式；坐公交车看到学生卡、老人卡、爱心卡，我会想到策略模式；等等。大家看到的这本书，很多地方都融入了这种形象思维。

为什么写书？

自 2019 年《Spring 5 核心原理与 30 个类手写实战》出版以来，深受广大读者喜爱，有些学员将此书作为学习的参考教材。为继续满足大家对技术的追求，借此机会将本人多年对 Netty

的研究笔记整理成书奉献给各位“汤粉”。

在此，特别感谢责任编辑董英、李秀梅及电子社的团队成员的付出，尤其在疫情期间，即使是远程办公也同样坚守岗位审稿至深夜。

谭勇德（Tom）
2020 年 3 月 于长沙

目　录

第1篇　I/O基础篇

第 2 篇 Netty 初体验

第 3 篇　Netty 核心篇

第 4 篇 Netty 实战篇

1

第 1 篇

I/O 基础篇

1

第 1 章 网络通信原理

1.1 网络基础架构

1.1.1 C/S 架构

有工作经验的人都知道，C 指的是 Client（客户端），S 指的是 Server（服务端），我们用 Socket 的目的就是实现 C/S 软件架构的服务端与客户端之间的网络通信。

1.1.2 C/S 信息传输流程

完成一次网络通信，大致要经过以下 5 个步骤。

（1）客户端产生数据，存放于客户端应用的内存中，然后调用接口将自己内存中的数据发送 / 拷贝给操作系统内存。

（2）客户端操作系统收到数据后，按照客户端应用指定的规则（即协议），调用网卡并发送数据。

（3）网络传输数据。

（4）服务端应用调用系统接口，想要将数据从操作系统内存拷贝到自己的内存中。

（5）服务端操作系统收到指令后，使用与客户端相同的规则（即协议）从网卡读取数据，然后拷贝给服务端应用。

1.2　TCP/IP 五层模型详解

计算机与计算机之间要有统一的连接标准才能够完成相互通信，这个标准被称为互联网协议，而网络就是物理链接介质+互联网协议。按照功能不同，人们将互联网协议从不同维度分为 OSI 七层、TCP/IP 五层或 TCP/IP 四层，如下图所示。

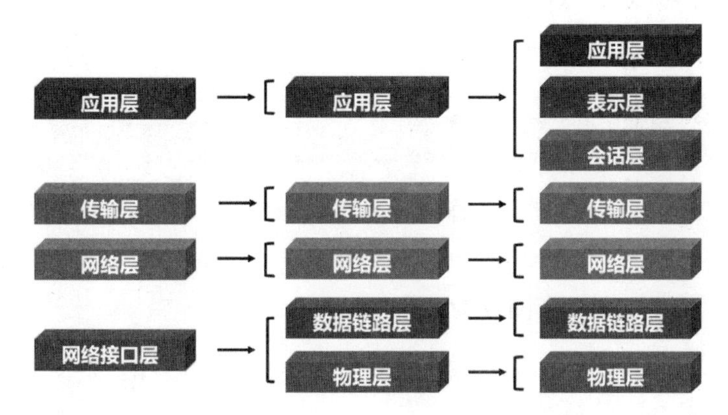

每层运行的常见设备如下图所示。

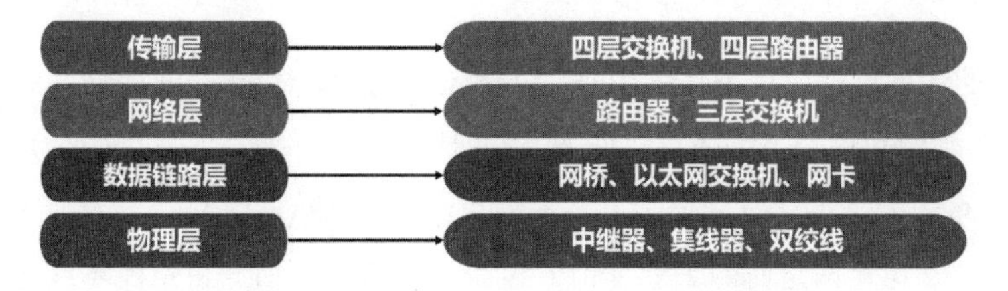

1.2.1　物理层

物理层主要是基于电器特性发送高低电平信号，电平即“电压平台”，指的是电路中某一点电压的高低状态，在网络信号中高电平用数字“1”表示，低电平用数字“0”表示。电平的高低是个相对概念，3V 对于 7V 是低电平，但对于 1V 就是高电平。

1.2.2 数据链路层

由于单纯的电平信号“0”和“1”没有任何意义，在实际应用中，我们会将电平信号进行分组处理，多少位一组、每组什么意思，这样数据才有具体的含义。数据链路层的功能就是定义电平信号的分组方式。

1. 以太网协议

数据链路层使用以太网协议进行数据传输，基于 MAC 地址的广播方式实现数据传输，只能在局域网内广播。早期各个公司都有自己的分组方式，后来形成了统一的标准，即以太网协议 Ethernet。

2. Ethernet 以太网

由一组电平信号构成一个数据包，叫作“帧”，每一数据帧由报头 Head 和数 Data 两部分组成，如下图所示。

Head：固定 18 字节，其中发送者 / 源地址 6 字节，接收者 / 目标地址 6 字节，数据类型 6 字节。

Data：最短 46 字节，最长 1 500 字节。

数据包的具体内容格式为：Head 长度＋Data 长度＝最短 64 字节，最长 1 518 字节（超过最大限制就分片发送）。

3. MAC 地址

Head 中包含的源地址和目标地址的由来：Ethernet 规定接入 Internet 的设备必须配有网卡，发送端和接收端的地址便是指网卡的地址，即 MAC 地址。

MAC 地址：每块网卡出厂时都被印上一个世界唯一的 MAC 地址，它是一个长度为 48 位的二进制数，通常用 12 位十六进制数表示（前 6 位是厂商编号，后 6 位是流水线号）。

4. Broadcast 广播

有了 MAC 地址，同一网络内的两台主机就可以通信了（一台主机通过 ARP 协议获取另外一台主机的 MAC 地址），下面是以太网通信数据帧的详细示意图。

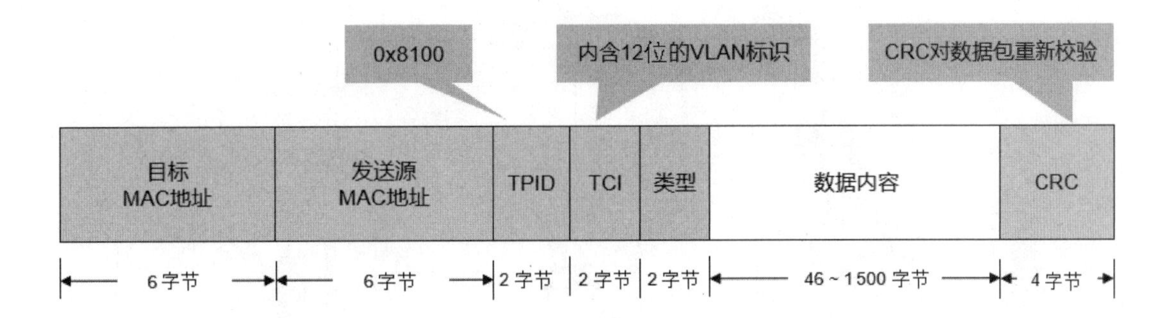

其实 Ethernet 采用非常原始的广播方式进行通信，也就是说计算机之间的通信基本靠“吼”。例如，有多台 PC 组成了一个网络，并通过硬件设施链接具备了通信条件，如下图所示。

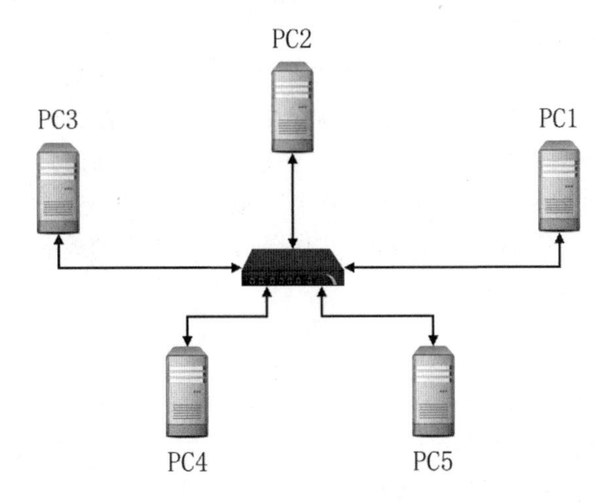

上图中，PC1 按照固定协议格式以广播的方式发送以太网包给 PC4，然而，PC2、PC3、PC5 都会收到 PC1 发来的数据包，拆开后如果发现目标 MAC 地址不是自己就会丢弃，如果是自己就响应。

1.2.3　网络层

有了 Ethernet、MAC 地址、广播的发送方式，世界上的计算机就可以彼此进行通信了，问题是世界范围的互联网是由一个个彼此隔离的小的局域网组成的（如下图所示），如果所有的通信都采用以太网的广播方式，那么一台机器发送的数据包全世界都会收到，这就不仅仅是效率低的问题了，这会是一种灾难。

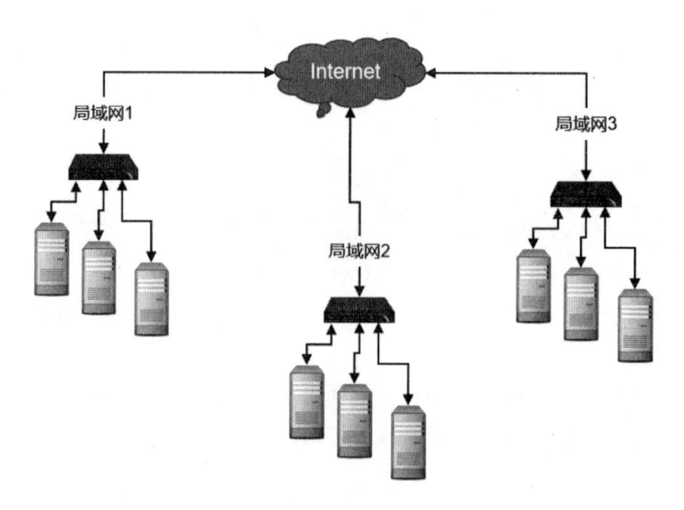

全世界的大网络由一个个小的彼此隔离的局域网组成，以太网包只能在一个局域网内发送，一个局域网是一个广播域，跨广播域通信只能通过路由转发。由此得出结论：必须找出一种方法来区分哪些计算机属于同一广播域，哪些不是。如果是就采用广播的方式发送，如果不是就采用路由的方式发送（向不同广播域 / 子网分发数据包），MAC 地址是无法区分的，它只跟厂商有关，网络层就是用来解决这一问题的。网络层的作用就是引入一套新的地址来区分不同的广播域 / 子网，这套地址叫作网络地址。

1. IP

规定网络地址的协议叫作 IP（Internet Protocol，网际互联协议），它定义的地址称为 IP 地址。广泛采用 v4 版本即 IPv4，规定网络地址由 32 位二进制数表示。一个 IP 地址通常写成四段十进制数，例如 172.16.10.1，其取值范围为：0.0.0.0~255.255.255.255。

IP 地址由两部分组成：网络部分（用来标识子网）和主机部分（用来标识主机）。

> **注意：**单纯的 IP 地址段只是标识了 IP 地址的种类，从网络部分或主机部分都无法辨识一个 IP 地址所处的子网。
>
> 例如，并不能确定 172.16.10.1 与 172.16.10.2 处于同一子网。因此，就需要子网掩码。

2. 子网掩码

所谓“子网掩码”，就是表示子网络特征的一个参数。它在形式上等同于 IP 地址，也是一个 32 位二进制数字，它的网络部分全部为 1，主机部分全部为 0。比如，IP 地址 172.16.10.1，如果已知网络部分是前 24 位，主机部分是后 8 位，那么子网络掩码就是 11111111.11111111.11111111.00000000，写成十进制就是 255.255.255.0。

我们根据“子网掩码”就能判断任意两个 IP 地址是否处于同一个子网络。方法是将两个 IP 地址与子网掩码分别进行&运算（两个数位都为 1，运算结果为 1，否则为 0），然后比较结果是否相同，如果相同，就表明它们在同一个子网络中，否则就不在。比如，已知 IP 地址 172.16.10.1 和 172.16.10.2 的子网掩码都是 255.255.255.0，请问它们是否在同一个子网络中？

我们将二者 IP 地址与子网掩码分别进行&运算，具体规则如下图所示。

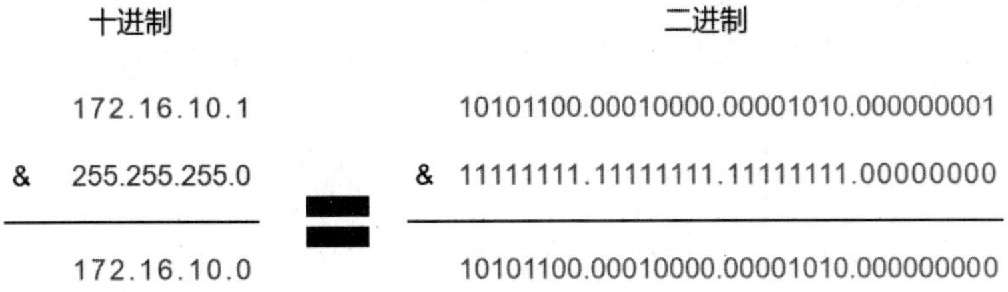

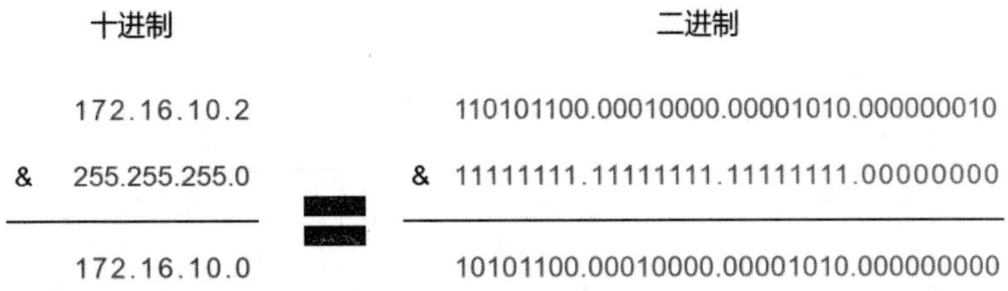

十进制		二进制
172.16.10.1		10101100.00010000.00001010.000000001
& 255.255.255.0	=	& 11111111.11111111.11111111.00000000
172.16.10.0		10101100.00010000.00001010.000000000

十进制		二进制
172.16.10.2		110101100.00010000.00001010.000000010
& 255.255.255.0	=	& 11111111.11111111.11111111.00000000
172.16.10.0		10101100.00010000.00001010.000000000

运算结果都是 172.16.10.0，因此它们在同一个子网络中。

总结一下，IP 的作用主要有两个，一个是为每一台计算机分配 IP 地址，另一个是确定哪些地址在同一个子网络中。

3. IP 数据包

IP 数据包也分为 Head 和 Data 两部分，无须为 IP 数据包定义单独的栏位，直接放入以太网包的 Data 部分即可。

Head（IP 头部）：长度为 20~60 字节。

Data（IP 数据）：最长为 65 515 字节。

而以太网数据包的 Data 部分，最长只有 1 500 字节。因此，如果 IP 数据包超过 1 500 字节，它就需要分割成几个以太网数据包，分开发送。其具体结构如下图所示。

以太网头部	IP头部	IP数据

4. ARP

我们已经知道计算机通信方式基本靠“吼”，也就是广播的方式。所有上层的数据包到最后都要封装到以太网头，然后通过以太网协议发送。在谈及以太网协议的时候，我们已经了解到，通信基于MAC地址的广播方式实现的，计算机在发送数据包时，获取自身的MAC地址是容易的，获取目标主机的MAC地址，需要通过ARP（Address Resolution Protocol，地址解析协议）来实现。

ARP用于实现从IP地址到MAC地址的映射，即询问目标IP地址对应的MAC地址，以广播的方式发送数据包，获取目标主机的 MAC 地址。我们通过一个案例来说明其具体通信原理，假设主机IP地址都已知。

- 主机A的IP地址为10.1.20.64，MAC地址为00:08:ca:xx:xx:xx；
- 主机B的IP地址为10.1.20.109，MAC地址为44:6d:57:xx:xx:xx。

当主机A要与主机B通信时，ARP可以将主机B的IP地址（10.1.20.109）解析成主机B的MAC地址，以下为工作流程。

第一步：通过IP地址和子网掩码计算出自己所处的子网，得出如下表所示的结果。

场　景	数据包地址
同一子网	目标主机MAC地址，目标主机IP地址
不同子网	网关MAC地址，目标主机IP地址

第二步：分析主机A和B是否处于同一网络，如果不是同一网络，那么下表中目标IP地址为10.1.20.109（访问路由器的路由表），通过ARP获取的是网关的MAC地址。

数据报文格式	源MAC地址	目标MAC地址	源IP地址	目标IP地址	数据部分
数据报文内容	发送端MAC地址	FF:FF:FF:FF:FF:FF	10.1.20.64	10.1.20.109	数据

第三步：根据主机A上的路由表内容，确定用于访问主机B的转发IP地址是10.1.20.109。然后主机A在自己的本地ARP缓存中检查主机B的匹配MAC地址。

第四步：如果主机A在ARP缓存中没有找到映射，它将询问10.1.20.64的硬件地址，从而将ARP请求帧广播到本地网络上的所有主机。源主机A的IP地址和MAC地址都包括在ARP请求中。本地网络上的每台主机都接收到ARP请求并且检查是否与自己的IP地址匹配。如果主机发现请求的IP地址与自己的IP地址不匹配，它将丢弃ARP请求。

第五步：主机B确定ARP请求中的IP地址与自己的IP地址匹配，将主机A的IP地址和MAC地址映射添加到本地ARP缓存中。

第六步：主机B将包含其MAC地址的ARP回复消息直接发送回主机A。

第七步：当主机 A 接收到从主机 B 发来的 ARP 回复消息时，会用主机 B 的 IP 地址和 MAC 地址映射更新 ARP 缓存。本机缓存是有生存期的，生存期结束后，将再次重复上面的过程。主机 B 的 MAC 地址一旦确定，主机 A 就能向主机 B 发送 IP 地址了。

为了让大家更好地理解 ARP 以及广播和单播的概念，我们可以利用网络抓包工具 Wireshark 来看一下抓取到的真实网络中的 ARP 过程，通过数据包的方式来呈现，部分 MAC 地址隐藏部分用 xx 代替。

假如，主机 A↔主机 B 通信：

- 主机 A：IP 地址为 10.1.20.64，MAC 地址为 00:08:ca:xx:xx:xx。
- 主机 B：IP 地址为 10.1.20.109，MAC 地址为 44:6d:57:xx:xx:xx。

ARP 请求数据包的内容如下。

```
    Frame 1490: 42 bytes on wire (336 bits), 42 bytes captured (336 bits) on interface 0
    Ethernet II, Src: Microsoft_xx:xx:xx (00:08:ca:xx:xx:xx, Dst: Broadcast
(ff:ff:ff:ff:ff:ff)
    Address Resolution Protocol (request)
    Hardware type: Ethernet (1)
    Protocol type: IP (0x080O)
    Hardware size: 6
    Protocol size: 4
    Opcode: request (1)
    Sender MAC address: Microsoft_xx:xx:xx (00:08:ca:xx:xx:xx)
    Sender IP address: 10.1.20.64 (10.1.20.64)
    Target MAC address: 00:00:00_00:00:00(00:00:00:00:00:00)
    Target IP address: 10.1.20.109 (10.1.20.109)
```

上面请求数据的含义是，我是主机 A，IP 地址为 10.1.20.64，MAC 地址为 00:08:ca:xx:xx:xx。

请问主机 B，IP 地址为 10.1.20.109，你的 MAC 地址是多少？

00:00:00_00:00:00 是置空位（留坑），表示询问者不知道，等待接收方回应（填坑）。

ARP 回应数据包的内容如下。

```
    Frame 1: 342 bytes on wire (2736 bits), 342 bytes captured (2736 bits) on interface 0
    Ethernet II, Src: Microsoft_xx:xx:xx (44:6d:57:xx:xx:xx), Dst: Microsoft_xx:xx:xx
(00:08:ca:xx:xx:xx)
    Address Resolution Protocol (reply)
     Hardware type: Ethernet (1)
     Protocol type: IP (0x0800)
```

```
    Hardware size: 6
  Protocol size: 4
  Opcode: reply (2)
    Sender MAC address: Microsoft_xx:xx:xx (44:6d:57:xx:xx:xx)
  Sender IP address: 10.1.20.109 (10.1.20.109)
    Target MAC address: Microsoft_xx:xx:xx (00:08:ca:xx:xx:xx)
  Target IP address: 10.1.20.64 (10.1.20.64)
```

上面请求数据的含义是，我是主机 B，IP 地址为 10.1.20.109，MAC 地址为 44:6d:57:xx:xx:xx。

你好，主机 A，IP 地址为 10.1.20.64，MAC 地址为 00:08:ca:xx:xx:xx。

下表是 ARP 中每个属性的具体含义详解。

属　性	含　义
Hardware type	硬件类型，标识链路层协议
Protocol type	协议类型，标识网络层协议
Hardware size	硬件地址大小，标识MAC地址长度，这里是6字节（48位）
Protocol size	协议地址大小，标识IP地址长度，这里是4字节（32位）
Opcode	操作代码，标识ARP数据包类型，1表示请求，2表示回应
Sender MAC address	发送者MAC地址
Sender IP address	发送者IP地址
Target MAC address	目标MAC地址，此处全为0，表示在请求
Target IP address	目标IP地址

1.2.4 传输层

现在我们已经知道，网络层的 IP 地址帮我们区分子网，以太网层的 MAC 地址帮我们找到主机。大家使用的都是应用程序，你的计算机上可能同时开启 QQ、微信等多个应用程序，那么我们通过 IP 地址和 MAC 地址找到了一台特定的主机，如何标识这台主机上的应用程序？答案就是端口，端口就是应用程序与网卡关联的编号。那么传输层就是用来建立端口到端口的通信机制的。

补充： 主机端口的取值范围为 0~65 535，其中 0~1 023 为系统保留端口的取值范围，也叫作 BSD 保留端口。用户可注册端口的取值范围为 1 024~49 152，还有随机动态端口的取值范围为 49 152~65 535。

为什么取值范围只能为 0~65 535，多一个都不行？从协议来讲，在 TCP 头部留给存储端口的空间只有 2 字节，最大值就是 65 535。

1. TCP

TCP（Transmission Control Protocol，传输控制协议）是一种可靠传输协议，TCP 数据包没有长度限制，理论上可以无限长，但是为了保证网络的效率，通常 TCP 数据包的长度不会超过 IP 数据包的长度，以确保单个 TCP 数据包不必再分割，其数据结构如下图所示。

以太网头部	IP头部	TCP头部	数据

2. UDP

UDP（User Datagram Protocol，用户数据报协议）是一种不可靠传输协议，“报头”部分总共有 8 字节，总长度不超过 65 535 字节，正好放进一个 IP 数据包，其数据结构如下图所示。

以太网头部	IP头部	UDP头部	数据

3. TCP 报文结构

TCP 报文是 TCP 层传输的数据单元，也叫作报文段。TCP 报文结构如下图所示。

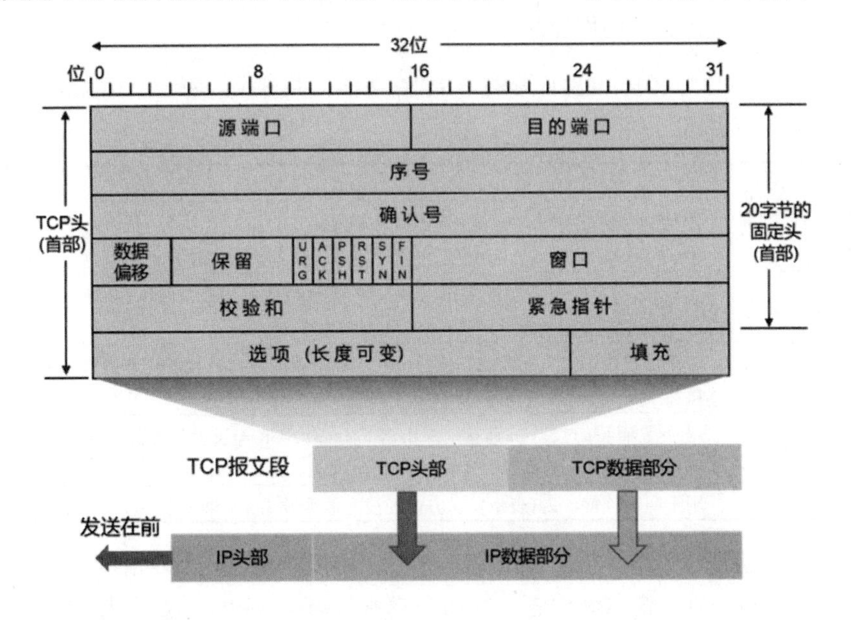

下面对报文内容做详细介绍。

（1）端口号：用来标识同一台计算机的不同应用进程。

- 源端口：源端口和 IP 地址的作用是标识报文的返回地址。

- 目的端口：目的端口指明接收方计算机上的应用程序接口。

TCP 报头中的源端口号和目的端口号同 IP 数据包中的源 IP 地址和目的 IP 地址唯一确定一条 TCP 连接。

（2）序号和确认号：TCP 可靠传输的关键部分。序号是本报文段发送的数据组的第一个字节的序号。在 TCP 传送的流中，每一个字节都有一个序号。例如：一个报文段的序号为 300，此报文段数据部分共有 100 字节，则下一个报文段的序号为 400。所以序号确保了 TCP 传输的有序性。

确认号，即 ACK，指明下一个期待收到的字节序号，表明该序号之前的所有数据已经正确无误地收到。确认号只有当 ACK 标志为 1 时才有效。比如建立连接时，SYN 报文的 ACK 标志为 0。

（3）数据偏移 / 头部长度：4 位。由于头部可能含有可选项内容，TCP 报头的长度是不确定的，报头不包含任何任选属性则长度为 20 字节，4 位头部长度属性所能表示的最大值为 1 111，转化成十进制为 15，15×32/8 = 60，故报头最大长度为 60 字节。头部长度也叫数据偏移，是因为头部长度实际上指示了数据区在报文段中的起始偏移值。

（4）保留：为将来定义新的用途保留，现在一般设置为 0。

（5）标志位：URG、ACK、PSH、RST、SYN、FIN，共 6 个，每一个标志位都表示一个控制功能，具体含义如下表所示。

字　段	中文名称	含　　义
URG	紧急指针标志	为1表示紧急指针有效，为0则忽略紧急指针
ACK	确认序号标志	为1表示确认号有效，为0表示报文中不含确认信息，忽略确认号属性
PSH	接收信号标志	为1表示带有Push标志的数据，指示接收方在接收到该报文段以后，应尽快将这个报文段交给应用程序，而不是在缓冲区排队
RST	重置连接标志	用于重置由于主机崩溃或其他原因而出现错误的连接，或者用于拒绝非法的报文段和拒绝连接请求
SYN	同步序号标志	用于建立连接过程，在连接请求中，SYN=1和ACK=0表示该数据段没有使用捎带的确认域，而连接应答捎带一个确认，即SYN=1和ACK=1
FIN	完成标志	用于释放连接，为1表示发送方已经没有数据发送了，即关闭本方数据流

（6）窗口：滑动窗口大小，用来告知发送端接收端的缓存大小，以此控制发送端发送数据的速率，从而达到流量控制。窗口大小是一个 16 位属性，因而窗口大小最大为 65 535。

（7）校验和：奇偶校验，此校验和针对整个 TCP 报文段，包括 TCP 头部和 TCP 数据，以 16 位属性进行计算所得。由发送端计算和存储，并由接收端进行验证。

（8）紧急指针：只有当 URG 标志为 1 时紧急指针才有效。紧急指针是一个正的偏移量，和顺序号属性中的值相加表示紧急数据最后一个字节的序号。TCP 的紧急方式是发送端向另一端发送紧急数据的一种方式。

（9）选项和填充：最常见的可选属性是最长报文大小，又称为 MSS（Maximum Segment Size），每个连接方通常都在通信的第一个报文段（为建立连接而设置 SYN 标志为 1 的那个段）中指明这个选项，它表示本端所能接收的最大报文段的长度。选项长度不一定是 32 位的整数倍，所以要加填充位，即在这个属性中加入额外的零，以保证 TCP 头部长度是 32 位的整数倍。

（10）数据部分：TCP 报文段中的数据部分是可选的。在一个连接建立和一个连接终止时，双方交换的报文段仅有 TCP 头部。如果一方没有数据要发送，也使用没有任何数据的头部来确认收到的数据。在处理超时的许多情况中，也会发送不带任何数据的报文段。

4. TCP 交互流程

传输连接包括三个阶段：连接建立、数据传送和连接释放。传输连接管理就是对连接建立和连接释放过程的管控，使其能正常运行，以达到这些目的：使通信双方能够确知对方的存在、可以允许通信双方协商一些参数（最大报文段长度、最大窗口大小等）、能够对运输实体资源进行分配（缓存大小等）。TCP 连接的建立采用客户端-服务器模式：主动发起连接建立的应用进程叫作客户端，被动等待连接建立的应用进程叫作服务器。接下来，介绍 TCP 完成数据传输的三次握手和四次挥手的详细过程。

第一次握手：建立连接时，客户端发送 SYN 包（syn=1）到服务器，并进入 SYN_SENT 状态，等待服务器确认。

第二次握手：服务器收到 SYN 包，必须确认客户端的 SYN 包（ack=x+1），同时自己也发送一个 SYN 包（syn=1），即 SYN+ACK 包，此时服务器进入 SYN_RECV 状态。

第三次握手：客户端收到服务器的 SYN+ACK 包，向服务器发送确认包 ACK（ack=y+1），此包发送完毕，客户端和服务器进入 ESTABLISHED（TCP 连接成功）状态，完成三次握手。

至此，TCP 连接就建立了，客户端和服务器可以愉快地“玩耍”了。只要通信双方没有一方发出连接释放的请求，连接就将一直保持。如果有一方释放连接，就会发起挥手操作。

第一次挥手：客户端进程发出连接释放报文，并且停止发送数据。释放数据报文头部，FIN=1，其序列号为 seq=u（等于前面已经传送过来的数据的最后一个字节的序号加 1），此时，客户端进入 FIN_WAIT_1（终止等待 1）状态。TCP 规定，FIN 报文段即使不携带数据，也要消耗一个序号。

第二次挥手：服务器收到连接释放报文，发出确认报文，ACK=1，ack=u+1，并且带上自己的序列号 seq=v，此时，服务器就进入了 CLOSE_WAIT（关闭等待）状态。TCP 服务器通知高层的应用进程，客户端向服务器方向的连接就被释放了，这时候处于半关闭状态，即客户端已经没有数据要发送了，但是服务器若发送数据，客户端依然要接收。这个状态还要持续一段时间，也就是整个 CLOSE_WAIT 状态持续的时间。

客户端收到服务器的确认请求后，客户端就进入 FIN_WAIT_2（终止等待 2）状态，等待服务器发送连接释放报文（在这之前还需要接收服务器发送的最后的数据）。

第三次挥手：服务器将最后的数据发送完毕后，就向客户端发送连接释放报文，FIN=1，ack=u+1，由于在半关闭状态，服务器很可能又发送了一些数据，假定此时的序列号为 seq=w，此时，服务器就进入了 LAST_ACK（最后确认）状态，等待客户端的确认。

第四次挥手：客户端收到服务器的连接释放报文后，必须发出确认，ACK=1，ack=w+1，而自己的序列号是 seq=u+1，此时，客户端就进入了 TIME_WAIT（时间等待）状态。注意此时 TCP 连接还没有释放，必须经过 2×MSL（最长报文段寿命）的时间，当客户端撤销相应的 TCB（Transmit Control Block，传输控制模块）后，才进入 CLOSED 状态。

最后，服务器只要收到了客户端发出的确认，就立即进入 CLOSED 状态。同样，撤销 TCB 后，就结束了这次 TCP 连接。可以看到，服务器结束 TCP 连接的时间要比客户端早一些。

TCP 交互的详细过程如下图所示。

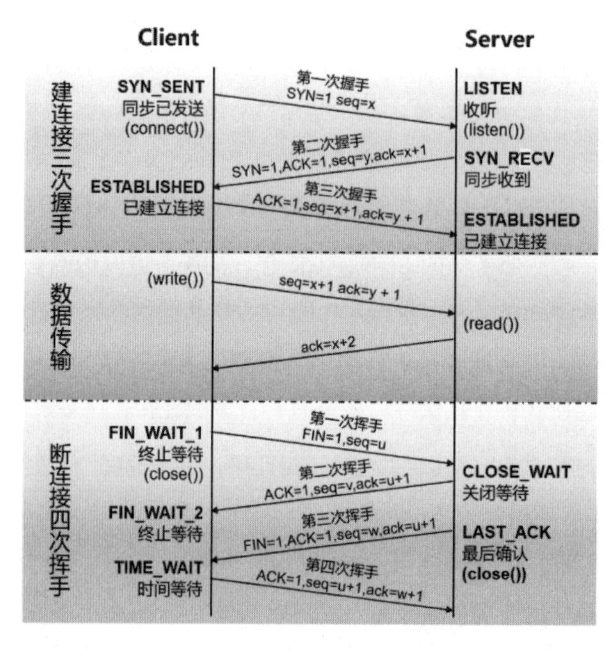

1.2.5　应用层

在日常操作中，用户使用的都是应用程序，应用程序都工作在应用层。互联网是开放的，大家都可以开发自己的应用程序，数据多种多样，必须规定好数据的组织形式。应用层的功能就是规定应用程序的数据格式。

例如：TCP 可以为各种各样的程序传递数据，比如 SMTP、HTTP、FTP、POP3 等，那么，必须有不同的协议规定电子邮件、网页、FTP 数据的格式，这些应用程序协议就构成了“应用层”。如下图所示是应用层协议的基本组成结构示意。

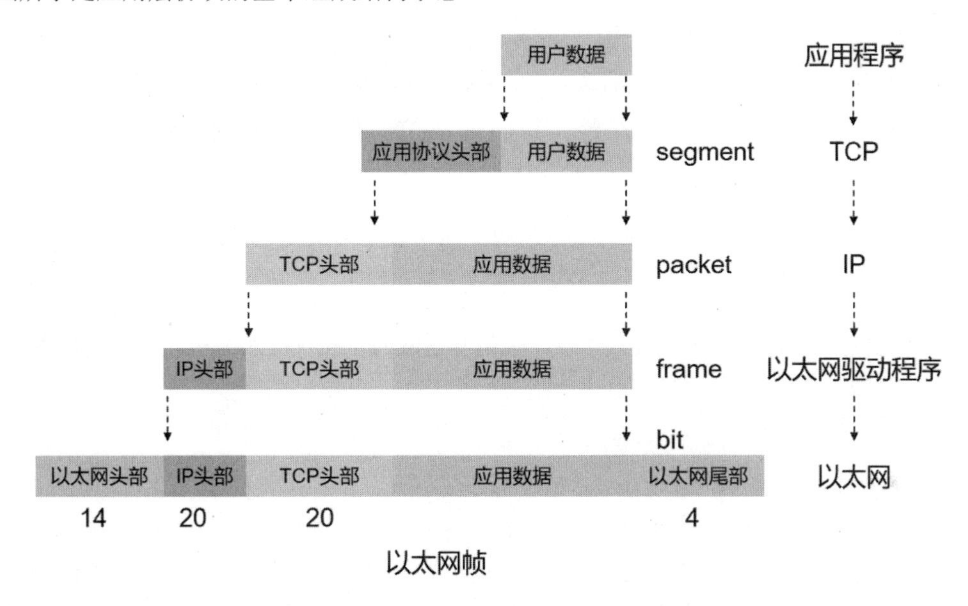

用户从应用程序中发送数据是一个对数据封装的过程，而接收数据则是一个解封装的动作。下面简单介绍一下常用的应用层协议。

1. DNS 协议

DNS 是英文 Domain Name System（域名系统）的缩写，用来把便于人们使用的机器名字转换为 IP 地址。现在顶级域名 TLD（Totel Lead Domination）分为三大类：国家顶级域名 nTLD、通用顶级域名 gTLD 和基础结构域名。域名服务器分为四种类型：根域名服务器、顶级域名服务器、本地域名服务器和权限域名服务器。DNS 使用 TCP 和 UDP 端口 53。当前，对于每一级域名长度的限制是 63 个字符，域名总长度则不能超过 253 个字符。

2. HTTP

HTTP（HyperText Transfer Protocol，超文本传输协议）是面向事务的应用层协议。它是互联网上能够可靠地交换信息的重要基础。HTTP 使用面向连接的 TCP 作为运输层协议，保证了数据的可靠传输。

3. FTP

FTP（File Transfer Protocol，文件传输协议）是互联网上使用最广泛的文件传送协议。FTP 提供交互式的访问，允许客户指明文件类型与格式，并允许文件具有存取权限。FTP 基于 TCP 工作。

4. SMTP

SMTP（Simple Mail Transfer Protocol，简单邮件传输协议）规定了在两个相互通信的 SMTP 进程之间应如何交换信息。SMTP 通信的三个阶段：建立连接、邮件传送和连接释放。

5. POP3

POP3（Post Office Protocol 3，邮件读取协议）通常被用来接收电子邮件。

6. Telnet 协议

Telnet 协议是一个简单的远程终端协议，也是互联网的正式标准，又称为终端仿真协议。

1.2.6 小结

总结一下 OSI 七层模型，它为开放互联信息系统提供了一种结构框架。建立七层模型的主要目的是解决异种网络互联时所遇到的兼容性问题。它的最大优点是将服务、接口和协议这三个概念明确地区分开来：服务说明某一层为上一层提供一些什么功能，接口说明上一层如何使用下一层的服务，而协议涉及如何实现本层的服务；这样各层之间具有很强的独立性，互联网络中各实体采用什么样的协议是没有限制的，只要向上提供相同的服务并且不改变相邻层的接口就可以。其详细结构如下图所示。

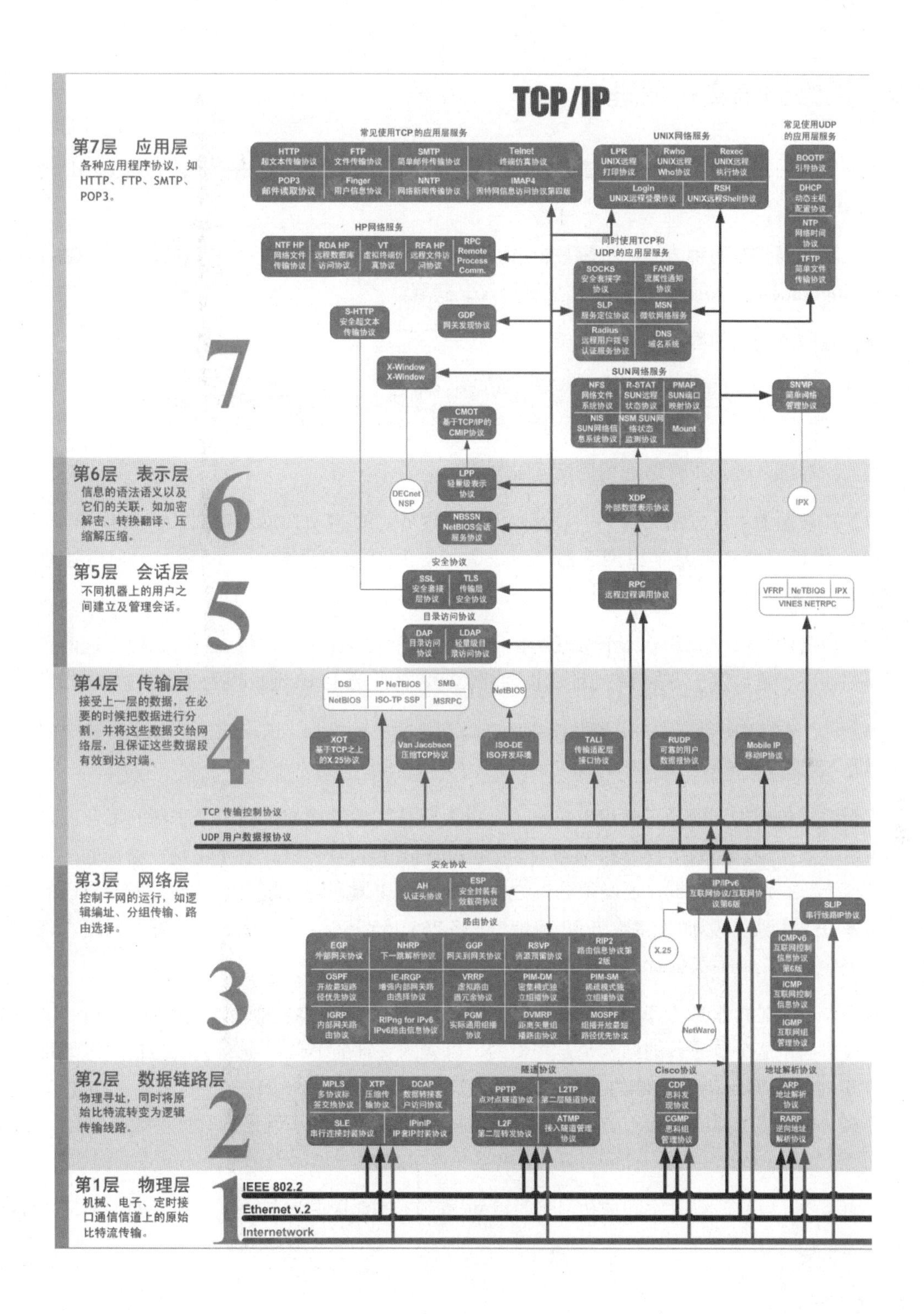

TCP/IP
第7层　应用层
各种应用程序协议，如HTTP、FTP、SMTP、POP3。
7
常见使用TCP的应用层服务
HTTP 超文本传输协议
FTP 文件传输协议
SMTP 简单邮件传输协议
Telnet 终端仿真协议
POP3 邮件读取协议
Finger 用户信息协议
NNTP 网络新闻传输协议
IMAP4 因特网信息访问协议第四版
UNIX网络服务
LPR UNIX远程打印协议
Rwho UNIX远程Who协议
Rexec UNIX远程执行协议
Login UNIX远程登录协议
RSH UNIX远程Shell协议
常见使用UDP的应用层服务
BOOTP 引导协议
DHCP 动态主机配置协议
NTP 网络时间协议
TFTP 简单文件传输协议
HP网络服务
NTF HP 网络文件传输协议
RDA HP 远程数据库访问协议
VT 虚拟终端仿真协议
RFA HP 远程文件访问协议
RPC Remote Process Comm.
同时使用TCP和UDP的应用层服务
SOCKS 安全套接字协议
FANP 流属性通知协议
SLP 服务定位协议
MSN 微软网络服务
Radius 远程用户拨号认证服务协议
DNS 域名系统
S-HTTP 安全超文本传输协议
GDP 网关发现协议
X-Window X-Window
SUN网络服务
NFS 网络文件系统协议
R-STAT SUN远程状态协议
PMAP SUN端口映射协议
NIS SUN网络信息系统协议
NSM SUN网络状态监测协议
Mount
CMOT 基于TCP/IP的CMIP协议
SNMP 简单网络管理协议
第6层　表示层
信息的语法语义以及它们的关联，如加密解密、转换翻译、压缩解压缩。
6
DECnet NSP
LPP 轻量级表示协议
NBSSN NetBIOS会话服务协议
XDP 外部数据表示协议
IPX
第5层　会话层
不同机器上的用户之间建立及管理会话。
5
安全协议
SSL 安全套接层协议
TLS 传输层安全协议
目录访问协议
DAP 目录访问协议
LDAP 轻量级目录访问协议
RPC 远程过程调用协议
VFRP NeTBIOS IPX
VINES NETRPC
第4层　传输层
接受上一层的数据，在必要的时候把数据进行分割，并将这些数据交给网络层，且保证这些数据段有效到达对端。
4
DSI IP NeTBIOS SMB
NetBIOS ISO-TP SSP MSRPC
NetBIOS
XOT 基于TCP之上的X.25协议
Van Jacobson 压缩TCP协议
ISO-DE ISO开发环境
TALI 传输适配层接口协议
RUDP 可靠的用户数据报协议
Mobile IP 移动IP协议
TCP 传输控制协议
UDP 用户数据报协议
第3层　网络层
控制子网的运行，如逻辑编址、分组传输、路由选择。
3
安全协议
AH 认证头协议
ESP 安全封装有效载荷协议
IP/IPv6 互联网协议/互联网协议第6版
SLIP 串行线路IP协议
路由协议
EGP 外部网关协议
NHRP 下一跳解析协议
GGP 网关到网关协议
RSVP 资源预留协议
RIP2 路由信息协议第2版
OSPF 开放最短路径优先协议
IE-IRGP 增强内部网关路由选择协议
VRRP 虚拟路由器冗余协议
PIM-DM 密集模式独立组播协议
PIM-SM 稀疏模式独立组播协议
IGRP 内部网关路由协议
RIPng for IPv6 IPv6路由信息协议
PGM 实际通用组播协议
DVMRP 距离矢量组播路由协议
MOSPF 组播开放最短路径优先协议
X.25
NetWare
ICMPv6 互联网控制信息协议第6版
ICMP 互联网控制信息协议
IGMP 互联网组管理协议
第2层　数据链路层
物理寻址，同时将原始比特流转变为逻辑传输线路。
2
MPLS 多协议标签交换协议
XTP 压缩传输协议
DCAP 数据转接客户访问协议
SLE 串行连接封装协议
IPinIP IP套IP封装协议
隧道协议
PPTP 点对点隧道协议
L2TP 第二层隧道协议
L2F 第二层转发协议
ATMP 接入隧道管理协议
Cisco协议
CDP 思科发现协议
CGMP 思科组管理协议
地址解析协议
ARP 地址解析协议
RARP 逆向地址解析协议
第1层　物理层
机械、电子、定时接口通信信道上的原始比特流传输。
1
IEEE 802.2
Ethernet v.2
Internetwork

1.3 网络通信实现原理

要想实现网络通信，每台主机需具备四要素：本机的 IP 地址、子网掩码、网关的 IP 地址和 DNS 的 IP 地址。

获取这四要素有两种方式：一是静态获取，即手动配置；二是动态获取，即通过 DHCP（Dynamic Host Configuration Protocol，动态主机配置协议）获取。

下图是网络通信的数据结构示意图。

以太网头部	IP头部	UDP头部	DHCP数据包

我们来详细分析一下网络通信的交互过程。

（1）最前面的“以太网头部”，设置发出方（本机）的 MAC 地址和接收方（DHCP 服务器）的 MAC 地址。前者就是本机网卡的 MAC 地址，后者这时不知道，就填入一个广播地址：FF-FF-FF-FF-FF-FF。

（2）后面的“IP 头部”，设置发出方的 IP 地址和接收方的 IP 地址。这时，对于这两者，本机都不知道。于是，发出方的 IP 地址就设为 0.0.0.0，接收方的 IP 地址设为 255.255.255.255。

（3）最后的“UDP 头部”，设置发出方的端口和接收方的端口。这一部分是 DHCP 规定好的，发出方是 68 端口，接收方是 67 端口。

这个数据包构造完成后，就可以发出了。以太网是广播发送的，同一个子网的每台计算机都收到了这个数据包。因为接收方的 MAC 地址是 FF-FF-FF-FF-FF-FF，看不出是发给谁的，所以每台收到这个数据包的计算机，还必须分析这个数据包的 IP 地址，才能确定是不是发给自己的。看到发出方 IP 地址是 0.0.0.0，接收方 IP 地址是 255.255.255.255，于是 DHCP 服务器知道“这个数据包是发给我的”，而其他计算机就可以丢弃这个数据包。

接下来，DHCP 服务器读出这个数据包的数据内容，分配好 IP 地址，发送回去一个“DHCP 响应”数据包。这个响应包的结构也是类似的，以太网头部的 MAC 地址是双方的网卡地址，IP 头部的 IP 地址是 DHCP 服务器的 IP 地址（发出方）和 255.255.255.255（接收方），UDP 头部的端口是 67（发出方）和 68（接收方），分配给请求端的 IP 地址和本网络的具体参数则包含在 Data 部分。

新加入的计算机收到这个响应包，于是就知道了自己的 IP 地址、子网掩码、网关地址、DNS 服务器等参数。

1.4　向浏览器输入 URL 后发生了什么

当在浏览器地址栏中输入网址后，浏览器是怎么把最终的页面呈现出来的呢？这个过程大致可以分为两个部分：网络通信和页面渲染。下面详细分析完整的通信过程。

第一步，本机设置以下信息。

本机IP地址	子网掩码	网关IP地址	DNS地址
192.168.1.100	255.255.255.0	192.168.1.1	8.8.8.8

第二步，打开浏览器，想要访问咕泡官网，在地址栏中输入网址 www.gupaoedu.com。

第三步，通过访问 DNS 域名系统服务器（基于 UDP）获得 IP 地址。

下图完整地说明了一次网络请求如何获取目标服务器 IP 地址的全过程。

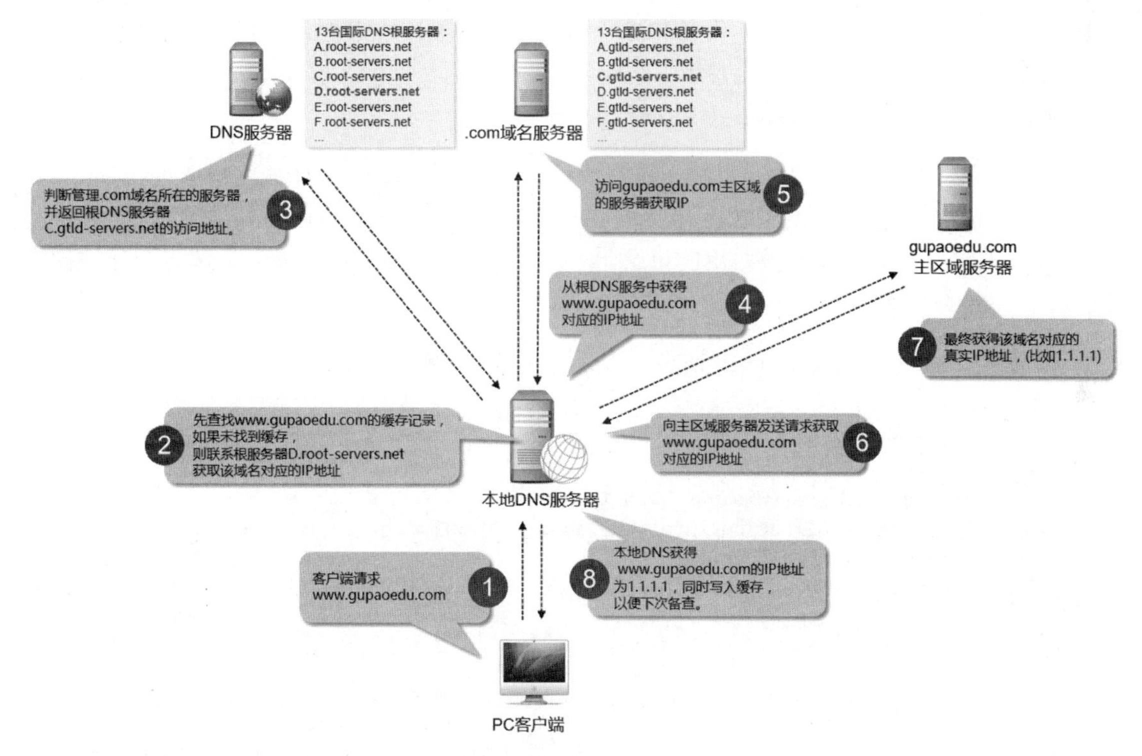

图中 13 台国际 DNS 根服务器 IP 地址具体如下。

```
A.root-servers.net 198.41.0.4 美国
B.root-servers.net 192.228.79.201 美国（另支持 IPv6）
```

```
C.root-servers.net 192.33.4.12 法国
D.root-servers.net 128.8.10.90 美国
E.root-servers.net 192.203.230.10 美国
F.root-servers.net 192.5.5.241 美国（另支持 IPv6）
G.root-servers.net 192.112.36.4 美国
H.root-servers.net 128.63.2.53 美国（另支持 IPv6）
I.root-servers.net 192.36.148.17 瑞典
J.root-servers.net 192.58.128.30 美国
K.root-servers.net 193.0.14.129 英国（另支持 IPv6）
L.root-servers.net 198.32.64.12 美国
M.root-servers.net 202.12.27.33 日本（另支持 IPv6）
```

通过域名寻找到目标机器所在位置。下面简单科普一下域名知识，域名有顶级域名和二级域名。顶级域名如.com、.net、.org、.cn 等属于国际顶级域名。根据目前的国际互联网域名体系，国际顶级域名分为两类：类别顶级域名（gTLD）和地理顶级域名（ccTLD）。类别顶级域名是以“com”“net”“org”“biz”“info”等结尾的域名，均由国外公司负责管理。地理顶级域名是以国家或地区代码为结尾的域名，如“cn”代表中国，“uk”代表英国。地理顶级域名一般由各个国家或地区负责管理。在不同的地域还会使用二级域名，二级域名是以顶级域名为基础的地理域名，例如中国的二级域有.com.cn、.net.cn、.org.cn、.gd.cn 等。

在实际的网站应用中，通常会使用子域名。比如父域名是 abc.com，子域名就是 www.abc.com 或者*.abc.com。一般来说，子域名是域名的一条记录，比如 gupaoedu.com 是一个域名，www.gupaoedu.com 是其中比较常用的记录，一般默认类似*.gupaoedu.com 的域名全部被称作 gupaoedu.com 的子域名。

第四步，向目标机器发起 HTTP 请求，获得如下格式的数据内容。

```
GET / HTTP/1.1
Host: www.gupaoedu.com
Connection: keep-alive
User-Agent: Mozilla/5.0 (Windows NT 6.1) ……
Accept: text/html,application/xhtml+xml,application/xml;q=0.9,*/*;q=0.8
Accept-Encoding: gzip,deflate,sdch
Accept-Language: zh-CN,zh;q=0.8
Accept-Charset: GBK,utf-8;q=0.7,*;q=0.3
Cookie: … …
```

我们假定这个部分的长度为 4 960 字节，它会被嵌在 TCP 数据包中。

第五步，TCP。TCP 数据包需要设置端口，接收方（咕泡官网）的 HTTP 端口默认是 80，发送方（本机）的端口是一个随机生成的 1 024~65 535 之间的整数，假定为 51 775。TCP 数据包的头部长度为 20 字节，加上嵌入 HTTP 的数据包，总长度变为 4 980 字节。

第六步，IP。TCP 数据包再嵌入 IP 数据包。IP 数据包需要设置双方的 IP 地址，这是已知的。IP 数据包的头部长度为 20 字节，加上嵌入的 TCP 数据包，总长度变为 5 000 字节。

第七步，以太网协议。IP 数据包嵌入以太网数据包。以太网数据包需要设置双方的 MAC 地址，发送方为本机的网卡 MAC 地址，接收方为网关 192.168.1.1 的 MAC 地址（通过 ARP 得到）。以太网数据包的数据部分最大长度为 1 500 字节，而现在的 IP 数据包长度为 5 000 字节。因此，IP 数据包必须分割成四个包。因为每个包都有自己的 IP 头部（20 字节），所以四个包的长度分别为 1 500 字节、1 500 字节、1 500 字节、560 字节。如下图所示是以太网数据包示意图。

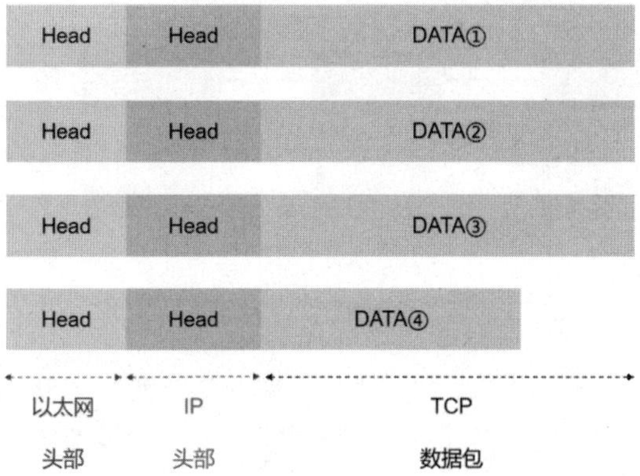

Head Head DATA①
Head Head DATA②
Head Head DATA③
Head Head DATA④
以太网头部　IP 头部　TCP 数据包

第八步，服务器响应。经过多个网关的转发，咕泡官网的服务器收到了这四个以太网数据包。

根据 IP 头部的序号，咕泡官网将四个包拼起来，取出完整的 TCP 数据包，然后读出里面的"HTTP 请求"，接着做出"HTTP 响应"，再用 TCP 发回来。本机收到 HTTP 响应以后，就可以将网页显示出来，完成一次网络通信。

1.5　网络通信之"魂"——Socket

我们知道两个进程进行通信一个最基本的前提是能够唯一地标识一个进程。在本地进程通信中，我们可以使用 PID 来唯一标识一个进程，但 PID 只在本地唯一，网络中的两个进程 PID 冲突概率很大，这时候就需要另辟蹊径了。我们知道 IP 层的 IP 地址可以唯一标识主机，而 TCP 层的协议和端口号可以唯一标识主机的一个进程，可以用 IP 地址＋协议＋端口号唯一标识网络中的一个进程。

能够唯一标识网络中的进程后，它们就可以利用 Socket 进行通信了。那么，什么是 Socket 呢？我们经常把 Socket 翻译为套接字，Socket 是在应用层和传输层之间的一个抽象层，它把 TCP/IP

层复杂的操作抽象为几个简单的接口供应用层调用，以实现进程在网络中的通信，具体结构如下图所示。

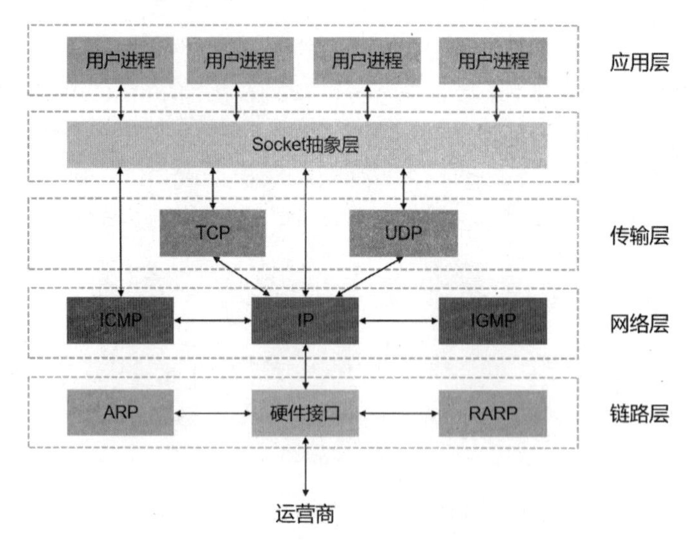

Socket 起源于 UNIX，在 UNIX“一切皆文件”的哲学思想下，Socket 是一种从打开，到完成读、写操作，最后关闭的模式，服务器和客户端各自维护一个“文件”，在建立连接打开文件后，可以向自己的文件写入内容供对方读取或者读取对方的内容，通信结束时关闭文件。

2

第 2 章 Java I/O 演进之路

2.1 I/O 的问世

2.1.1 什么是 I/O

我们都知道在 UNIX 世界里一切皆文件，而文件是什么呢？文件就是一串二进制流而已，其实不管是 Socket，还是 FIFO（First Input First Output，先进先出队列））、管道、终端。对计算机来说，一切都是文件，一切都是流。在信息交换的过程中，计算机都是对这些流进行数据的收发操作，简称为 I/O 操作（Input and Output），包括往流中读出数据、系统调用 Read、写入数据、系统调用 Write。不过计算机里有那么多流，怎么知道要操作哪个流呢？实际上是由操作系统内核创建文件描述符（File Descriptor，FD）来标识的，一个 FD 就是一个非负整数，所以对这个整数的操作就是对这个文件（流）的操作。我们创建一个 Socket，通过系统调用会返回一个 FD，那么剩下的对 Socket 的操作就会转化为对这个描述符的操作，这又是一种分层和抽象的思想。

2.1.2 I/O 交互流程

通常用户进程中的一次完整 I/O 交互流程分为两阶段，首先是经过内核空间，也就是由操作系统处理；紧接着就是到用户空间，也就是交由应用程序。具体交互流程如下图所示。

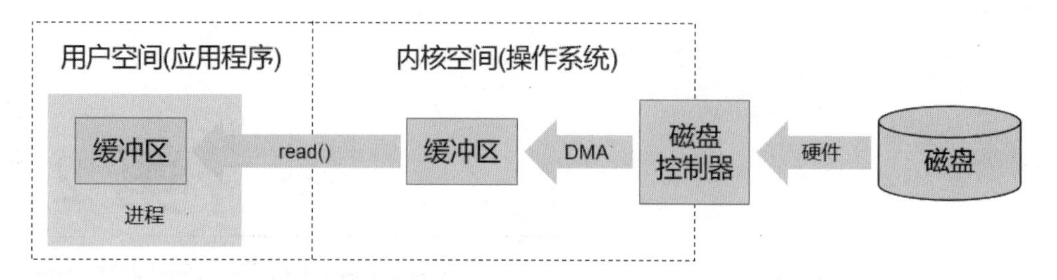

内核空间中存放的是内核代码和数据，而进程的用户空间中存放的是用户程序的代码和数据。不管是内核空间还是用户空间，它们都处于虚拟空间中，Linux 使用两级保护机制：0 级供内核（Kernel）使用，3 级供用户程序使用。每个进程都有各自的私有用户空间（0~3G），这个空间对系统中的其他进程是不可见的。最高的 1G 字节虚拟内核空间则为所有进程及内核共享。

操作系统和驱动程序运行在内核空间，应用程序运行在用户空间，两者不能简单地使用指针传递数据。因为 Linux 使用的虚拟内存机制，必须通过系统调用请求 Kernel 来协助完成 I/O 操作，内核会为每个 I/O 设备维护一个缓冲区，用户空间的数据可能被换出，所以当内核空间使用用户空间的指针时，对应的数据可能不在内存中。

对于一个输入操作来说，进程 I/O 系统调用后，内核会先看缓冲区中有没有相应的缓存数据，如果没有再到设备中读取。因为设备 I/O 一般速度较慢，需要等待，内核缓冲区有数据则直接复制到进程空间。所以，一个网络输入操作通常包括两个不同阶段。

（1）等待网络数据到达网卡，然后将数据读取到内核缓冲区。

（2）从内核缓冲区复制数据，然后拷贝到用户空间。

I/O 有内存 I/O、网络 I/O 和磁盘 I/O 三种，通常我们说的 I/O 指的是后两者。如下图所示是 I/O 通信过程的调度示意。

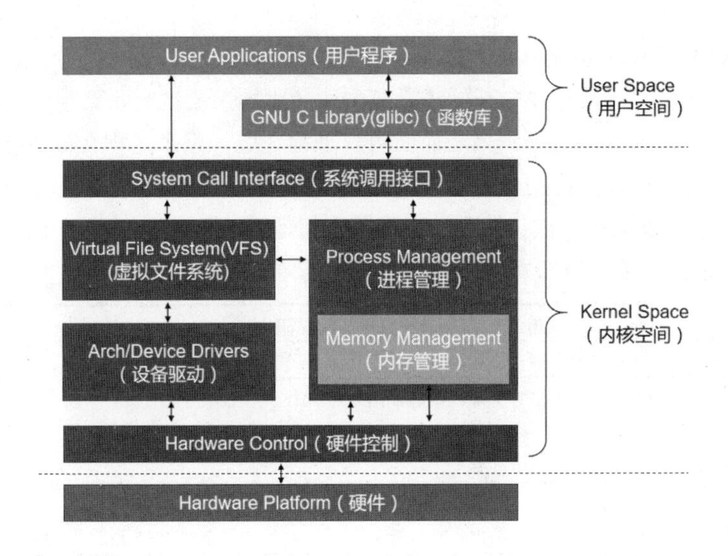

2.2　五种 I/O 通信模型

在网络环境下，通俗地讲，将 I/O 分为两步：第一步是等待；第二步是数据搬迁。

如果想要提高 I/O 效率，需要将等待时间降低。因此发展出来五种 I/O 模型，分别是：阻塞 I/O 模型、非阻塞 I/O 模型、多路复用 I/O 模型、信号驱动 I/O 模型、异步 I/O 模型。其中，前四种被称为同步 I/O，下面对每一种 I/O 模型进行详细分析。

2.2.1　阻塞 I/O 模型

阻塞 I/O 模型的通信过程示意如下图所示。

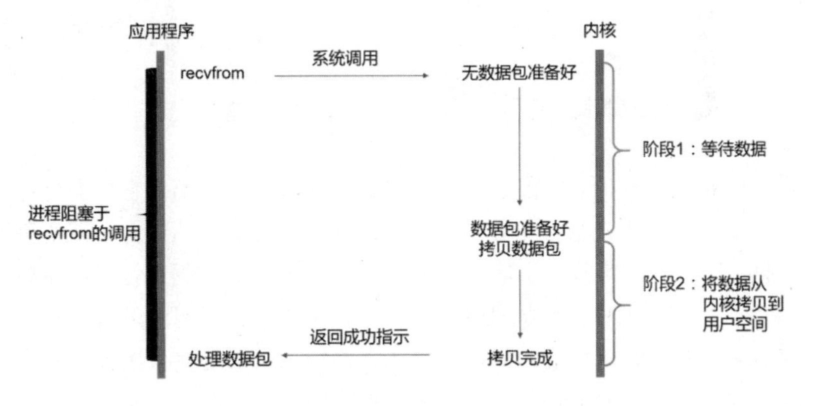

当用户进程调用了 recvfrom 这个系统调用，内核就开始了 I/O 的第一个阶段：准备数据。对于网络 I/O 来说，很多时候数据在一开始还没有到达（比如，还没有收到一个完整的 UDP 包），这个时候内核就要等待足够的数据到来。而在用户进程这边，整个进程会被阻塞，当数据准备好时，它就会将数据从内核拷贝到用户内存，然后返回结果，用户进程才解除阻塞的状态，重新运行起来。几乎所有的开发者第一次接触到的网络编程都是从 listen()、send()、recv()等接口开始的，这些接口都是阻塞型的。阻塞 I/O 模型的特性总结如下表所示。

特　　点	在I/O执行的两个阶段（等待数据和拷贝数据）都被阻塞
典型应用	阻塞Socket，Java BIO
优　　点	• 进程阻塞挂起不消耗CPU资源，及时响应每个操作 • 实现难度低，开发应用较容易 • 适合并发量小的网络应用开发
缺　　点	• 不适合并发量大的应用，因为一个请求I/O会阻塞进程 • 需要为每个请求分配一个处理进程（线程）以及时响应，系统开销大

2.2.2 非阻塞 I/O 模型

非阻塞 I/O 模型的通信过程示意如下图所示。

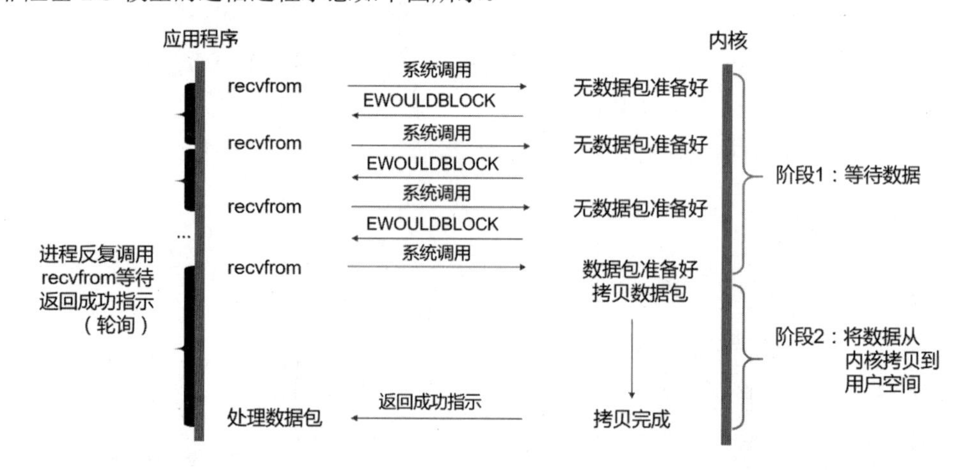

当用户进程发出 read 操作时，如果内核中的数据还没有准备好，那么它并不会阻塞用户进程，而是立刻返回一个 error。从用户进程角度讲，它发起一个 read 操作后，并不需要等待，而是马上就得到了一个结果，用户进程判断结果是一个 error 时，它就知道数据还没有准备好。于是它可以再次发送 read 操作，一旦内核中的数据准备好了，并且再次收到了用户进程的系统调用，那么它会马上将数据拷贝到用户内存，然后返回，非阻塞型接口相比于阻塞型接口的显著差异在于，在被调用之后立即返回。非阻塞 I/O 模型的特性总结如下表所示。

特　　点	用户进程需要不断地主动询问内核（Kernel）数据准备好了没有
典型应用	Socket设置NON_BLOCK
优　　点	实现难度低，开发应用相对阻塞I/O模型较难
缺　　点	• 进程轮询（重复）调用，消耗CPU的资源 • 适合并发量较小且不需要及时响应的网络应用开发

非阻塞模式套接字与阻塞模式套接字相比，不容易使用。使用非阻塞模式套接字，需要编写更多的代码，但是，非阻塞模式套接字在控制建立多个连接、数据的收发量不均、时间不定时，具有明显优势。

2.2.3　多路复用 I/O 模型

多路复用 I/O 模型的通信过程示意如下图所示。

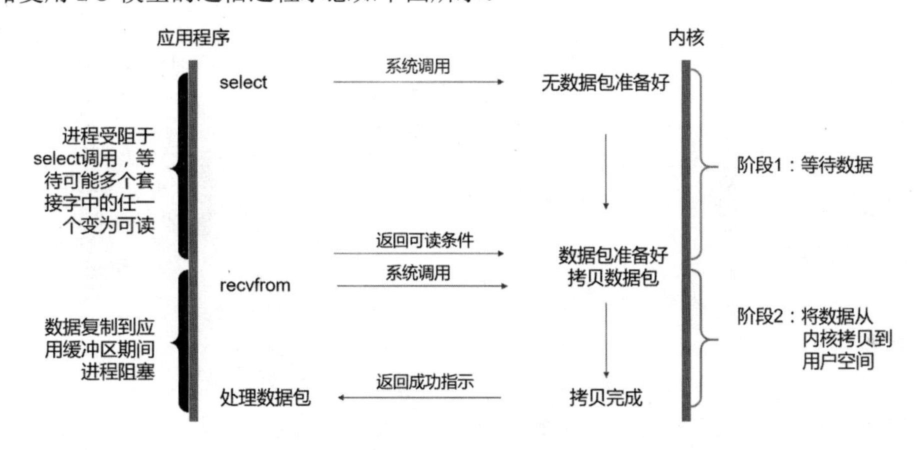

多个进程的 I/O 可以注册到一个复用器（Selector）上，当用户进程调用该 Selector，Selector 会监听注册进来的所有 I/O，如果 Selector 监听的所有 I/O 在内核缓冲区都没有可读数据，select 调用进程会被阻塞，而当任一 I/O 在内核缓冲区中有可读数据时，select 调用就会返回，而后 select 调用进程可以自己或通知另外的进程（注册进程）再次发起读取 I/O，读取内核中准备好的数据，多个进程注册 I/O 后，只有一个 select 调用进程被阻塞。

多路复用 I/O 相对阻塞和非阻塞更难简单说明，所以额外解释一段，其实多路复用 I/O 模型和阻塞 I/O 模型并没有太大的不同，事实上，还更差一些，因为这里需要使用两个系统调用（select 和 recvfrom），而阻塞 I/O 模型只有一次系统调用（recvfrom）。但是，用 Selector 的优势在于它可以同时处理多个连接，所以如果处理的连接数不是很多，使用 select/epoll 的 Web Server 不一定比使用多线程加阻塞 I/O 的 Web Server 性能更好，可能延迟还更大，select/epoll 的优势并不是对于单个连接能处理得更快，而是能处理更多的连接。多路复用 I/O 模型的特性总结如下表所示。

特　　点	对于每一个Socket，一般都设置成非阻塞，但是整个用户的进程其实是一直被阻塞的，只不过进程是被select函数阻塞，而不是被Socket I/O阻塞
典型应用	Java NIO，Nginx（epoll，poll，select）
优　　点	• 专一进程解决多个进程I/O的阻塞问题，性能好，Reactor模式 • 适合高并发服务应用开发，一个进程/线程响应多个请求
缺　　点	实现和开发应用难度较大

2.2.4　信号驱动 I/O 模型

信号驱动 I/O 模型的通信过程示意如下图所示。

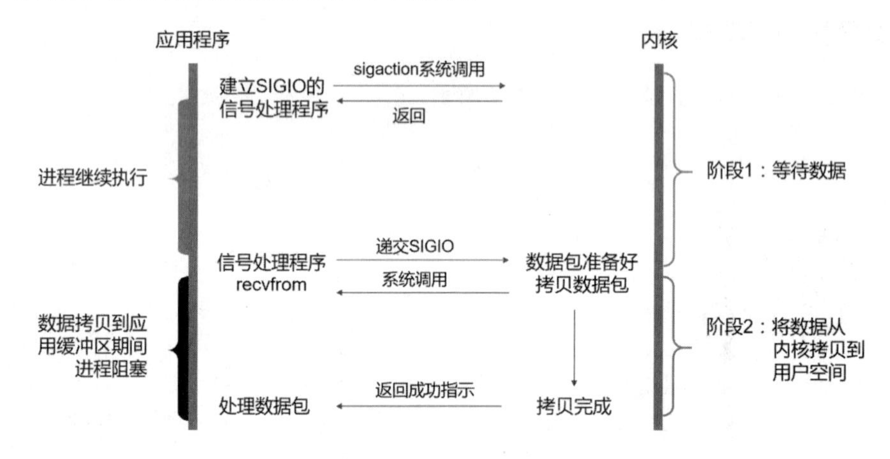

信号驱动 I/O 是指进程预先告知内核，向内核注册一个信号处理函数，然后用户进程返回不阻塞，当内核数据就绪时会发送一个信号给进程，用户进程便在信号处理函数中调用 I/O 读取数据。从上图可以看出，实际上 I/O 内核拷贝到用户进程的过程还是阻塞的，信号驱动 I/O 并没有实现真正的异步，因为通知到进程之后，依然由进程来完成 I/O 操作。这和后面的异步 I/O 模型很容易混淆，需要理解 I/O 交互并结合五种 I/O 模型进行比较阅读。信号驱动 I/O 模型的特性总结如下表所示。

特　　点	并不符合异步I/O要求，只能算是伪异步，并且实际中并不常用
典型应用	应用场景较少
优　　点	应用较少，不做详细总结
缺　　点	实现和开发应用难度大

2.2.5　异步 I/O 模型

异步 I/O 模型的通信过程示意如下图所示。

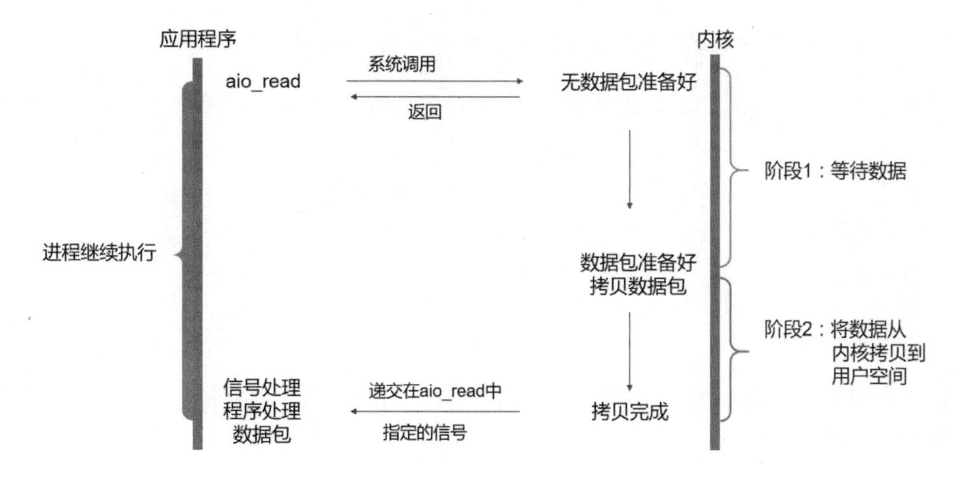

用户进程发起 aio_read 操作后，给内核传递与 read 相同的描述符、缓冲区指针、缓冲区大小三个参数及文件偏移，告诉内核当整个操作完成时，如何通知我们立刻就可以开始去做其他的事；而另一方面，从内核的角度，当它收到一个 aio_read 之后，首先它会立刻返回，所以不会对用户进程产生任何阻塞，内核会等待数据准备完成，然后将数据拷贝到用户内存，当这一切都完成之后，内核会给用户进程发送一个信号，告诉它 aio_read 操作完成。

异步 I/O 的工作机制是：告知内核启动某个操作，并让内核在整个操作完成后通知我们，这种模型与信号驱动 I/O 模型的区别在于，信号驱动 I/O 模型是由内核通知我们何时可以启动一个 I/O 操作，这个 I/O 操作由用户自定义的信号函数来实现，而异步 I/O 模型由内核告知我们 I/O 操作何时完成。

异步 I/O 模型的特性总结如下表所示。

特　点	真正实现了异步I/O，是五种I/O模型中唯一的异步模型
典型应用	Java 7 AIO，高性能服务器应用
优　点	• 不阻塞，数据一步到位，采用Proactor模式 • 非常适合高性能、高并发应用
缺　点	• 需要操作系统的底层支持，Linux 2.5内核首现，Linux 2.6产品的内核标准特性 • 实现和开发应用难度大

2.2.6　易混淆的概念澄清

在实际开发中，我们经常会听到同步、异步、阻塞、非阻塞这些概念，每次遇到的时候都会“蒙圈”，然后就查网上各种资料，结果越查越迷糊。大部分文章都千篇一律，没有说到本质上的区别，所以下次再碰到这些概念，印象还是比较模糊，尤其是在一些场景下觉得同步与阻塞、异步与非阻塞没什么区别，但其实这四个术语描述的还真不是一回事。

下面我们来慢慢探讨它们之间的区别与联系，在这之前，我们还会经常看到下面的组合术语。

（1）同步阻塞。

（2）同步非阻塞。

（3）异步阻塞。

（4）异步非阻塞。

在什么是同步和异步、阻塞和非阻塞的概念还没弄清楚之前，更别提上面这些组合术语了，只会让你更加困惑。

1. 同步和异步

同步和异步其实是指CPU时间片的利用，主要看请求发起方对消息结果的获取是主动发起的，还是被动通知的，如下图所示。如果是请求方主动发起的，一直在等待应答结果（同步阻塞），或者可以先去处理其他事情，但要不断轮询查看发起的请求是否有应答结果（同步非阻塞），因为不管如何都要发起方主动获取消息结果，所以形式上还是同步操作。如果是由服务方通知的，也就是请求方发出请求后，要么一直等待通知（异步阻塞），要么先去干自己的事（异步非阻塞）。当事情处理完成后，服务方会主动通知请求方，它的请求已经完成，这就是异步。异步通知的方式一般通过状态改变、消息通知或者回调函数来完成，大多数时候采用的都是回调函数。

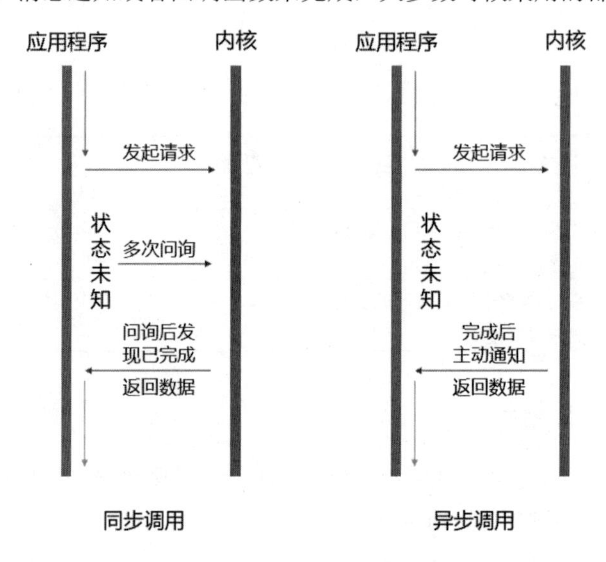

2. 阻塞和非阻塞

阻塞和非阻塞在计算机的世界里，通常指针对 I/O 的操作，如网络 I/O 和磁盘 I/O 等。那么什么是阻塞和非阻塞呢？简单地说，就是我们调用了一个函数后，在等待这个函数返回结果之前，当前的线程是处于挂起状态还是运行状态。如果是挂起状态，就意味着当前线程什么都不能干，就等着获取结果，这就是同步阻塞；如果仍然是运行状态，就意味着当前线程是可以继续处理其他任务的，但要时不时地看一下是否有结果了，这就是同步非阻塞。具体如下图所示。

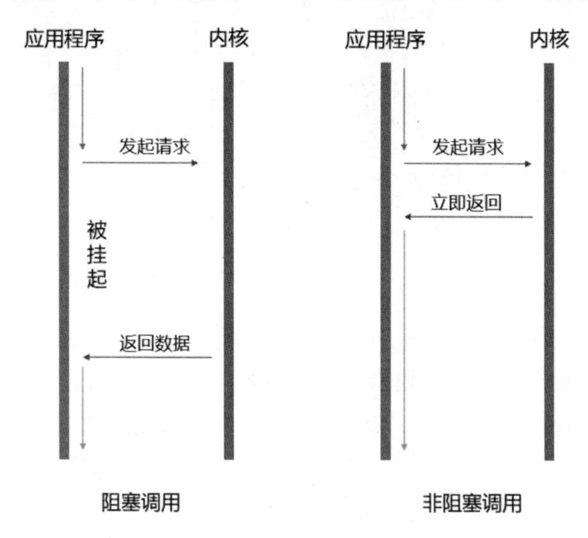

3. 实际生活场景

同步、异步、阻塞和非阻塞可以组合成上面提到过的四种结果。

举个例子，比如我们去照相馆拍照，拍完照片之后，商家说需要 30min 左右才能洗出来照片。

（1）这个时候，如果我们一直在店里面什么都不干，一直等待直到洗完照片，这个过程就叫同步阻塞。

（2）当然，大部分人很少这么干，更多的是大家拿起手机开始看电视，看一会儿就会问老板洗完没，老板说没洗完，然后接着看，再过一会儿接着问，直到照片洗完，这个过程就叫同步非阻塞。

（3）由于店里生意太好了，越来越多的人过来拍，店里面快没地方坐了，老板说你把手机号留下，我一会儿洗好了就打电话告诉你过来取，然后你去外面找了一个长凳开始躺着睡觉等待老板打电话，什么都不干，这个过程就叫异步阻塞（实际不应用）。

（4）当然实际情况是，大家可能会先去逛街或者吃饭，或者做其他活动，这样一来，两不耽误，这个过程就叫异步非阻塞（效率最高）。

4. 小结

从上面的描述中，我们能够看到阻塞和非阻塞通常是指在客户端发出请求后，在服务端处理这个请求的过程中，客户端本身是直接挂起等待结果，还是继续做其他的任务。而异步和同步则是对于请求结果的获取是客户端主动获取结果，还是由服务端来通知结果。从这一点来看，同步和阻塞其实描述的是两个不同角度的事情，阻塞和非阻塞指的是客户端等待消息处理时本身的状态，是挂起还是继续干别的。同步和异步指的是对于消息结果是客户端主动获取的，还是由服务端间接推送的。记住这两点关键的区别将有助于我们更好地区分和理解它们。

2.2.7 各 I/O 模型的对比与总结

其实前四种 I/O 模型都是同步 I/O 操作，它们的区别在于第一阶段，而第二阶段是一样的：在数据从内核拷贝到应用缓冲区期间（用户空间），进程阻塞于 recvfrom 调用。

有人可能会说，NIO（Non-Blocking I/O）并没有被阻塞。这里有个非常“狡猾”的地方，定义中所指的“I/O Operation”是指真实的 I/O 操作。NIO 在执行 recvfrom 的时候，如果内核（Kernel）的数据没有准备好，这时候不会阻塞进程。但是，当内核（Kernel）中数据准备好的时候，recvfrom 会将数据从内核（Kernel）拷贝到用户内存中，这个时候进程就被阻塞了。在这段时间内，进程是被阻塞的。下图是各 I/O 模型的阻塞状态对比。

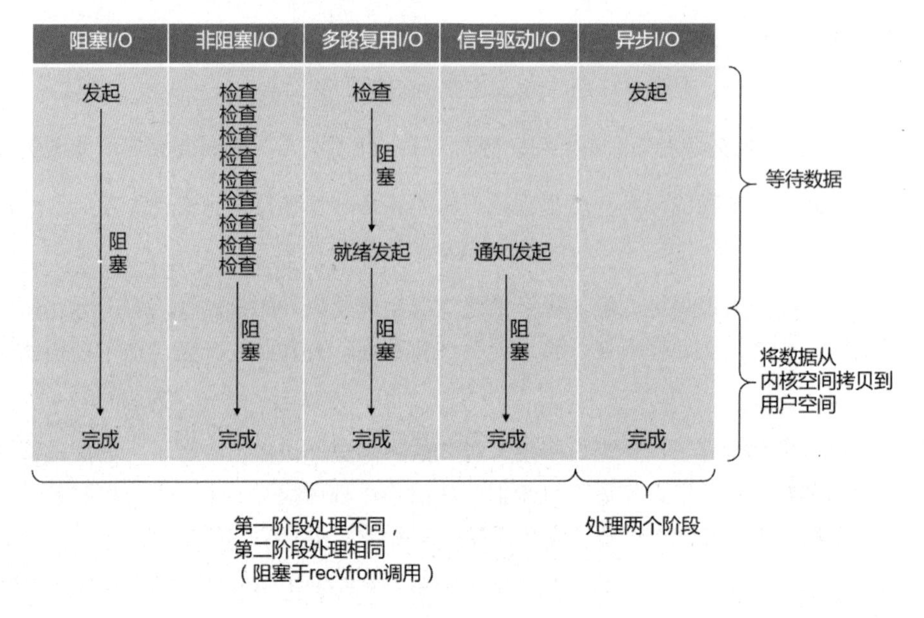

从上图可以看出，阻塞程度：阻塞 I/O>非阻塞 I/O>多路复用 I/O>信号驱动 I/O>异步 I/O，效率是由低到高的。最后，再看一下下表，从多维度总结了各 I/O 模型之间的差异，可以加深理解。

属　性	同步阻塞I/O	伪异步I/O	非阻塞I/O（NIO）	异步I/O
客户端数:I/O线程数	1:1	M:N（M>=N）	M:1	M:0
阻塞类型	阻塞	阻塞	非阻塞	非阻塞
同步	同步	同步	同步（多路复用）	异步
API使用难度	简单	简单	复杂	一般
调试难度	简单	简单	复杂	复杂
可靠性	非常差	差	高	高
吞吐量	低	中	高	高

2.3　从 BIO 到 NIO 的演进

下表总结了 Java BIO（Blocking I/O）和 NIO（Non-Blocking I/O）之间的主要差异。

I/O模型	BIO	NIO
通　信	面向流	面向缓冲区
处　理	阻塞I/O	非阻塞I/O
触　发	无	选择器

2.3.1　面向流与面向缓冲

Java NIO 和 BIO 之间第一个最大的区别是，BIO 是面向流的，NIO 是面向缓冲区的。Java BIO 面向流意味着每次从流中读一个或多个字节，直至读取所有字节，它们没有被缓存在任何地方。此外，不能前后移动流中的数据。如果需要前后移动从流中读取的数据，需要先将它缓存到一个缓冲区。Java NIO 的缓冲导向方法略有不同。数据读取到一个它稍后处理的缓冲区，需要时可在缓冲区中前后移动。这就增加了处理过程的灵活性。但是，还需要检查该缓冲区是否包含所有需要处理的数据。而且，要确保当更多的数据读入缓冲区时，不能覆盖缓冲区里尚未处理的数据。

2.3.2　阻塞与非阻塞

Java BIO 的各种流是阻塞的。这意味着，当一个线程调用 read()或 write()时，该线程被阻塞，直到有一些数据被读取，或数据完全写入。该线程在此期间不能再干任何事情。Java NIO 的非阻塞模式，是一个线程从某通道（Channel）发送请求读取数据，但是它仅能得到目前可用的数据，如果目前没有数据可用，就什么都不会获取，而不是保持线程阻塞，所以直到数据变成可以读取之前，该线程可以继续做其他的事情。非阻塞写也是如此。一个线程请求写入某通道一些数据，但不需要等待它完全写入，这个线程同时可以去做别的事情。线程通常将非阻塞 I/O 的空闲时间用于在其他通道上执行 I/O 操作，所以一个单独的线程现在可以管理多个 I/O 通道。

2.3.3 选择器在 I/O 中的应用

Java NIO 的选择器（Selector）允许一个单独的线程监视多个输入通道，可以注册多个通道使用一个选择器，然后使用一个单独的线程来“选择”通道：这些通道里已经有可以处理的输入，或者选择已准备写入的通道。这种选择机制使一个单独的线程很容易管理多个通道。

2.3.4 NIO 和 BIO 如何影响应用程序的设计

无论选择 BIO 还是 NIO 工具箱，都可能会影响应用程序设计的以下几个方面。

（1）对 NIO 或 BIO 类的 API 调用。

（2）数据处理逻辑。

（3）用来处理数据的线程数。

1. API 调用

当然，使用 NIO 的 API 调用看起来与使用 BIO 时有所不同，但这并不意外，因为并不是仅从一个 InputStream 逐字节读取，而是数据必须先读入缓冲区再处理。

2. 数据处理

使用纯粹的 NIO 设计相较 BIO 设计，数据处理也会受到影响。

在 BIO 设计中，我们从 InputStream 或 Reader 逐字节读取数据。假设你正在处理一个基于行的文本数据流，有如下一段文本。

```
Name:Tom
Age:18
Email: tom@qq.com
Phone:13888888888
```

该文本行的流可以这样处理。

```
FileInputStream input = new FileInputStream("d://info.txt");
BufferedReader reader = new BufferedReader(new InputStreamReader(input));
String nameLine = reader.readLine();
String ageLine = reader.readLine();
String emailLine = reader.readLine();
```

```
String phoneLine = reader.readLine();
```

请注意处理状态由程序执行多久决定。换句话说，一旦 reader.readLine()方法返回，你就知道文本行肯定已读完，readline()阻塞直到整行读完，这就是原因。你也知道此行包含名称；同样，第二个 readline()调用返回的时候，你知道这行包含年龄。正如你可以看到，该处理程序仅在有新数据读入时运行，并知道每步的数据是什么。一旦正在运行的线程已处理过读入的某些数据，该线程不会再回退数据（大多如此）。下图也说明了这条原则。

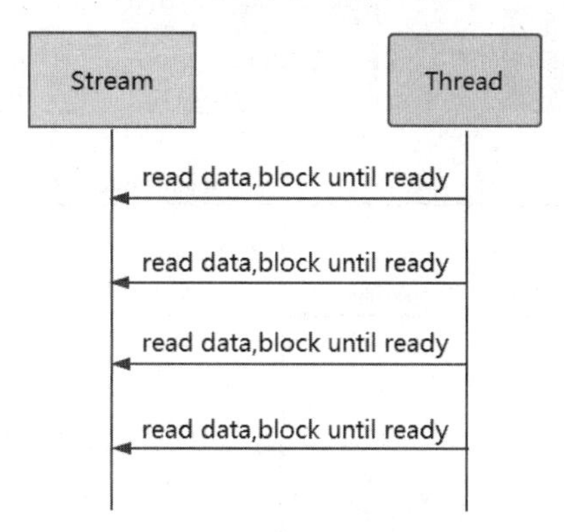

Java BIO 从一个阻塞的流中读数据，而一个 NIO 的实现会有所不同，下面是一个简单的例子。

```
ByteBuffer buffer = ByteBuffer.allocate(48);
int bytesRead = inChannel.read(buffer);
```

注意第二行，从通道读取字节到 ByteBuffer。当这个方法调用返回时，你不知道你所需的所有数据是否在缓冲区内。你所知道的是，该缓冲区包含一些字节，这使得处理有点困难。

假设第一次 read(buffer)调用后，读入缓冲区的数据只有半行，例如，“Name:An”，你能处理数据吗？显然不能，需要等待，直到整行数据读入缓存。在此之前，对数据的任何处理都毫无意义。

所以，你怎么知道是否该缓冲区包含足够的数据可以处理呢？好了，你不知道。发现的方法只能查看缓冲区中的数据。其结果是，在你知道所有数据都在缓冲区里之前，你必须检查几次缓冲区的数据。这不仅效率低下，而且会使程序设计方案杂乱不堪。例如：

```
ByteBuffer buffer = ByteBuffer.allocate(48);
    int bytesRead = inChannel.read(buffer);
    while(!bufferFull(bytesRead)) {
        bytesRead = inChannel.read(buffer);
    }
```

bufferFull()方法必须跟踪有多少数据读入缓冲区，并返回真或假，这取决于缓冲区是否已满。换句话说，如果缓冲区准备好被处理，那么表示缓冲区已满。

bufferFull()方法扫描缓冲区，但必须保持与 bufferFull()方法被调用之前状态相同。如果没有，下一个读入缓冲区的数据可能无法读到正确的位置。虽然这是不可能的，但却是需要注意的又一个问题。

如果缓冲区已满，它可以被处理。如果它不满，并且在实际案例中有意义，或许能处理其中的部分数据，但是许多情况下并非如此。下图展示了“缓冲区数据循环就绪”。

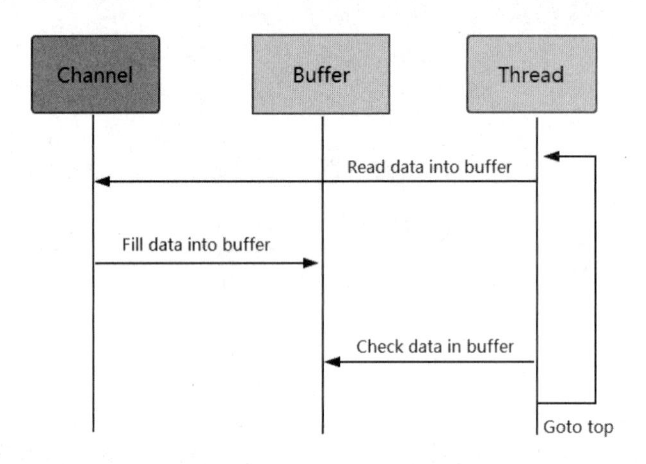

3. 设置处理线程数

NIO 可以只使用一个（或几个）单线程管理多个通道（网络连接或文件），但付出的代价是解析数据可能会比从一个阻塞流中读取数据更复杂。

如果需要管理同时打开的成千上万个连接，这些连接每次只是发送少量的数据，例如聊天服务器，实现 NIO 的服务器可能是一个优势。同样，如果需要维持许多打开的连接，如 P2P 网络中，使用一个单独的线程来管理所有出站连接，可能是一个优势。一个线程有多个连接的设计方案如下图所示。

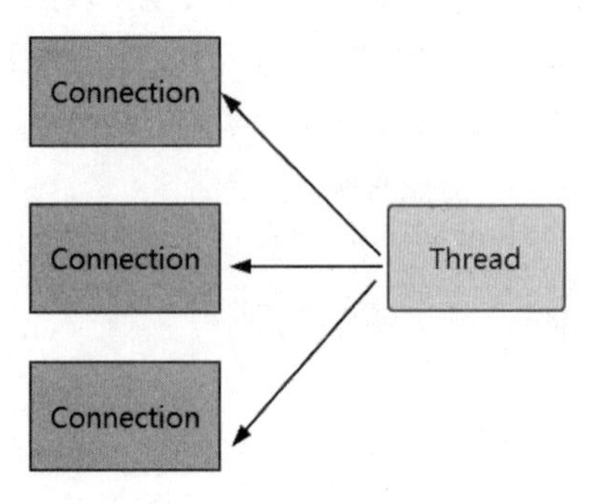

（1）Java NIO：单线程管理多个连接。如果有少量的连接使用非常高的带宽，一次发送大量的数据，也许用典型的 I/O 服务器实现可能非常契合。下图说明了一个典型的 I/O 服务器设计。

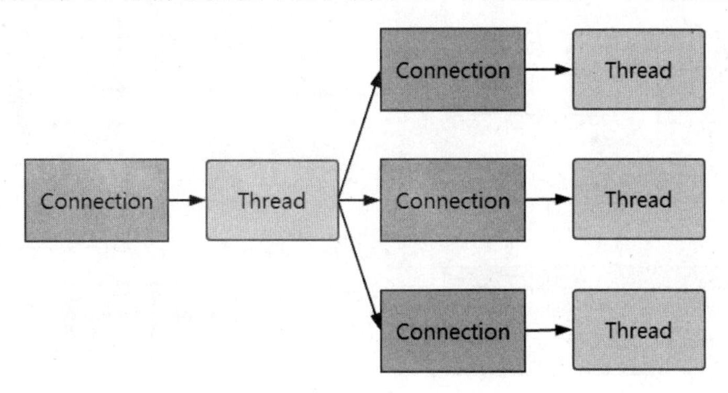

（2）Java BIO：一个典型的 I/O 服务器设计。一个连接只用一个线程来处理。

2.4　Java AIO 详解

JDK 1.7（NIO2）才是实现真正的异步 AIO(Asynchronous I/O)、把 I/O 读写操作完全交给操作系统，学习了 Linux Epoll 模式，下面我们来做一些演示。

2.4.1　AIO 基本原理

Java AIO 处理 API 中，重要的三个类分别是：AsynchronousServerSocketChannel（服务端）、AsynchronousSocketChannel（客户端）及 CompletionHandler（用户处理器）。CompletionHandler 接口实现应用程序向操作系统发起 I/O 请求，当完成后处理具体逻辑，否则做自己该做的事情，“真正”的异步 I/O 需要操作系统更强的支持。

在多路复用 I/O 模型中，事件循环将文件句柄的状态事件通知给用户线程，由用户线程自行读取数据、处理数据。而在异步 I/O 模型中，当用户线程收到通知时，数据已经被内核读取完毕，并放在了用户线程指定的缓冲区内，内核在 I/O 完成后通知用户线程直接使用即可。异步 I/O 模型使用 Proactor 设计模式实现这一机制，如下图所示。

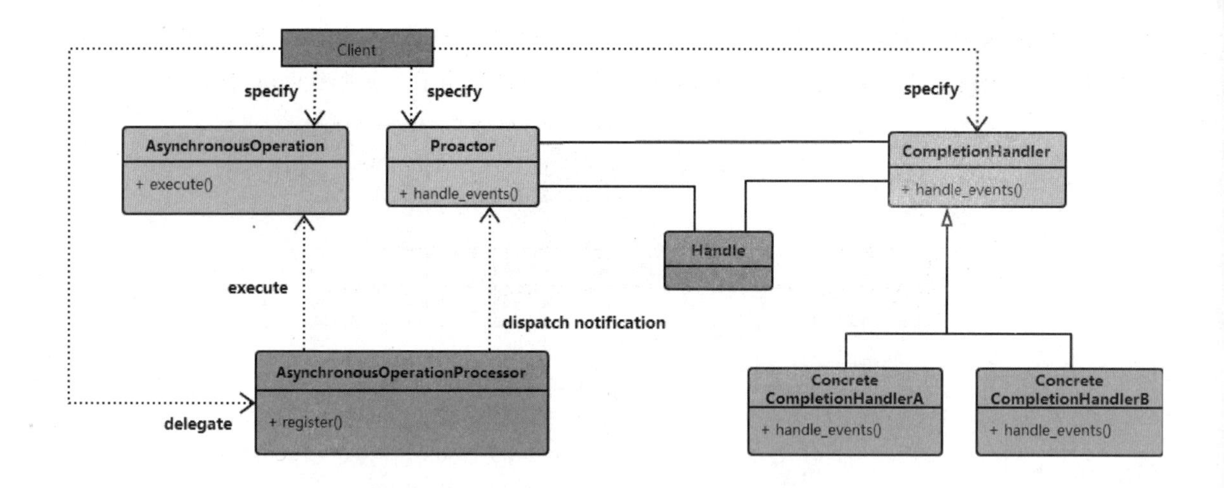

2.4.2 AIO 初体验

我们基于 AIO 先写一段简单的代码，来感受一下服务端和客户端的交互过程，同时也体验一下 API 的使用。先来看服务端代码。

```
package com.gupaoedu.vip.netty.io.aio;
import java.io.IOException;
import java.net.InetSocketAddress;
import java.nio.ByteBuffer;
import java.nio.channels.AsynchronousChannelGroup;
import java.nio.channels.AsynchronousServerSocketChannel;
import java.nio.channels.AsynchronousSocketChannel;
import java.nio.channels.CompletionHandler;
import java.util.concurrent.ExecutorService;
import java.util.concurrent.Executors;

/**
 * AIO 服务端
 */
public class AIOServer {

    private final int port;

    public static void main(String args[]) {
        int port = 8000;
        new AIOServer(port);
    }
```

```
    public AIOServer(int port) {
        this.port = port;
        listen();
    }

    private void listen() {
        try {
            ExecutorService executorService = Executors.newCachedThreadPool();
            AsynchronousChannelGroup threadGroup =
AsynchronousChannelGroup.withCachedThreadPool(executorService, 1);
            final AsynchronousServerSocketChannel server =
AsynchronousServerSocketChannel.open(threadGroup);
            server.bind(new InetSocketAddress(port));
            System.out.println("服务已启动，监听端口" + port);

            server.accept(null, new CompletionHandler<AsynchronousSocketChannel, Object>(){
                final ByteBuffer buffer = ByteBuffer.allocateDirect(1024);
                public void completed(AsynchronousSocketChannel result, Object attachment){
                    System.out.println("I/O 操作成功，开始获取数据");
                    try {
                        buffer.clear();
                        result.read(buffer).get();
                        buffer.flip();
                        result.write(buffer);
                        buffer.flip();
                    } catch (Exception e) {
                        System.out.println(e.toString());
                    } finally {
                        try {
                            result.close();
                            server.accept(null, this);
                        } catch (Exception e) {
                            System.out.println(e.toString());
                        }
                    }

                    System.out.println("操作完成");
                }

                @Override
                public void failed(Throwable exc, Object attachment) {
                    System.out.println("I/O 操作失败：" + exc);
                }
            });

            try {
```

```
                Thread.sleep(Integer.MAX_VALUE);
            } catch (InterruptedException ex) {
                System.out.println(ex);
            }
        } catch (IOException e) {
            System.out.println(e);
        }
    }
}
```

上述代码的主要功能就是开启一个监听端口，然后在 CompletionHandler 中处理接收到消息以后的逻辑，将接收到的信息再输出到客户端。下面来看客户端的代码。

```
package com.gupaoedu.vip.netty.io.aio;
import java.net.InetSocketAddress;
import java.nio.ByteBuffer;
import java.nio.channels.AsynchronousSocketChannel;
import java.nio.channels.CompletionHandler;

/**
 * AIO 客户端
 */
public class AIOClient {
    private final AsynchronousSocketChannel client;

    public AIOClient() throws Exception{
        client = AsynchronousSocketChannel.open();
    }

    public void connect(String host,int port)throws Exception{
        client.connect(new InetSocketAddress(host,port),null,new
CompletionHandler<Void,Void>() {
            @Override
            public void completed(Void result, Void attachment) {
                try {
                    client.write(ByteBuffer.wrap("这是一条测试数据".getBytes())).get();
                    System.out.println("已发送至服务器");
                } catch (Exception ex) {
                    ex.printStackTrace();
                }
            }

            @Override
            public void failed(Throwable exc, Void attachment) {
                exc.printStackTrace();
            }
```

```
        });
        final ByteBuffer bb = ByteBuffer.allocate(1024);
        client.read(bb, null, new CompletionHandler<Integer,Object>(){

                    @Override
                    public void completed(Integer result, Object attachment) {
                        System.out.println("I/O 操作完成" + result);
                        System.out.println("获取反馈结果" + new String(bb.array()));
                    }

                    @Override
                    public void failed(Throwable exc, Object attachment) {
                        exc.printStackTrace();
                    }
                }
        );

        try {
            Thread.sleep(Integer.MAX_VALUE);
        } catch (InterruptedException ex) {
            System.out.println(ex);
        }

    }

    public static void main(String args[])throws Exception{
        new AIOClient().connect("localhost",8000);
    }
}
```

客户端的代码的主要功能是发送一串字符到服务端。同时，在 CompletionHandler 接口处理服务端发送过来的结果。

服务端执行结果如下图所示。

客户端执行结果如下图所示。

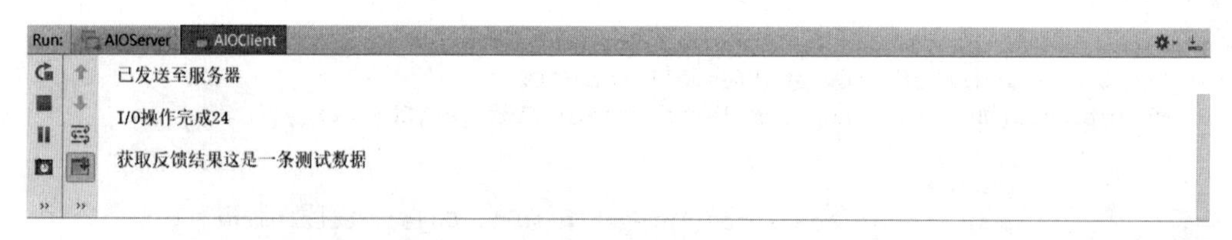

运行代码后，我们会发现不管是客户端还是服务端，其处理接收消息的逻辑都是异步操作，和 BIO、NIO 的 API 使用有根本上的区别。

2

第 2 篇

Netty 初体验

第 3 章 Netty 与 NIO 之前世今生

3.1 Java NIO 三件套

在 NIO 中有三个核心对象需要掌握：缓冲区（Buffer）、选择器（Selector）和通道（Channel）。

3.1.1 缓冲区

1. Buffer 操作基本 API

缓冲区实际上是一个容器对象，更直接地说，其实就是一个数组，在 NIO 库中，所有数据都是用缓冲区处理的。在读取数据时，它是直接读到缓冲区中的；在写入数据时，它也是写入缓冲区的；任何时候访问 NIO 中的数据，都是将它放到缓冲区中。而在面向流 I/O 系统中，所有数据都是直接写入或者直接将数据读取到 Stream 对象中。

在 NIO 中，所有的缓冲区类型都继承于抽象类 Buffer，最常用的就是 ByteBuffer，对于 Java 中的基本类型，基本都有一个具体 Buffer 类型与之相对应，它们之间的继承关系如下图所示。

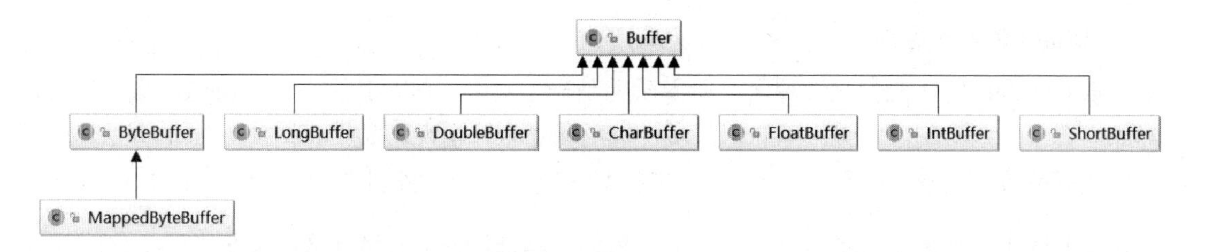

下面是一个简单的使用 IntBuffer 的例子。

```
package com.gupaoedu.vip.netty.io.nio.buffer;
import java.nio.IntBuffer;

public class IntBufferDemo {
    public static void main(String[] args) {
        //分配新的 int 缓冲区，参数为缓冲区容量
        //新缓冲区的当前位置将为 0，其界限（限制位置）为其容量。它具有一个底层实现数组，其数组偏移量
为 0
        IntBuffer buffer = IntBuffer.allocate(8);

        for (int i = 0; i < buffer.capacity(); ++i) {
            int j = 2 * (i + 1);
            //将给定整数写入此缓冲区的当前位置，当前位置递增
            buffer.put(j);
        }
        //重设此缓冲区，将限制位置设置为当前位置，然后将当前位置设置为 0
        buffer.flip();
        //查看在当前位置和限制位置之间是否有元素
        while (buffer.hasRemaining()) {
            //读取此缓冲区当前位置的整数，然后当前位置递增
            int j = buffer.get();
            System.out.print(j + "  ");
        }
    }
}
```

运行后可以看到如下图所示的结果。

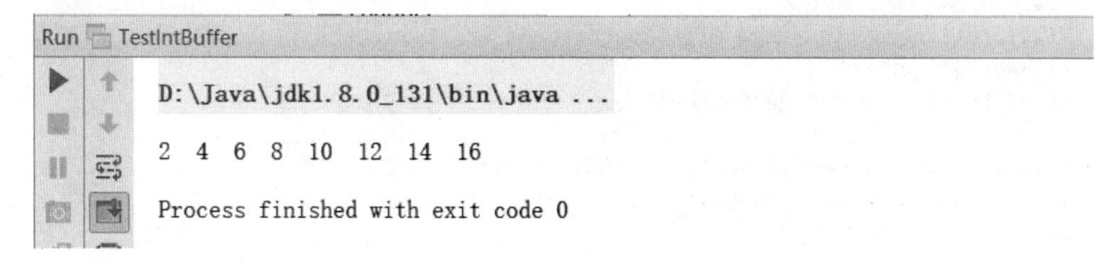

2. Buffer 的基本原理

在谈到缓冲区时，我们说缓冲区对象本质上是一个数组，但它其实是一个特殊的数组，缓冲区对象内置了一些机制，能够跟踪和记录缓冲区的状态变化情况，如果我们使用 get()方法从缓冲区获取数据或者使用 put()方法把数据写入缓冲区，都会引起缓冲区状态的变化。

在缓冲区中，最重要的属性有下面三个，它们一起合作完成对缓冲区内部状态的变化跟踪。

- position：指定下一个将要被写入或者读取的元素索引，它的值由 get()/put()方法自动更新，在新创建一个 Buffer 对象时，position 被初始化为 0。
- limit：指定还有多少数据需要取出（在从缓冲区写入通道时），或者还有多少空间可以放入数据（在从通道读入缓冲区时）。
- capacity：指定了可以存储在缓冲区中的最大数据容量，实际上，它指定了底层数组的大小，或者至少是指定了准许我们使用的底层数组的容量。

以上三个属性值之间有一些相对大小的关系：0 <= position <= limit <= capacity。如果我们创建一个新的容量大小为 10 的 ByteBuffer 对象，在初始化的时候，position 设置为 0，limit 和 capacity 设置为 10，在以后使用 ByteBuffer 对象过程中，capacity 的值不会再发生变化，而其他两个将会随着使用而变化。

准备一个 txt 文档，存放在 E 盘，输入以下内容。

```
Tom.
```

我们用一段代码来验证 position、limit 和 capacity 这三个值的变化过程，代码如下。

```java
package com.gupaoedu.vip.netty.io.nio.buffer;
import java.io.FileInputStream;
import java.nio.*;
import java.nio.channels.*;

public class BufferDemo {
    public static void main(String args[]) throws Exception {
        //这里用的是文件 I/O 处理
        FileInputStream fin = new FileInputStream("E://test.txt");
        //创建文件的操作管道
        FileChannel fc = fin.getChannel();

        //分配一个 10 个大小的缓冲区，其实就是分配一个 10 个大小的 Byte 数组
        ByteBuffer buffer = ByteBuffer.allocate(10);
        output("初始化", buffer);
```

```
        //先读一下
        fc.read(buffer);
        output("调用 read()", buffer);

        //准备操作之前，先锁定操作范围
        buffer.flip();
        output("调用 flip()", buffer);

        //判断有没有可读数据
        while (buffer.remaining() > 0) {
            byte b = buffer.get();
            //System.out.print(((char)b));
        }
        output("调用 get()", buffer);

        //可以理解为解锁
        buffer.clear();
        output("调用 clear()", buffer);

        //最后把管道关闭
        fin.close();
    }

    //把这个缓冲区里的实时状态打印出来
    public static void output(String step, Buffer buffer) {
        System.out.println(step + " : ");
        //容量，数组大小
        System.out.print("capacity: " + buffer.capacity() + ", ");
        //当前操作数据所在的位置，也可以叫作游标
        System.out.print("position: " + buffer.position() + ", ");
        //锁定值，flip，数据操作范围索引只能在 position - limit 之间
        System.out.println("limit: " + buffer.limit());
        System.out.println();
    }
}
```

完成后的输出结果如下图所示。

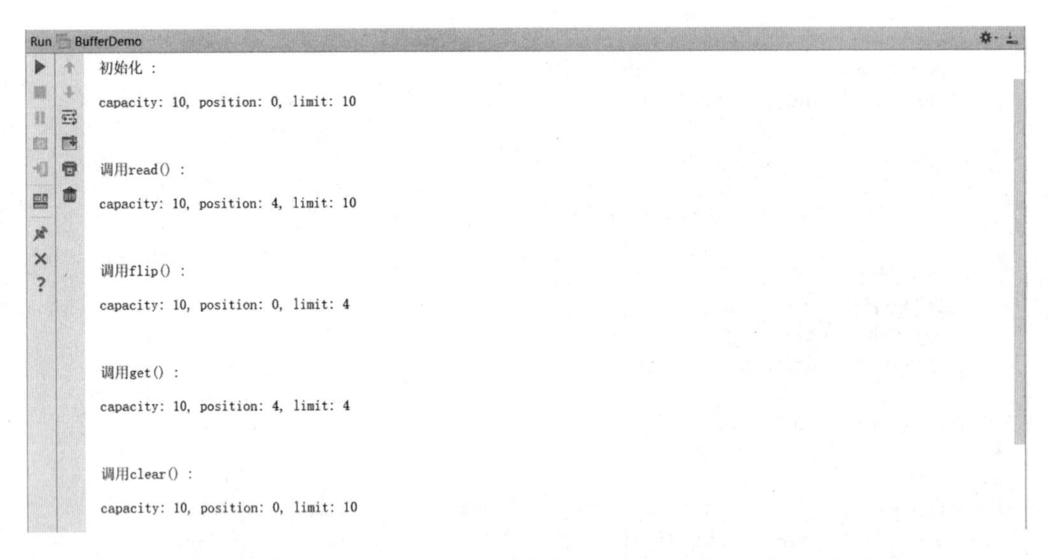

我们已经看到运行结果，下面对以上结果进行图解，三个属性值分别如下图所示。

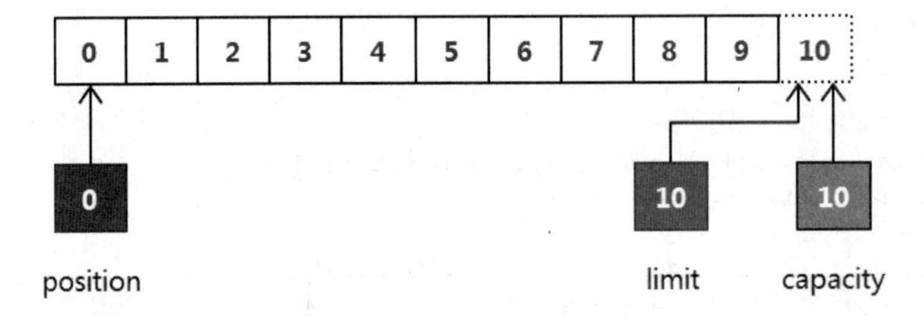

我们可以从通道中读取一些数据到缓冲区中，注意从通道读取数据，相当于往缓冲区写入数据。如果读取 4 个自己的数据，则此时 position 的值为 4，即下一个将要被写入的字节索引为 4，而 limit 仍然是 10，如下图所示。

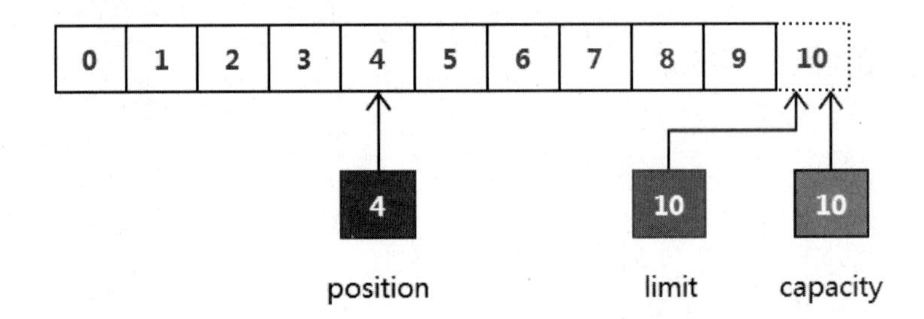

下一步把读取的数据写入输出通道，相当于从缓冲区中读取数据，在此之前，必须调用 flip() 方法。该方法将会完成以下两件事情：一是把 limit 设置为当前的 position 值。二是把 position 设置为 0。

由于 position 被设置为 0，所以可以保证在下一步输出时读取的是缓冲区的第一个字节，而 limit 被设置为当前的 position，可以保证读取的数据正好是之前写入缓冲区的数据，如下图所示。

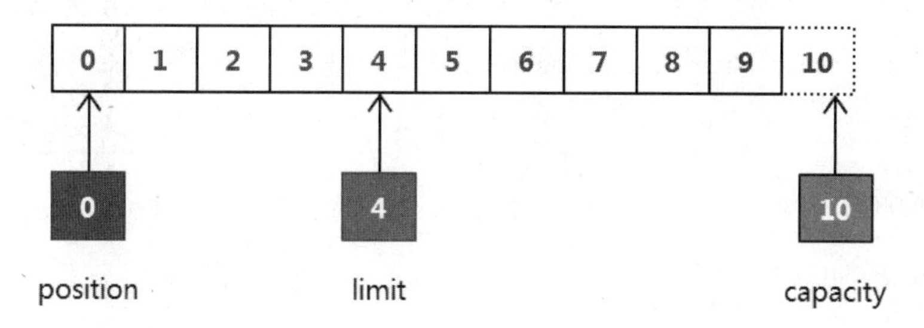

现在调用 get()方法从缓冲区中读取数据写入输出通道，这会导致 position 的增加而 limit 保持不变，但 position 不会超过 limit 的值，所以在读取之前写入缓冲区的 4 字节之后，position 和 limit 的值都为 4，如下图所示。

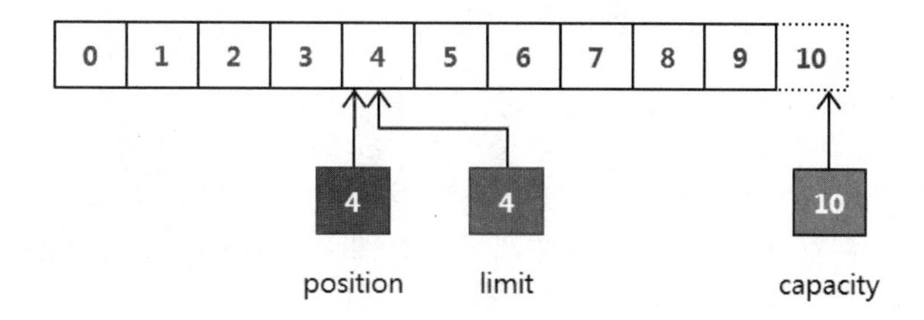

在从缓冲区中读取数据完毕后，limit 的值仍然保持在调用 flip()方法时的值，调用 clear()方法能够把所有的状态变化设置为初始化时的值，如下图所示。

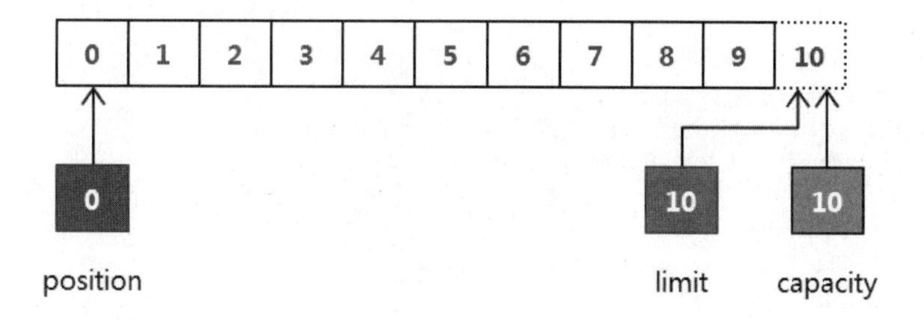

3. 缓冲区的分配

在前面的几个例子中，我们已经看到，在创建一个缓冲区对象时，会调用静态方法 allocate()来指定缓冲区的容量，其实调用 allocate()方法相当于创建了一个指定大小的数组，并把它包装为缓冲区对象。或者我们也可以直接将一个现有的数组包装为缓冲区对象，示例代码如下。

```
package com.gupaoedu.vip.netty.io.nio.buffer;
import java.nio.ByteBuffer;

/** 手动分配缓冲区 */
public class BufferWrap {

    public void myMethod() {
        //分配指定大小的缓冲区
        ByteBuffer buffer1 = ByteBuffer.allocate(10);

        //包装一个现有的数组
        byte array[] = new byte[10];
        ByteBuffer buffer2 = ByteBuffer.wrap( array );
    }
}
```

4. 缓冲区分片

在 NIO 中，除了可以分配或者包装一个缓冲区对象，还可以根据现有的缓冲区对象创建一个子缓冲区，即在现有缓冲区上切出一片作为一个新的缓冲区，但现有的缓冲区与创建的子缓冲区在底层数组层面上是数据共享的，也就是说，子缓冲区相当于现有缓冲区的一个视图窗口。调用 slice()方法可以创建一个子缓冲区，下面我们通过例子来看一下。

```
package com.gupaoedu.vip.netty.io.nio.buffer;
import java.nio.ByteBuffer;

/**
 * 缓冲区分片
 */
public class BufferSlice {
    static public void main( String args[] ) throws Exception {
        ByteBuffer buffer = ByteBuffer.allocate( 10 );

        //缓冲区中的数据 0~9
        for (int i=0; i<buffer.capacity(); ++i) {
            buffer.put( (byte)i );
        }

        //创建子缓冲区
        buffer.position( 3 );
```

```
        buffer.limit( 7 );
        ByteBuffer slice = buffer.slice();

        //改变子缓冲区的内容
        for (int i=0; i<slice.capacity(); ++i) {
            byte b = slice.get( i );
            b *= 10;
            slice.put( i, b );
        }

        buffer.position( 0 );
        buffer.limit( buffer.capacity() );

        while (buffer.remaining()>0) {
            System.out.println( buffer.get() );
        }
    }
}
```

在该示例中，分配了一个容量大小为 10 的缓冲区，并在其中放入了数据 0~9，而在该缓冲区基础上又创建了一个子缓冲区，并改变子缓冲区中的内容，从最后输出的结果来看，只有子缓冲区“可见的”那部分数据发生了变化，并且说明子缓冲区与原缓冲区是数据共享的，输出结果如下图所示。

5. 只读缓冲区

只读缓冲区非常简单，可以读取它们，但是不能向它们写入数据。可以通过调用缓冲区的 asReadOnlyBuffer()方法，将任何常规缓冲区转换为只读缓冲区，这个方法返回一个与原缓冲区完全相同的缓冲区，并与原缓冲区共享数据，只不过它是只读的。如果原缓冲区的内容发生了变化，只读缓冲区的内容也随之发生变化。具体代码如下。

```java
package com.gupaoedu.vip.netty.io.nio.buffer;
import java.nio.*;

/** 只读缓冲区 */
public class ReadOnlyBuffer {
    static public void main( String args[] ) throws Exception {
        ByteBuffer buffer = ByteBuffer.allocate( 10 );

        //缓冲区中的数据 0~9
        for (int i=0; i<buffer.capacity(); ++i) {
            buffer.put( (byte)i );
        }

        //创建只读缓冲区
        ByteBuffer readonly = buffer.asReadOnlyBuffer();

        //改变原缓冲区的内容
        for (int i=0; i<buffer.capacity(); ++i) {
            byte b = buffer.get( i );
            b *= 10;
            buffer.put( i, b );
        }

        readonly.position(0);
        readonly.limit(buffer.capacity());

        //只读缓冲区的内容也随之改变
        while (readonly.remaining()>0) {
            System.out.println( readonly.get());
        }
    }
}
```

如果尝试修改只读缓冲区的内容，则会报 ReadOnlyBufferException 异常。只读缓冲区对于保护数据很有用。在将缓冲区传递给某个对象的方法时，无法知道这个方法是否会修改缓冲区中的数据。创建一个只读缓冲区可以保证该缓冲区不会被修改。只可以把常规缓冲区转换为只读缓冲区，而不能将只读缓冲区转换为可写的缓冲区。

6. 直接缓冲区

直接缓冲区是为加快 I/O 速度，使用一种特殊方式为其分配内存的缓冲区，JDK 文档中的描述为：给定一个直接字节缓冲区，Java 虚拟机将尽最大努力直接对它执行本机 I/O 操作。也就是说，它会在每一次调用底层操作系统的本机 I/O 操作之前（或之后），尝试避免将缓冲区的内容拷贝到一个中间缓冲区或者从一个中间缓冲区拷贝数据。要分配直接缓冲区，需要调用

allocateDirect()方法，而不是 allocate()方法，使用方式与普通缓冲区并无区别，如下面的文件所示。

```
package com.gupaoedu.vip.netty.io.nio.buffer;
import java.io.*;
import java.nio.*;
import java.nio.channels.*;
/**
 * 直接缓冲区
  */
public class DirectBuffer {
    static public void main( String args[] ) throws Exception {

     //首先从磁盘上读取之前写出的文件内容
        String infile = "E://test.txt";
        FileInputStream fin = new FileInputStream( infile );
        FileChannel fcin = fin.getChannel();

        //把读取的内容写入一个新的文件
        String outfile = String.format("E://testcopy.txt");
        FileOutputStream fout = new FileOutputStream( outfile );
        FileChannel fcout = fout.getChannel();

        //使用 allocateDirect，而不是 allocate
        ByteBuffer buffer = ByteBuffer.allocateDirect(1024);

        while (true) {
            buffer.clear();

            int r = fcin.read(buffer);

            if (r==-1) {
                break;
            }

            buffer.flip();

            fcout.write(buffer);
        }
    }
}
```

7. 内存映射

内存映射是一种读和写文件数据的方法，可以比常规的基于流或者基于通道的 I/O 快得多。内存映射文件 I/O 通过使文件中的数据表现为内存数组的内容来完成，这初听起来似乎不过就是将整个文件读到内存中，但事实上并不是这样的。一般来说，只有文件中实际读取或写入的部分才会映射到内存中。来看下面的示例代码。

```
package com.gupaoedu.vip.netty.io.nio.buffer;
import java.io.*;
import java.nio.*;
import java.nio.channels.*;

/**
 * I/O 映射缓冲区
 */
public class MappedBuffer {
    static private final int start = 0;
    static private final int size = 1024;

    static public void main( String args[] ) throws Exception {
        RandomAccessFile raf = new RandomAccessFile( "E://test.txt", "rw" );
        FileChannel fc = raf.getChannel();

        //把缓冲区跟文件系统进行一个映射关联
        //只要操作缓冲区里面的内容，文件内容也会跟着改变
        MappedByteBuffer mbb = fc.map( FileChannel.MapMode.READ_WRITE,start, size );

        mbb.put( 0, (byte)97 );
        mbb.put( 1023, (byte)122 );

        raf.close();
    }
}
```

3.1.2 选择器

传统的 Client/Server 模式会基于 TPR（Thread per Request），服务器会为每个客户端请求建立一个线程，由该线程单独负责处理一个客户请求。这种模式带来的一个问题就是线程数量的剧增，大量的线程会增大服务器的开销。大多数的实现为了避免这个问题，都采用了线程池模型，并设置线程池中线程的最大数量，这又带来了新的问题，如果线程池中有 200 个线程，而有 200 个用户都在进行大文件下载，会导致第 201 个用户的请求无法及时处理，即便第 201 个用户只想请求一个几 KB 大小的页面。传统的 Client/Server 模式如下图所示。

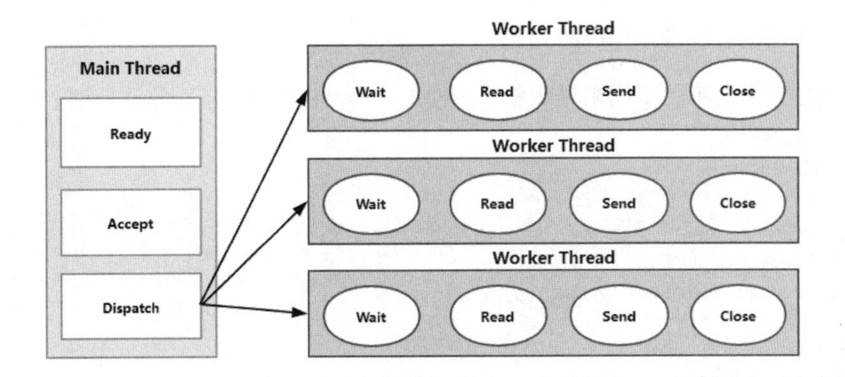

NIO 中非阻塞 I/O 采用了基于 Reactor 模式的工作方式，I/O 调用不会被阻塞，而是注册感兴趣的特定 I/O 事件，如可读数据到达、新的套接字连接等，在发生特定事件时，系统再通知我们。NIO 中实现非阻塞 I/O 的核心对象是 Selector，Selector 是注册各种 I/O 事件的地方，而且当那些事件发生时，就是 Seleetor 告诉我们所发生的事件，如下图所示。

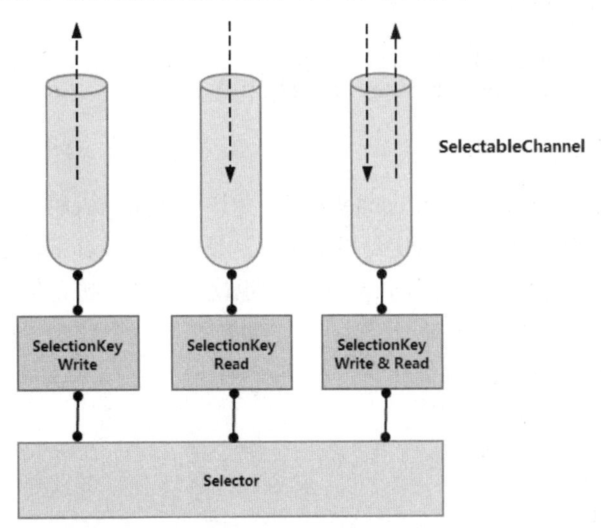

从图中可以看出，当有读或写等任何注册的事件发生时，可以从 Selector 中获得相应的 SelectionKey，同时从 SelectionKey 中可以找到发生的事件和该事件所发生的具体的 SelectableChannel，以获得客户端发送过来的数据。

使用 NIO 中非阻塞 I/O 编写服务器处理程序，大体上可以分为下面三个步骤。

（1）向 Selector 对象注册感兴趣的事件。

（2）从 Selector 中获取感兴趣的事件。

（3）根据不同的事件进行相应的处理。

下面我们用一个简单的示例来说明整个过程。首先是向 Selector 对象注册感兴趣的事件。

```
/*
     * 注册事件
     */
    private Selector getSelector() throws IOException {
        //创建 Selector 对象
        Selector sel = Selector.open();

        //创建可选择通道，并配置为非阻塞模式
        ServerSocketChannel server = ServerSocketChannel.open();
        server.configureBlocking(false);

        //绑定通道到指定端口
        ServerSocket socket = server.socket();
        InetSocketAddress address = new InetSocketAddress(port);
        socket.bind(address);

        //向 Selector 注册感兴趣的事件
        server.register(sel, SelectionKey.OP_ACCEPT);
        return sel;
    }
```

上述代码中先创建了 ServerSocketChannel 对象，并调用 configureBlocking()方法，配置为非阻塞模式。接下来的三行代码把该通道绑定到指定端口，最后向 Selector 注册事件。此处指定的参数是 OP_ACCEPT，即指定想要监听 accept 事件，也就是新的连接发生时所产生的事件。对于 ServerSocketChannel 通道来说，我们唯一可以指定的参数就是 OP_ACCEPT。从 Selector 中获取感兴趣的事件，即开始监听，进入内部循环。

```
/*
* 开始监听
*/
public void listen() {
    System.out.println("listen on " + port);
    try {
        while(true) {
            //该调用会阻塞，直到至少有一个事件发生
            selector.select();
            Set<SelectionKey> keys = selector.selectedKeys();
            Iterator<SelectionKey> iter = keys.iterator();
            while (iter.hasNext()) {
                SelectionKey key = (SelectionKey) iter.next();
                iter.remove();
                process(key);
```

```
            }
        }
    } catch (IOException e) {
        e.printStackTrace();
    }
}
```

在非阻塞 I/O 中，内部循环模式基本都遵循这种方式。首先调用 select()方法，该方法会阻塞，直到至少有一个事件发生，然后使用 selectedKeys()方法获取发生事件的 SelectionKey，再使用迭代器进行循环。

最后一步就是根据不同的事件，编写相应的处理代码。

```
/*
 * 根据不同的事件做处理
 */
private void process(SelectionKey key) throws IOException{

    //接受请求
    if (key.isAcceptable()) {
        ServerSocketChannel server = (ServerSocketChannel) key.channel();
        SocketChannel channel = server.accept();
        channel.configureBlocking(false);
        channel.register(selector, SelectionKey.OP_READ);
}

    //读数据
    else if (key.isReadable()) {
        SocketChannel channel = (SocketChannel) key.channel();
        int len = channel.read(buffer);
        if (len > 0) {
            buffer.flip();
            content = new String(buffer.array(),0,len);
            SelectionKey sKey = channel.register(selector, SelectionKey.OP_WRITE);
            sKey.attach(content);
        } else {
            channel.close();
        }
        buffer.clear();
}

    //写事件
    else if (key.isWritable()) {
        SocketChannel channel = (SocketChannel) key.channel();
        String content = (String) key.attachment();
        ByteBuffer block = ByteBuffer.wrap(("输出内容：" + content).getBytes());
```

```
        if(block != null){
            channel.write(block);
        }else{
            channel.close();
        }
    }
}
```

此处判断是接受请求、读数据还是写事件，分别做不同的处理。在 Java 1.4 之前的 I/O 系统中，提供的都是面向流的 I/O 系统，系统一次一个字节地处理数据，一个输入流产生一个字节的数据，一个输出流消费一个字节的数据，面向流的 I/O 速度非常慢；而在 Java 1.4 中推出了 NIO，这是一个面向块的 I/O 系统，系统以块的方式处理数据，每一个操作在一步中都产生或者消费一个数据库，按块处理数据要比按字节处理数据快得多。

3.1.3 通道

通道是一个对象，通过它可以读取和写入数据，当然所有数据都通过 Buffer 对象来处理。我们永远不会将字节直接写入通道，而是将数据写入包含一个或者多个字节的缓冲区。同样也不会直接从通道中读取字节，而是将数据从通道读入缓冲区，再从缓冲区获取这个字节。

NIO 提供了多种通道对象，所有的通道对象都实现了 Channel 接口。它们之间的继承关系如下图所示。

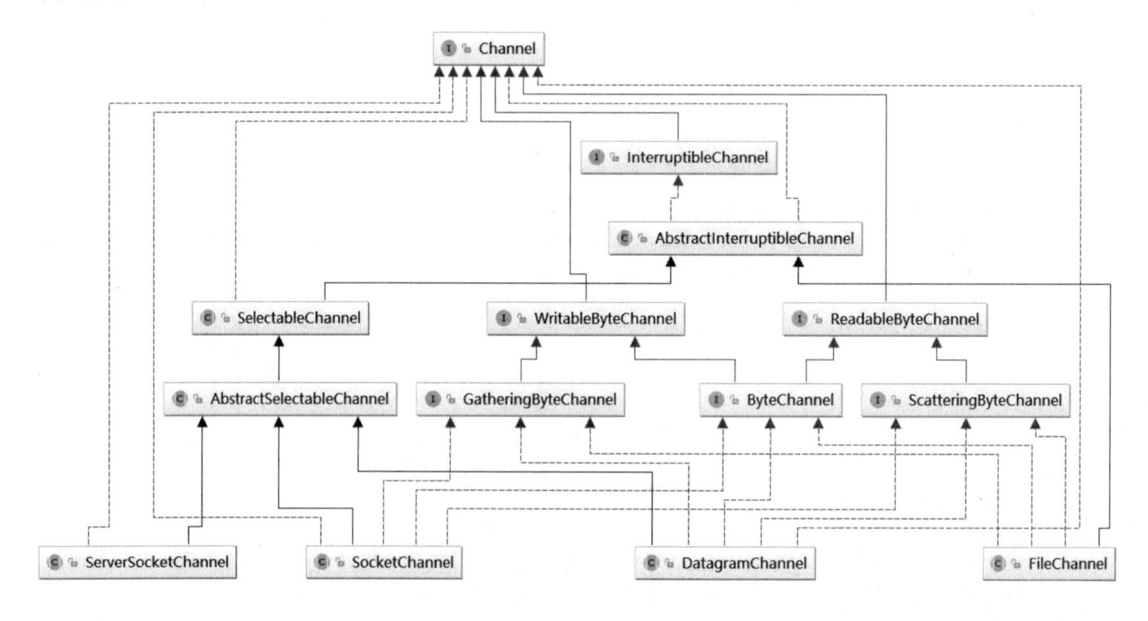

1. 使用 NIO 读取数据

前面我们说过，任何时候读取数据，都不是直接从通道读取的，而是从通道读取到缓冲区的。所以使用 NIO 读取数据可以分为下面三个步骤。

（1）从 FileInputStream 获取 Channel。

（2）创建 Buffer。

（3）将数据从 Channel 读取到 Buffer 中。

下面是一个简单的使用 NIO 从文件中读取数据的例子。

```
package com.gupaoedu.vip.netty.io.nio.channel;
import java.io.*;
import java.nio.*;
import java.nio.channels.*;

public class FileOutputDemo {
    static private final byte message[] = { 83, 111, 109, 101, 32, 98, 121, 116, 101, 115, 46 };

    static public void main( String args[] ) throws Exception {
        FileOutputStream fout = new FileOutputStream( "E://test.txt" );

        FileChannel fc = fout.getChannel();

        ByteBuffer buffer = ByteBuffer.allocate( 1024 );

        for (int i=0; i<message.length; ++i) {
            buffer.put( message[i] );
        }

        buffer.flip();

        fc.write( buffer );

        fout.close();
    }
}
```

2. 使用 NIO 写入数据

使用 NIO 写入数据与读取数据的过程类似，数据同样不是直接写入通道的，而是写入缓冲区，可以分为下面三个步骤。

（1）从 FileInputStream 获取 Channel。

（2）创建 Buffer。

（3）将数据从 Channel 写入 Buffer。

```
public class FileInputDemo {
    static public void main( String args[] ) throws Exception {
        FileInputStream fin = new FileInputStream("E://test.txt");

        //从 FileInputStream 获取 Channel
        FileChannel fc = fin.getChannel();

        //创建 Buffer
        ByteBuffer buffer = ByteBuffer.allocate(1024);

        //将数据从 Channel 写入 Buffer
        fc.read(buffer);

        buffer.flip();

        while (buffer.remaining() > 0) {
            byte b = buffer.get();
            System.out.print(((char)b));
        }

        fin.close();
    }
}
```

下面是一个简单的使用 NIO 向文件中写入数据的例子。

```
package com.gupaoedu.vip.netty.io.nio.channel;
import java.io.*;
import java.nio.*;
import java.nio.channels.*;

public class FileOutputDemo {
    static private final byte message[] = { 83, 111, 109, 101, 32,
        98, 121, 116, 101, 115, 46 };

    static public void main( String args[] ) throws Exception {
        FileOutputStream fout = new FileOutputStream( "E://test.txt" );

        FileChannel fc = fout.getChannel();

        ByteBuffer buffer = ByteBuffer.allocate( 1024 );

        for (int i=0; i<message.length; ++i) {
            buffer.put( message[i] );
        }
```

```
        buffer.flip();
        fc.write( buffer );
        fout.close();
    }
}
```

3. 多路复用 I/O

我们试想一下这样的现实场景。

一个餐厅同时有 100 位客人到店，到店后要做的第一件事情就是点菜。但是问题来了，餐厅老板为了节约人力成本，目前只有一名大堂服务员拿着唯一的一本菜单给客人提供服务。

那么最笨（但是最简单）的方法（方法 A）是，无论有多少客人等待点餐，服务员都把仅有的一份菜单递给其中一位客人，然后站在客人身旁等待这位客人完成点菜过程。在记录客人的点菜内容后，把点菜记录交给后堂厨师。然后是第二位客人、第三位客人……很明显，没有老板会这样设置服务流程。因为随后的 80 位客人，在等待超时后就会离店（还会给差评）。

于是还有一种办法（方法 B），老板马上新雇佣 99 名服务员，同时印制 99 本新的菜单。每一名服务员手持一本菜单负责一位客人（关键不只在于服务员，还在于菜单。因为没有菜单，客人也无法点菜）。在客人点完菜后，服务员记录点菜内容交给后堂厨师（当然为了更高效，后堂厨师最好也有 100 名），如下图所示。这样每一位客人享受的都是 VIP 服务，当然客人也不会走，但是人力成本却很高（必亏无疑）。

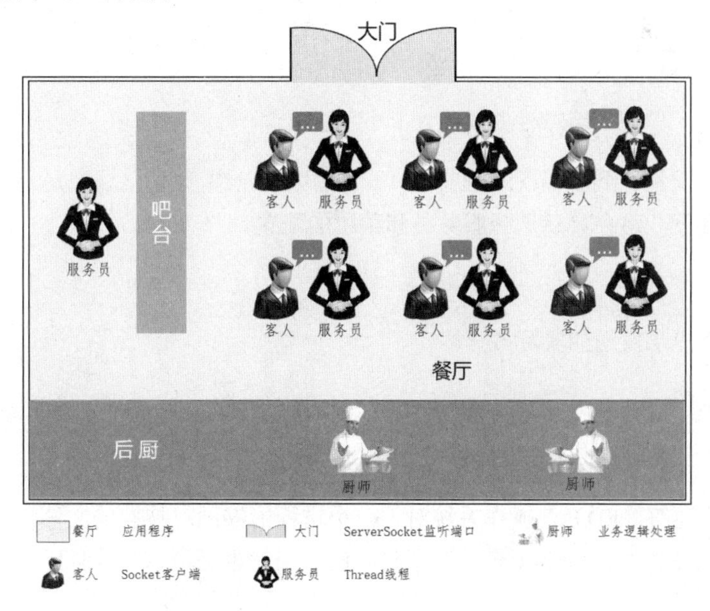

另外一种办法（方法 C），就是改进点菜的方式，当客人到店后，自己申请一本菜单。想好自己要点的菜后，就呼叫服务员。服务员站在自己身边记录客人的菜单内容。将菜单递给厨师的过程也要进行改进，并不是每一份菜单记录好以后，都要交给后堂厨师。服务员可以记录好多份菜单后，同时交给厨师就行了，如下图所示。那么这种方式，对于老板来说人力成本是最低的；对于客人来说，虽然不再享受 VIP 服务，并且要进行一段时间的等待，但是这些都是可以接受的；对于服务员来说，基本上她的时间都没有浪费，最大程度地提高了时间利用率。

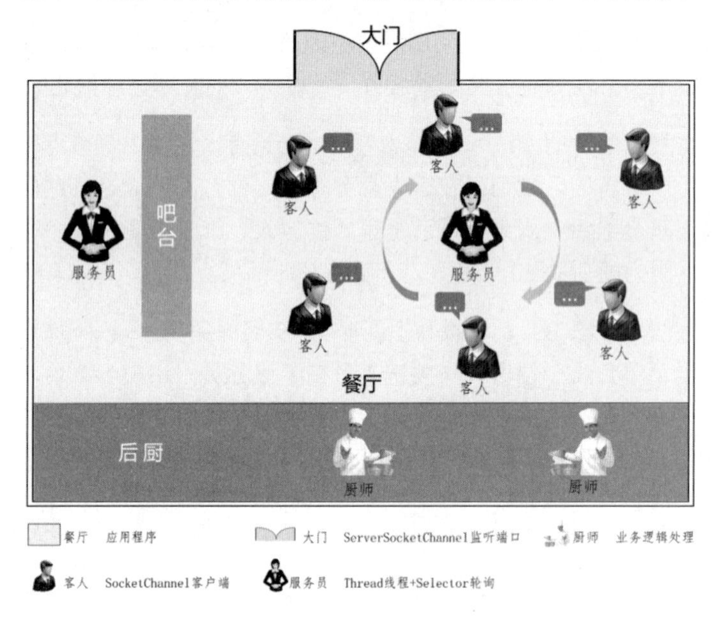

如果你是老板，你会采用哪种方式呢？

到店情况：并发量。到店情况不理想时，一个服务员一本菜单，当然是足够的。所以不同的老板在不同的场合下，将会灵活选择服务员和菜单的配置。

客人：客户端请求。

点餐内容：客户端发送的实际数据。

老板：操作系统。

人力成本：系统资源。

菜单：文件描述符（FD）。操作系统对于一个进程能够同时持有的文件描述符的个数是有限制的，在 Linux 系统中用$ulimit-n 查看这个限制值，当然也是可以（并且应该）进行内核参数调整的。

服务员：操作系统内核用于 I/O 操作的线程（内核线程）。

厨师：应用程序线程（当然厨房就是应用程序进程）。

方法 A：同步 I/O。

方法 B：同步 I/O。

方法 C：多路复用 I/O。

目前流行的多路复用 I/O 的实现主要包括四种：select、poll、epoll、kqueue。如下表所示是它们的一些重要特性的比较。

复用模型	相对性能	关键思路	操作系统	Java支持
select	较高	Reactor	Windows/Linux	支持，Reactor模式（反应器设计模式）。Linux Kernel 2.4之前，默认使用select；而目前Windows下对同步I/O的支持，都是select模型
poll	较高	Reactor	Linux	Linux下的Java NIO框架，Linux Kernel 2.6之前使用poll进行支持，也是使用Reactor模式
epoll	高	Reactor/Proactor	Linux	Linux Kernel 2.6及以后使用epoll进行支持；Linux Kernel 2.6之前使用poll进行支持；另外一定注意，由于Linux下没有Windows下的IOCP技术提供真正的异步I/O 支持，所以Linux下使用epoll模拟异步I/O
kqueue	高	Proactor	Linux	目前Java的版本不支持

多路复用 I/O 技术最适用的是“高并发”场景，所谓“高并发”是指 1ms 内至少同时有上千个连接请求准备好。其他情况下多路复用 I/O 技术发挥不出它的优势。另外，使用 Java NIO 进行功能实现，相对于传统的套接字实现要复杂一些，所以实际应用中，需要根据自己的业务需求进行技术选择。

3.2　NIO 源码初探

说到源码，先得从 Selector 的 open 方法开始看，我们看 java.nio.channels.Selector 类的源码。

```
public static Selector open() throws IOException {
    return SelectorProvider.provider().openSelector();
}
```

看 SelectorProvider.provider()的具体代码。

```
public static SelectorProvider provider() {
        synchronized (lock) {
            if (provider != null)
                return provider;
            return AccessController.doPrivileged(
                new PrivilegedAction<SelectorProvider>() {
                    public SelectorProvider run() {
                            if (loadProviderFromProperty())
                                return provider;
                             if (loadProviderAsService())

                                return provider;

                            provider = sun.nio.ch.DefaultSelectorProvider.create();
                            return provider;
                        }
                    });
        }
}
```

其中 provider = sun.nio.ch.DefaultSelectorProvider.create()会根据操作系统来返回不同的实现类，Windows 平台返回 WindowsSelectorProvider；而 if (provider != null) return provider 保证了整个 Server 程序中只有一个 WindowsSelectorProvider 对象；看 WindowsSelectorProvider.openSelector()代码。

```
public AbstractSelector openSelector() throws IOException {
    return new WindowsSelectorImpl(this);
}
```

new WindowsSelectorImpl(SelectorProvider)的代码如下。

```
WindowsSelectorImpl(SelectorProvider sp) throws IOException {
        super(sp);
        pollWrapper = new PollArrayWrapper(INIT_CAP);
        wakeupPipe = Pipe.open();
        wakeupSourceFd = ((SelChImpl)wakeupPipe.source()).getFDVal();

         //Disable the Nagle algorithm so that the wakeup is more immediate
        SinkChannelImpl sink = (SinkChannelImpl)wakeupPipe.sink();
        (sink.sc).socket().setTcpNoDelay(true);
        wakeupSinkFd = ((SelChImpl)sink).getFDVal();

        pollWrapper.addWakeupSocket(wakeupSourceFd, 0);
}
```

其中 Pipe. open()是关键，这个方法的调用过程如下。

```
public static Pipe open() throws IOException {
```

```
    return SelectorProvider.provider().openPipe();
}
```

在 SelectorProvider 中，代码如下。

```
public Pipe openPipe() throws IOException {
    return new PipeImpl(this);
}
```

再看一下 PipeImpl()的代码。

```
PipeImpl(SelectorProvider sp) {
        long pipeFds = IOUtil.makePipe(true);
        int readFd = (int) (pipeFds >>> 32);
        int writeFd = (int) pipeFds;
        FileDescriptor sourcefd = new FileDescriptor();
        IOUtil.setfdVal(sourcefd, readFd);
        source = new SourceChannelImpl(sp, sourcefd);
        FileDescriptor sinkfd = new FileDescriptor();
        IOUtil.setfdVal(sinkfd, writeFd);
        sink = new SinkChannelImpl(sp, sinkfd);
}
```

其中 IOUtil.makePipe(true)是一个本地方法。

```
/**
* Returns two file descriptors for a pipe encoded in a long.
* The read end of the pipe is returned in the high 32 bits,
* while the write end is returned in the low 32 bits.
*/
staticnativelong makePipe(boolean blocking);
```

具体实现代码如下。

```
JNIEXPORT jlong JNICALL
    Java_sun_nio_ch_IOUtil_makePipe(JNIEnv *env, jobject this, jboolean blocking)
    {
        int fd[2];

        if (pipe(fd) < 0) {
            JNU_ThrowIOExceptionWithLastError(env, "Pipe failed");
            return 0;
        }
        if (blocking == JNI_FALSE) {
            if ((configureBlocking(fd[0], JNI_FALSE) < 0)
                   || (configureBlocking(fd[1], JNI_FALSE) < 0)) {
                    JNU_ThrowIOExceptionWithLastError(env, "Configure blocking failed");
                    close(fd[0]);
```

```
                    close(fd[1]);
                    return 0;
                }
            }
            return ((jlong) fd[0] << 32) | (jlong) fd[1];
        }
        static int
        configureBlocking(int fd, jboolean blocking)
        {
            int flags = fcntl(fd, F_GETFL);
            int newflags = blocking ? (flags & ~O_NONBLOCK) : (flags | O_NONBLOCK);

            return (flags == newflags) ? 0 : fcntl(fd, F_SETFL, newflags);
        }
```

正如下面这段注释所描述的内容。

```
  /**
    * Returns two file descriptors for a pipe encoded in a long.
    * The read end of the pipe is returned in the high 32 bits,
    * while the write end is returned in the low 32 bits.
    */
```

高位存放的是通道 read 端的文件描述符，低 32 位存放的是 write 端的文件描述符。所以取得 makepipe()返回值后要做移位处理。

```
pollWrapper.addWakeupSocket(wakeupSourceFd, 0);
```

这行代码把返回的 pipe 的 write 端的 FD 放在 pollWrapper 中（后面会发现这么做是为了实现 Selector 的 wakeup()）。

ServerSocketChannel.open()的实现代码如下。

```
    public static ServerSocketChannel open() throws IOException {
        return SelectorProvider.provider().openServerSocketChannel();
    }
```

SelectorProvider 的实现代码如下。

```
public ServerSocketChannel openServerSocketChannel() throws IOException {
            return new ServerSocketChannelImpl(this);
}
```

可见 ServerSocketChannelImpl 也有 WindowsSelectorImpl 的引用。

```
public ServerSocketChannelImpl(SelectorProvider sp) throws IOException {
            super(sp);
            this.fd =  Net.serverSocket(true);
```

```
    this.fdVal = IOUtil.fdVal(fd);
            this.state = ST_INUSE;
}
```

然后通过 serverChannel1.register(selector, SelectionKey.OP_ACCEPT)；把 Selector 和 Channel 绑定在一起，也就是把新建 ServerSocketChannel 时创建的 FD 与 Selector 绑定在一起。

到此，Server 端已启动完成，主要创建了以下对象。

（1）WindowsSelectorProvider：为单例对象，实际上是调用操作系统的 API。。

（2）WindowsSelectorImpl 中包含如下内容。

- pollWrapper：保存 Selector 上注册的 FD，包括 pipe 的 write 端 FD 和 ServerSocketChannel 所用的 FD。
- wakeupPipe：通道（其实就是两个 FD，一个是 read 端的，一个是 write 端的）。

下面来看 Selector 中的 select()方法，selector.select()主要调用了 WindowsSelectorImpl 中的 doSelect()方法。

```
protected int doSelect(long timeout) throws IOException {
        if (channelArray == null)
            throw new ClosedSelectorException();
        this.timeout = timeout; // set selector timeout
        processDeregisterQueue();
        if (interruptTriggered) {
            resetWakeupSocket();
            return 0;
        }
        adjustThreadsCount();
        finishLock.reset();
        startLock.startThreads();
        try {
            begin();
            try {
                subSelector.poll();
            } catch (IOException e) {
                finishLock.setException(e);
            }
            if (threads.size() > 0)
                finishLock.waitForHelperThreads();
          } finally {
             end();
          }
            finishLock.checkForException();
        processDeregisterQueue();
```

```
        int updated = updateSelectedKeys();

        resetWakeupSocket();
        return updated;
    }
```

其中 subSelector.poll()是核心，也就是轮询 pollWrapper 中保存的 FD；具体实现是调用 native 方法 poll0()。

```
private int poll() throws IOException{ //poll for the main thread
            return poll0(pollWrapper.pollArrayAddress,
                    Math.min(totalChannels, MAX_SELECTABLE_FDS),
                    readFds, writeFds, exceptFds, timeout);
    }
    private native int poll0(long pollAddress, int numfds,
                int[] readFds, int[] writeFds, int[] exceptFds, long timeout);
        //These arrays will hold result of native select().
   //The first element of each array is the number of selected sockets.
   //Other elements are file descriptors of selected sockets.
   private final int[] readFds = new int [MAX_SELECTABLE_FDS + 1];//保存发生 read 的 FD
   private final int[] writeFds = new int [MAX_SELECTABLE_FDS + 1]; //保存发生 write 的 FD
   private final int[] exceptFds = new int [MAX_SELECTABLE_FDS + 1]; //保存发生 except 的 FD
```

poll0.()会监听 pollWrapper 中的 FD 有没有数据进出，这会造成 I/O 阻塞，直到有数据读写事件发生。比如，由于 pollWrapper 中保存的也有 ServerSocketChannel 的 FD，所以只要 ClientSocket 发一份数据到 ServerSocket，那么 poll0()就会返回；又由于 pollWrapper 中保存的也有 pipe 的 write 端的 FD，所以只要 pipe 的 write 端向 FD 发一份数据，也会造成 poll0()返回；如果这两种情况都没有发生，那么 poll0()就会一直阻塞，也就是 selector.select()会一直阻塞；如果有任何一种情况发生，那么 selector.select()就会返回，所以在 OperationServer 的 run()里要用 while (true)，这样可以保证在 Selector 接收数据并处理完后继续监听 poll()。

再来看 WindowsSelectorImpl.Wakeup()。

```
public Selector wakeup() {
          synchronized (interruptLock) {
              if (!interruptTriggered) {
                  setWakeupSocket();
                  interruptTriggered = true;
              }
          }
          return this;
      }
   private void setWakeupSocket() {
          setWakeupSocket0(wakeupSinkFd);
      }
```

```
private native void setWakeupSocket0(int wakeupSinkFd);
JNIEXPORT void JNICALL
Java_sun_nio_ch_WindowsSelectorImpl_setWakeupSocket0(JNIEnv *env, jclass this,
                                                    jint scoutFd)
{
    /* Write one byte into the pipe */
    const char byte = 1;
    send(scoutFd, &byte, 1, 0);
}
```

可见 wakeup()是通过 pipe 的 write 端 send(scoutFd, &byte, 1, 0)发送一个字节 1 来唤醒 poll()的，所以在需要的时候就可以调用 selector.wakeup()来唤醒 Selector。

3.3　反应堆

现在我们已经对阻塞 I/O 有了一定了解，知道阻塞 I/O 在调用 InputStream.read()方法时是阻塞的，它会一直等到数据到来（或超时）时才会返回；同样，在调用 ServerSocket.accept()方法时，也会一直阻塞到有客户端连接才会返回，每个客户端连接成功后，服务端都会启动一个线程去处理该客户端的请求。阻塞 I/O 的通信模型示意如下图所示。

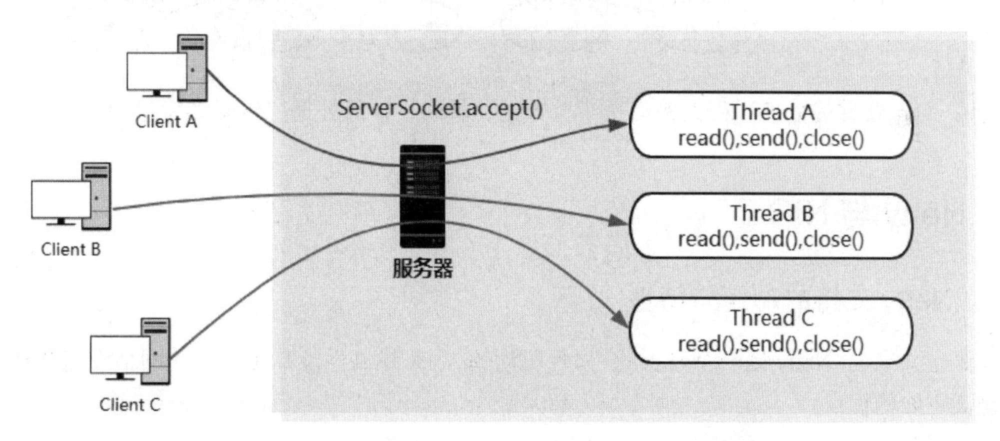

如果仔细分析，一定会发现阻塞 I/O 存在一些缺点。根据阻塞 I/O 通信模型，总结了它的两个缺点。

（1）当客户端多时，会创建大量的处理线程。且每个线程都要占用栈空间和一些 CPU 时间。

（2）阻塞可能带来频繁的上下文切换，且大部分上下文切换可能是无意义的。在这种情况下非阻塞 I/O 就有了它的应用前景。

Java NIO 是从 JDK 1.4 开始使用的，它既可以说成是“新 I/O”，也可以说成是“非阻塞 I/O”。下面是 Java NIO 的工作原理。

（1）由一个专门的线程来处理所有的 I/O 事件，并负责分发。

（2）事件驱动机制：事件到的时候触发，而不是同步地去监视事件。

（3）线程通信：线程之间通过 wait、notify 等方式通信。保证每次上下文切换都是有意义的，减少无谓的线程切换。

下面是笔者理解的 Java NIO 反应堆的工作原理图。

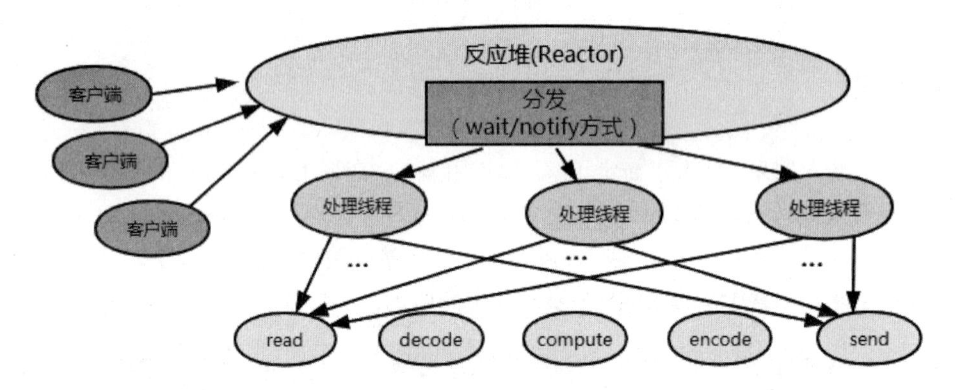

（注：每个线程的处理流程大概都是读取数据、解码、计算处理、编码和发送响应。）

3.4 Netty 与 NIO

3.4.1 Netty 支持的功能与特性

按照定义来说，Netty 是一个异步的、事件驱动的、用来做高性能高可靠性的网络应用的框架。下面是其主要的优点。

（1）框架设计优雅，底层模型随意切换，适应不同的网络协议要求。

（2）提供了很多标准的协议、安全、编解码的支持。

（3）解决了很多 NIO 不易用的问题。

（4）社区更为活跃，在很多开源框架中使用，如 Dubbo、RocketMQ、Spark 等。

Netty 支持的功能与特性如下图所示。

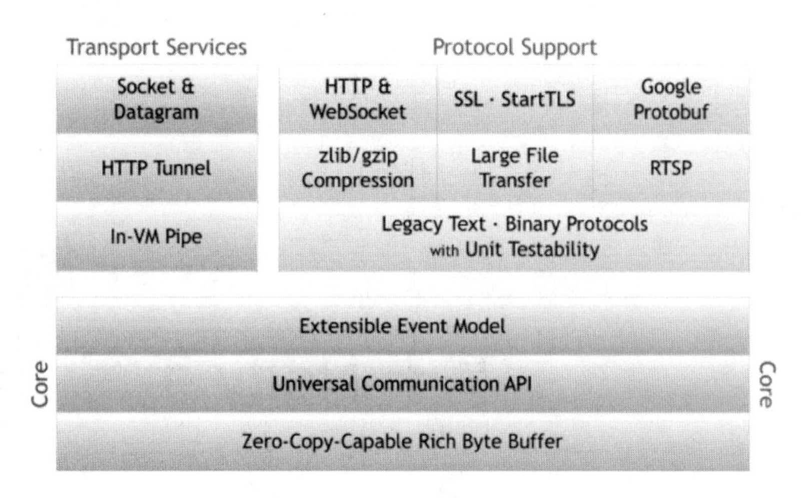

（1）底层核心有：Zero-Copy-Capable Buffer，非常易用的零拷贝 Buffer（这个内容很有意思，稍后专门来讲）；统一的 API；标准可扩展的事件模型。

（2）传输方面的支持有：管道通信；HTTP 隧道；TCP 与 UDP。

（3）协议方面的支持有：基于原始文本和二进制的协议；解压缩；大文件传输；流媒体传输；ProtoBuf 编解码；安全认证；HTTP 和 WebSocket。

3.4.2　Netty 采用 NIO 而非 AIO 的理由

（1）Netty 不看重 Windows 上的使用，在 Linux 系统上，AIO 的底层实现仍使用 epoll，没有很好地实现 AIO，因此在性能上没有明显的优势，而且被 JDK 封装了一层，不容易深度优化。

（2）Netty 整体架构采用 Reactor 模型，而 AIO 采用 Proactor 模型，混在一起会非常混乱，把 AIO 也改造成 Reactor 模型，看起来是把 Epoll 绕个弯又绕回来。

（3）AIO 还有个缺点是接收数据需要预先分配缓存，而 NIO 是需要接收时才分配缓存，所以对连接数量非常大但流量小的情况，AIO 浪费很多内存。

（4）Linux 上 AIO 不够成熟，处理回调结果的速度跟不上处理需求，比如外卖员太少，顾客太多，供不应求，造成处理速度有瓶颈。

4

第 4 章 基于 Netty 手写 Tomcat

Netty 作为底层通信框架，用来实现 Web 容器自然也不难，我们先介绍一下整体实现思路。我们知道，Tomcat 是基于 J2EE 规范的 Web 容器，主要入口是 web.xml 文件。web.xml 文件中主要配置 Servlet、Filter、Listener 等，而 Servlet、Filter、Listener 在 J2EE 中只是抽象的实现，具体业务逻辑由开发者来实现。本章内容，就以最常用的 Servlet 为例来详细展开。

4.1 环境准备

4.1.1 定义 GPServlet 抽象类

首先，我们创建 GPServlet 类。我们都知道 GPServlet 生命周期中最常用的方法是 doGet()方法和 doPost()方法，而 doGet()方法和 doPost()方法是 service()方法的分支实现，看下面的简易版 Servlet 源码实现。

```
package com.gupaoedu.vip.netty.tomcat.http;

public abstract class GPServlet {

    public void service(GPRequest request,GPResponse response) throws Exception{
```

```
        //由 service()方法决定是调用 doGet()还是调用 doPost()
        if("GET".equalsIgnoreCase(request.getMethod())){
            doGet(request, response);
        }else{
            doPost(request, response);
        }

    }

    public abstract void doGet(GPRequest request,GPResponse response) throws Exception;

    public abstract void doPost(GPRequest request,GPResponse response) throws Exception;

}
```

从上面的代码中，我们看到，doGet()方法和 doPost()方法中有两个参数 GPRequest 和 GPResponse 对象，这两个对象是由 Web 容器创建的，主要是对底层 Socket 的输入输出的封装。其中 GPRequest 是对 Input 的封装，GPResponse 是对 Output 的封装。

4.1.2　创建用户业务代码

下面基于 GPServlet 来实现两个业务逻辑 FirstServlet 和 SecondServlet。FirstServlet 类的实现代码如下。

```
package com.gupaoedu.vip.netty.io.bio.tomcat.servlet;

import com.gupaoedu.vip.netty.io.bio.tomcat.http.GPRequest;
import com.gupaoedu.vip.netty.io.bio.tomcat.http.GPResponse;
import com.gupaoedu.vip.netty.io.bio.tomcat.http.GPServlet;

public class FirstServlet extends GPServlet {

    public void doGet(GPRequest request, GPResponse response) throws Exception {
        this.doPost(request, response);
    }

    public void doPost(GPRequest request, GPResponse response) throws Exception {
        response.write("This is First Servlet");
    }

}
```

SecondServlet 类的实现代码如下。

```
package com.gupaoedu.vip.netty.io.bio.tomcat.servlet;

import com.gupaoedu.vip.netty.io.bio.tomcat.http.GPRequest;
import com.gupaoedu.vip.netty.io.bio.tomcat.http.GPResponse;
import com.gupaoedu.vip.netty.io.bio.tomcat.http.GPServlet;

public class SecondServlet extends GPServlet {

    public void doGet(GPRequest request, GPResponse response) throws Exception {
        this.doPost(request, response);
    }

    public void doPost(GPRequest request, GPResponse response) throws Exception {
        response.write("This is Second Servlet");
    }

}
```

4.1.3 完成 web.properties 配置

为了简化操作，我们用 web.properties 文件代替 web.xml 文件，具体内容如下。

```
servlet.one.url=/firstServlet.do
servlet.one.className=com.gupaoedu.vip.netty.io.bio.tomcat.servlet.FirstServlet

servlet.two.url=/secondServlet.do
servlet.two.className=com.gupaoedu.vip.netty.io.bio.tomcat.servlet.SecondServlet
```

上述代码分别给两个 Servlet 配置了/firstServlet.do 和/secondServlet.do 的 URL 映射。

4.2 基于传统 I/O 手写 Tomcat

下面我们来看 GPRequest 和 GPResponse 的基本实现。

4.2.1 创建 GPRequest 对象

GPRequest 主要就是对 HTTP 的请求头信息进行解析。我们从浏览器发送一个 HTTP 请求，如在浏览器地址栏中输入 http://localhost:8080，后台服务器获取的请求其实就是一串字符串，具体格式如下。

```
GET / HTTP/1.1
Host: localhost:8080
```

```
Connection: keep-alive
Upgrade-Insecure-Requests: 1
User-Agent: Mozilla/5.0 (Windows NT 10.0; Win64; x64) AppleWebKit/537.36 (KHTML, like Gecko)
Chrome/75.0.3770.142 Safari/537.36
Accept:
text/html,application/xhtml+xml,application/xml;q=0.9,image/webp,image/apng,*/*;q=0.8,appl
ication/signed-exchange;v=b3
Accept-Encoding: gzip, deflate, br
Accept-Language: zh-CN,zh;q=0.9
```

在 GPRequest 获得输入内容之后，对这一串满足 HTTP 的字符信息进行解析。我们来看 GPRequest 简单直接的代码实现。

```
package com.gupaoedu.vip.netty.io.bio.tomcat.http;

import java.io.InputStream;

/**
 * Created by Tom.
 */
public class GPRequest {

    private String method;
    private String url;

    public GPRequest(InputStream in){
        try {
            //获取 HTTP 内容
            String content = "";
            byte[] buff = new byte[1024];
            int len = 0;
            if ((len = in.read(buff)) > 0) {
                content = new String(buff,0,len);
            }

            String line = content.split("\\n")[0];
            String [] arr = line.split("\\s");

            this.method = arr[0];
            this.url = arr[1].split("\\?")[0];
        }catch (Exception e){
            e.printStackTrace();
        }
    }

    public String getUrl() {
        return url;
```

```
    }

    public String getMethod() {
        return method;
    }
}
```

在上面的代码中，GPRequest 主要提供了 getUrl()方法和 getMethod()方法。输入流 InputStream 作为 GPRequest 的构造参数传入，在构造函数中，用字符串切割的方法提取请求方式和 URL。

4.2.2 创建 GPResponse 对象

接下来看 GPResponse 的实现，与 GPRequest 的实现思路类似，就是按照 HTTP 规范从 Output 输出格式化的字符串，来看代码。

```
package com.gupaoedu.vip.netty.io.bio.tomcat.http;

import java.io.OutputStream;

/**
 * Created by Tom.
 */
public class GPResponse {
    private OutputStream out;
    public GPResponse(OutputStream out){
        this.out = out;
    }

    public void write(String s) throws Exception {
        //输出也要遵循 HTTP
        //状态码为 200
        StringBuilder sb = new StringBuilder();
        sb.append("HTTP/1.1 200 OK\n")
                .append("Content-Type: text/html;\n")
                .append("\r\n")
                .append(s);
        out.write(sb.toString().getBytes());
    }
}
```

上面的代码中，输出流 OutputStream 作为 GPResponse 的构造参数传入，主要提供了一个 write()方法。通过 write()方法按照 HTTP 规范输出字符串。

4.2.3　创建 GPTomcat 启动类

前面 4.2.1 和 4.2.2 两节只是对 J2EE 规范的再现，接下来就是真正 Web 容器的实现逻辑，分为三个阶段：初始化阶段、服务就绪阶段、接受请求阶段。

第一阶段：初始化阶段，主要是完成对 web.xml 文件的解析。

```
package com.gupaoedu.vip.netty.io.bio.tomcat;

import com.gupaoedu.vip.netty.io.bio.tomcat.http.GPRequest;
import com.gupaoedu.vip.netty.io.bio.tomcat.http.GPResponse;
import com.gupaoedu.vip.netty.io.bio.tomcat.http.GPServlet;

import java.io.FileInputStream;
import java.io.InputStream;
import java.io.OutputStream;
import java.net.ServerSocket;
import java.net.Socket;
import java.util.HashMap;
import java.util.Map;
import java.util.Properties;

/**
 * Created by Tom.
 */
public class GPTomcat {
    private int port = 8080;
    private ServerSocket server;
    private Map<String,GPServlet> servletMapping = new HashMap<String,GPServlet>();

    private Properties webxml = new Properties();


    private void init(){

        //加载 web.xml 文件，同时初始化 ServletMapping 对象
        try{
            String WEB_INF = this.getClass().getResource("/").getPath();
            FileInputStream fis = new FileInputStream(WEB_INF + "web.properties");

            webxml.load(fis);

            for (Object k : webxml.keySet()) {

                String key = k.toString();
                if(key.endsWith(".url")){
```

```
                String servletName = key.replaceAll("\\.url$", "");
                String url = webxml.getProperty(key);
                String className = webxml.getProperty(servletName + ".className");
                //单实例，多线程
                GPServlet obj = (GPServlet)Class.forName(className).newInstance();
                servletMapping.put(url, obj);
            }

        }


    }catch(Exception e){
        e.printStackTrace();
    }

  }
}
```

上面代码中，首先从 WEB-INF 读取 web.properties 文件并对其进行解析，然后将 URL 规则和 GPServlet 的对应关系保存到 servletMapping 中。

第二阶段：服务就绪阶段，完成 ServerSocket 的准备工作。在 GPTomcat 类中增加 start()方法。

```
  public void start(){

    //1.加载配置文件，初始化 ServletMapping
    init();

    try {
        server = new ServerSocket(this.port);

        System.out.println("GPTomcat 已启动，监听的端口是：" + this.port);

        //2.等待用户请求，用一个死循环来等待用户请求
        while (true) {
            Socket client = server.accept();
            //3.HTTP 请求，发送的数据就是字符串——有规律的字符串（HTTP）
            process(client);

        }

    } catch (Exception e) {
        e.printStackTrace();
    }
  }
```

第三阶段：接受请求阶段，完成每一次请求的处理。在 GPTomcat 中增加 process()方法的实现。

```
private void process(Socket client) throws Exception {

    InputStream is = client.getInputStream();
    OutputStream os = client.getOutputStream();

    //4.Request(InputStrean)/Response(OutputStrean)
    GPRequest request = new GPRequest(is);
    GPResponse response = new GPResponse(os);

    //5.从协议内容中获得 URL，把相应的 Servlet 用反射进行实例化
    String url = request.getUrl();

    if(servletMapping.containsKey(url)){
       //6.调用实例化对象的 service()方法，执行具体的逻辑 doGet()/doPost()方法
       servletMapping.get(url).service(request,response);
    }else{
       response.write("404 - Not Found");
    }


    os.flush();
    os.close();

    is.close();
    client.close();
}
```

每次客户端请求过来以后，从 servletMapping 中获取其对应的 Servlet 对象，同时实例化 GPRequest 和 GPResponse 对象，将 GPRequest 和 GPResponse 对象作为参数传入 service()方法，最终执行业务逻辑。最后，增加 main()方法。

```
public static void main(String[] args) {
    new GPTomcat().start();
}
```

服务启动后，运行效果如下图所示。

```
Run  GPTomcat
D:\Java\jdk1.8.0_131\bin\java ...
GP Tomcat 已启动，监听的端口是：8080
```

4.3 基于 Netty 重构 Tomcat 实现

了解了传统的 I/O 实现方式之后，我们发现 Netty 版本的实现就比较简单了，来看具体的代码实现。

4.3.1 重构 GPTomcat 逻辑

话不多说，直接看代码。

```
package com.gupaoedu.vip.netty.tomcat;

import com.gupaoedu.vip.netty.tomcat.http.GPRequest;
import com.gupaoedu.vip.netty.tomcat.http.GPResponse;
import com.gupaoedu.vip.netty.tomcat.http.GPServlet;
import io.netty.bootstrap.ServerBootstrap;
import io.netty.buffer.ByteBuf;
import io.netty.channel.*;
import io.netty.channel.nio.NioEventLoopGroup;
import io.netty.channel.socket.SocketChannel;
import io.netty.channel.socket.nio.NioServerSocketChannel;
import io.netty.handler.codec.http.HttpRequest;
import io.netty.handler.codec.http.HttpRequestDecoder;
import io.netty.handler.codec.http.HttpResponseEncoder;

import java.io.FileInputStream;
import java.nio.channels.Selector;
import java.util.HashMap;
import java.util.Map;
import java.util.Properties;


//Netty 就是一个同时支持多协议的网络通信框架
public class GPTomcat {
    //打开 Tomcat 源码，全局搜索 ServerSocket

    private int port = 8080;

    private Map<String,GPServlet> servletMapping = new HashMap<String,GPServlet>();

    private Properties webxml = new Properties();

    private void init(){

            //加载 web.xml 文件，同时初始化 ServletMapping 对象
            try{
```

```
            String WEB_INF = this.getClass().getResource("/").getPath();
            FileInputStream fis = new FileInputStream(WEB_INF + "web.properties");

            webxml.load(fis);

            for (Object k : webxml.keySet()) {

                String key = k.toString();
                if(key.endsWith(".url")){
                    String servletName = key.replaceAll("\\.url$", "");
                    String url = webxml.getProperty(key);
                    String className = webxml.getProperty(servletName +
".className");
                    GPServlet obj =
(GPServlet)Class.forName(className).newInstance();
                    servletMapping.put(url, obj);
                }
            }

        }catch(Exception e){
            e.printStackTrace();
        }
    }

    public void start(){

        init();


        //Netty 封装了 NIO 的 Reactor 模型，Boss，Worker
        //Boss 线程
        EventLoopGroup bossGroup = new NioEventLoopGroup();
        //Worker 线程
        EventLoopGroup workerGroup = new NioEventLoopGroup();
        try {

            //1.创建对象
            ServerBootstrap server = new ServerBootstrap();


            //2.配置参数
            //链路式编程
            server.group(bossGroup, workerGroup)
                //主线程处理类，看到这样的写法，底层就是用反射
                .channel(NioServerSocketChannel.class)
                //子线程处理类，Handler
                .childHandler(new ChannelInitializer<SocketChannel>() {
```

```
                    //客户端初始化处理
                        protected void initChannel(SocketChannel client) throws
Exception {
                            //无锁化串行编程
                            //Netty 对 HTTP 的封装，对顺序有要求
                            //HttpResponseEncoder 编码器
                            //责任链模式，双向链表 Inbound OutBound
                            client.pipeline().addLast(new HttpResponseEncoder());
                            //HttpRequestDecoder 解码器
                            client.pipeline().addLast(new HttpRequestDecoder());
                            //业务逻辑处理
                            client.pipeline().addLast(new GPTomcatHandler());
                    }
                })
                //针对主线程的配置 分配线程最大数量 128
                .option(ChannelOption.SO_BACKLOG, 128)
                //针对子线程的配置 保持长连接
                .childOption(ChannelOption.SO_KEEPALIVE, true);

                //3.启动服务器
                ChannelFuture f = server.bind(port).sync();
                System.out.println("GPTomcat 已启动，监听的端口是： " + port);
                f.channel().closeFuture().sync();
        }catch (Exception e){
                e.printStackTrace();
        }finally {
                //关闭线程池
                bossGroup.shutdownGracefully();
                workerGroup.shutdownGracefully();
        }
    }

    public class GPTomcatHandler extends ChannelInboundHandlerAdapter {
        @Override
        public void channelRead(ChannelHandlerContext ctx, Object msg) throws Exception
{

                if (msg instanceof HttpRequest){
                    System.out.println("hello");
                    HttpRequest req = (HttpRequest) msg;

                    //转交给我们自己的 Request 实现
                    GPRequest request = new GPRequest(ctx,req);
                    //转交给我们自己的 Response 实现
                    GPResponse response = new GPResponse(ctx,req);
                    //实际业务处理
                    String url = request.getUrl();
```

```
                    if(servletMapping.containsKey(url)){
                            servletMapping.get(url).service(request, response);
                    }else{
                            response.write("404 - Not Found");
                    }

                }
            }

            @Override
            public void exceptionCaught(ChannelHandlerContext ctx, Throwable cause) throws
Exception {

            }

    }

    public static void main(String[] args) {
            new GPTomcat().start();
    }

}
```

代码的基本思路和基于传统 I/O 手写的版本一致，不再赘述。

4.3.2　重构 GPRequest 逻辑

我们先来看代码。

```
package com.gupaoedu.vip.netty.tomcat.http;

import io.netty.channel.ChannelHandlerContext;
import io.netty.handler.codec.http.HttpRequest;
import io.netty.handler.codec.http.QueryStringDecoder;

import java.util.List;
import java.util.Map;

public class GPRequest {

    private ChannelHandlerContext ctx;

    private HttpRequest req;

    public GPRequest(ChannelHandlerContext ctx, HttpRequest req) {
            this.ctx = ctx;
```

```
        this.req = req;
    }

    public String getUrl() {
        return req.uri();
    }

    public String getMethod() {
        return req.method().name();
    }

    public Map<String, List<String>> getParameters() {
        QueryStringDecoder decoder = new QueryStringDecoder(req.uri());
        return decoder.parameters();
    }

    public String getParameter(String name) {
        Map<String, List<String>> params = getParameters();
        List<String> param = params.get(name);
        if (null == param) {
            return null;
        } else {
            return param.get(0);
        }
    }
}
```

和基于传统的 I/O 手写的版本一样，提供 getUrl()方法和 getMethod()方法。在 Netty 的版本中，我们增加了 getParameter()的实现，供大家参考。

4.3.3 重构 GPResponse 逻辑

还是继续看代码。

```
package com.gupaoedu.vip.netty.tomcat.http;

import io.netty.buffer.Unpooled;
import io.netty.channel.ChannelHandlerContext;
import io.netty.handler.codec.http.*;


public class GPResponse {

    //SocketChannel 的封装
    private ChannelHandlerContext ctx;
```

```
    private HttpRequest req;

    public GPResponse(ChannelHandlerContext ctx, HttpRequest req) {
        this.ctx = ctx;
        this.req = req;
    }

    public void write(String out) throws Exception {
        try {
            if (out == null || out.length() == 0) {
                return;
            }
            //设置 HTTP 及请求头信息
            FullHttpResponse response = new DefaultFullHttpResponse(
                //设置版本为 HTTP 1.1
                HttpVersion.HTTP_1_1,
                //设置响应状态码
                HttpResponseStatus.OK,
                //将输出内容编码格式设置为 UTF-8
                Unpooled.wrappedBuffer(out.getBytes("UTF-8")));

            response.headers().set("Content-Type", "text/html;");

            ctx.write(response);
        } finally {
            ctx.flush();
            ctx.close();
        }
    }
}
```

相对于基于传统的 I/O 手写的版本而言，主要变化就是利用 Netty 对 HTTP 的默认支持，可以使用现成的 API。

4.3.4　运行效果演示

启动容器，我们在浏览器地址栏中输入 http://localhost:8080/firstServlet.do，可以得到如下图所示的结果。

← → C ⓘ localhost:8080/firstServlet.do

This is First Serlvet

在浏览器地址栏中输入 http://localhost:8080/secondServlet.do，可以得到如下图所示的结果。

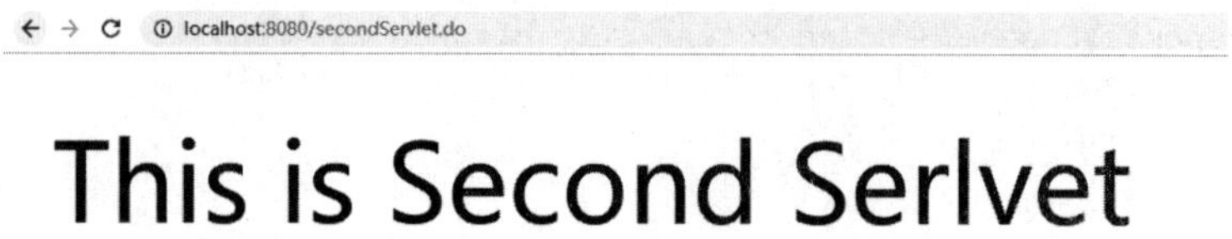

localhost:8080/secondServlet.do

This is Second Serlvet

5

第 5 章 基于 Netty 重构 RPC 框架

5.1 RPC 概述

大概很多小伙伴都见过下图，这是在 Dubbo 官网中找到的一张描述项目架构的演进过程的图。

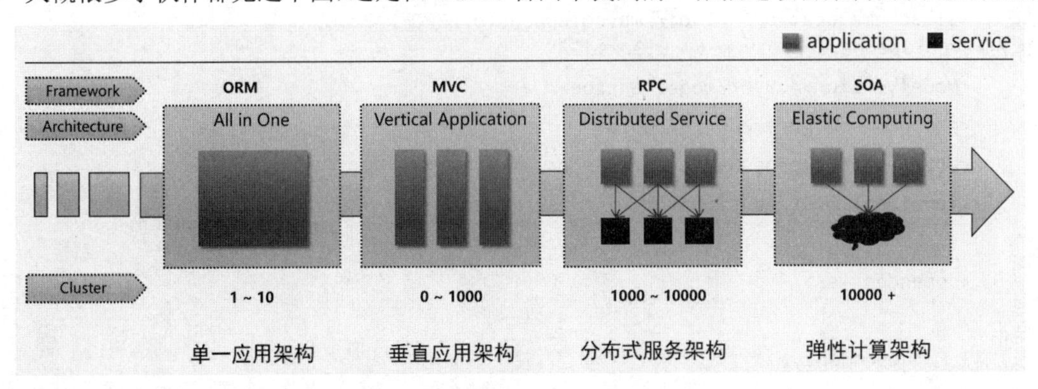

它描述了每一种架构需要的具体配置和组织形态。当网站流量很小时，只需一个应用，将所有功能都部署在一起，以减少节点部署和成本，我们通常会采用单一应用架构。之后才出现了 ORM 框架，该框架大大简化增删改查的操作流程，提高了开发者的工作效率。

随着用户量增加，当访问量逐渐增大，单一应用增加机器带来的加速度越来越小，我们需要将应用拆分成互不干扰的几个应用以提升效率，就出现了垂直应用架构。MVC（Model-View-Centroller，模型-视图-控制器）架构就是一种非常经典的用于加速前端页面开发的架构。

当垂直应用越来越多，应用之间交互不可避免，将核心业务抽取出来，作为独立的服务，逐渐形成稳定的服务中心，使前端应用能更快速地响应多变的市场需求，就出现了分布式服务架构。分布式架构下服务数量逐渐增加，为了提高管理效率，RPC（Memote Procedure Call，远程过程调用）框架应运而生。RPC 用于提高业务复用及整合，分布式服务框架下 RPC 是关键。

下一代框架，将会是流动计算架构占据主流。当服务越来越多，容量的评估、小服务的资源浪费等问题逐渐明显。此时，需要增加一个调度中心，基于访问压力实时管理集群容量，提高集群利用率。SOA（Service-Oriented Architecture，面向服务的架构）架构就是用于提高集群利用率的。在资源调度和治理中心方面 SOA 是关键。

Netty 基本上是作为架构的技术底层而存在的，主要完成高性能的网络通信。

5.2 环境预设

第一步，我们先将项目环境搭建起来，创建 pom.xml 配置文件。

```
    <project xmlns="http://maven.apache.org/POM/4.0.0"
xmlns:xsi="http://www.w3.org/2001/XMLSchema-instance"
        xsi:schemaLocation="http://maven.apache.org/POM/4.0.0
http://maven.apache.org/maven-v4_0_0.xsd">
        <modelVersion>4.0.0</modelVersion>
        <groupId>com.gupaoedu</groupId>
        <artifactId>gupaoedu-vip-netty-rpc</artifactId>
        <version>1.0.0</version>

        <dependencies>


            <dependency>
              <groupId>io.netty</groupId>
              <artifactId>netty-all</artifactId>
              <version>4.1.6.Final</version>
```

```
        </dependency>

        <dependency>
            <groupId>org.projectlombok</groupId>
            <artifactId>lombok</artifactId>
            <version>1.16.10</version>
        </dependency>

    </dependencies>

</project>
```

第二步，创建项目结构。

在没有 RPC 框架以前，我们的服务调用是这样的，如下图所示。

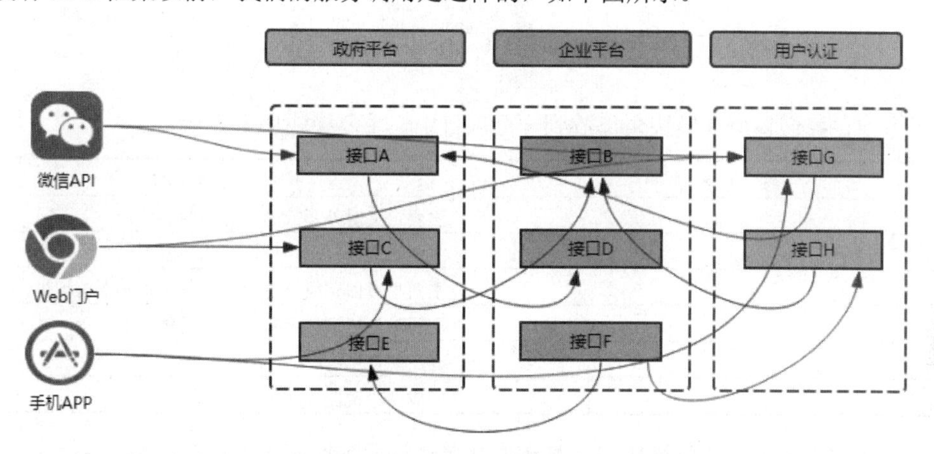

从上图可以看出，接口的调用完全没有规律可循，想怎么调用就怎么调用。这导致业务发展到一定阶段之后，对接口的维护变得非常困难。于是有人提出了服务治理的概念。所有服务间不允许直接调用，而是先到注册中心进行登记，再由注册中心统一协调和管理所有服务的状态并对外发布，调用者只需要记住服务名称，去注册中心获取服务即可。这样极大地规范了服务的管理，提高了所有服务端可控性。整个设计思想其实在我们生活中也能找到活生生的案例。例如：我们平时工作交流，大多都是用 IM 工具，而不是面对面“吼”。大家只需要相互记住运营商（也就是注册中心）提供的号码（如腾讯 QQ）即可。再比如：我们打电话，所有电话号码由运营商分配。我们需要和某一个人通话时，只需要拨通对方的号码，运营商（注册中心，如中国移动、中国联通、中国电信）就会帮我们将信号转接过去。

目前流行的 RPC 服务治理框架主要有 Dubbo 和 Spring Cloud，下面我们以比较经典的 Dubbo 为例。Dubbo 核心模块主要有四个：Registry（注册中心）、Provider（服务端）、Consumer（消费端）、Monitor（监控中心），如下图所示。

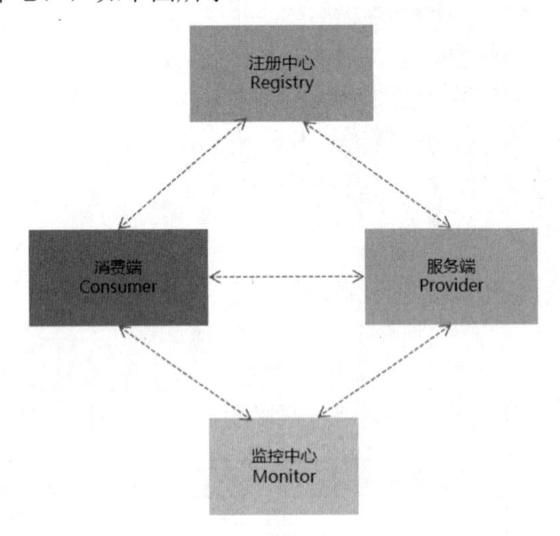

为了方便，我们将所有模块全部放到一个项目中，主要模块如下表所示。

模 块 名	主要功能
API	主要定义对外开放的功能与服务接口
Protocol	主要定义自定义传输协议的内容
Registry	主要负责保存所有可用的服务名称和服务地址
Provider	实现对外提供的所有服务的具体功能
Consumer	客户端调用
Monitor	完成调用链监控

下面，我们先把项目结构搭建好，具体的项目结构如下图所示。

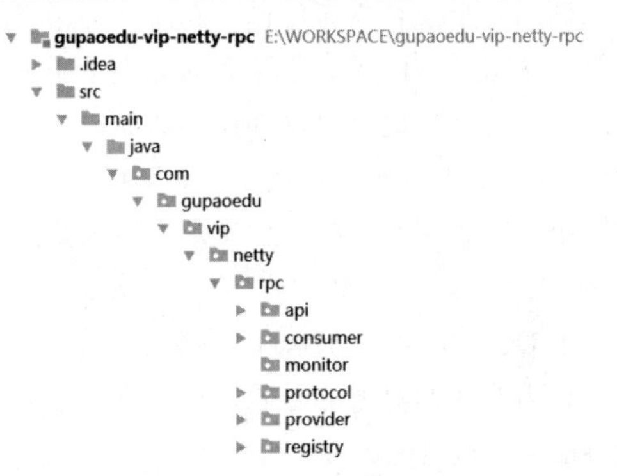

5.3　代码实战

5.3.1　创建 API 模块

首先创建 API 模块，Provider 和 Consumer 都遵循 API 模块的规范。为了简化，创建两个 Service 接口，分别如下。

（1）IRpcHelloService 接口，实现一个 hello()方法，主要目的是用来确认服务是否可用。

```
package com.gupaoedu.vip.netty.rpc.api;

public interface IRpcHelloService {
    String hello(String name);
}
```

（2）创建 IRpcService 接口，完成模拟业务的加、减、乘、除运算。

```
package com.gupaoedu.vip.netty.rpc.api;

public interface IRpcService {

    /** 加 */
    public int add(int a,int b);

    /** 减 */
    public int sub(int a,int b);

    /** 乘 */
    public int mult(int a,int b);

    /** 除 */
    public int div(int a,int b);

}
```

至此，API 模块就定义完成了，非常简单。接下来，我们要确定传输规则，也就是传输协议，协议内容当然要自定义才能体现出 Netty 的优势。

5.3.2　创建自定义协议

通过前面章节的初体验我们了解到，Netty 中的 HTTP 处理要 Netty 内置的 HTTP 的编解码器来完成解析。现在我们来看自定义协议如何设定。

在 Netty 中完成一个自定义协议其实非常简单，只需要定义一个普通的 Java 类即可。我们现在手写 RPC 主要是为了完成对 Java 代码的远程调用，类似于 RMI（Remote Method Invocation，远程方法调用），大家应该都很熟悉了吧。在远程调用 Java 代码时，哪些内容是必须由网络来传输的呢？譬如，服务名称？需要调用该服务的哪个方法？方法的实参是什么？这些信息都需要通过客户端传送到服务端。

下面我们来看具体的代码实现，定义 InvokerProtocol 类。

```
package com.gupaoedu.vip.netty.rpc.protocol;

import lombok.Data;
import java.io.Serializable;

/**
 * 自定义传输协议
 */
@Data
public class InvokerProtocol implements Serializable {

    private String className;//类名
    private String methodName;//函数名称
    private Class<?>[] parames;//参数类型
    private Object[] values;//参数列表

}
```

从上面的代码可以看出，协议中主要包含的信息有类名、函数名称、参数类型和参数列表，通过这些信息就可以定位一个具体的业务逻辑实现。

5.3.3 实现 Provider 业务逻辑

我们将 API 中定义的所有功能在 Provider 模块中实现，分别创建两个实现类。

（1）RpcHelloServiceImpl 类。

```
package com.gupaoedu.vip.netty.rpc.provider;

import com.gupaoedu.vip.netty.rpc.api.IRpcHelloService;

public class RpcHelloServiceImpl implements IRpcHelloService {

    public String hello(String name) {
        return "Hello " + name + "!";
    }

}
```

（2）RpcServiceImpl 类。

```
package com.gupaoedu.vip.netty.rpc.provider;

import com.gupaoedu.vip.netty.rpc.api.IRpcService;

public class RpcServiceImpl implements IRpcService {

    public int add(int a, int b) {
            return a + b;
    }

    public int sub(int a, int b) {
            return a - b;
    }

    public int mult(int a, int b) {
            return a * b;
    }

    public int div(int a, int b) {
            return a / b;
    }

}
```

5.3.4　完成 Registry 服务注册

Registry（注册中心）的主要功能就是负责将所有 Provider 的服务名称和服务引用地址注册到一个容器中，并对外发布。Registry 要启动一个对外的服务，很显然应该作为服务端，并提供一个对外可以访问的端口。先启动一个 Netty 服务，创建 RpcRegistry 类，具体代码如下。

```
package com.gupaoedu.vip.netty.rpc.registry;

import io.netty.bootstrap.ServerBootstrap;
import io.netty.channel.ChannelFuture;
import io.netty.channel.ChannelInitializer;
import io.netty.channel.ChannelOption;
import io.netty.channel.ChannelPipeline;
import io.netty.channel.EventLoopGroup;
import io.netty.channel.nio.NioEventLoopGroup;
import io.netty.channel.socket.SocketChannel;
import io.netty.channel.socket.nio.NioServerSocketChannel;
import io.netty.handler.codec.LengthFieldBasedFrameDecoder;
import io.netty.handler.codec.LengthFieldPrepender;
import io.netty.handler.codec.serialization.ClassResolvers;
```

```
import io.netty.handler.codec.serialization.ObjectDecoder;
import io.netty.handler.codec.serialization.ObjectEncoder;

public class RpcRegistry {
    private int port;
    public RpcRegistry(int port){
        this.port = port;
    }
    public void start(){
        EventLoopGroup bossGroup = new NioEventLoopGroup();
        EventLoopGroup workerGroup = new NioEventLoopGroup();

        try {
            ServerBootstrap b = new ServerBootstrap();
            b.group(bossGroup, workerGroup)
                  .channel(NioServerSocketChannel.class)
                   .childHandler(new ChannelInitializer<SocketChannel>() {

                     @Override
                     protected void initChannel(SocketChannel ch) throws Exception {
                         ChannelPipeline pipeline = ch.pipeline();
                         //自定义协议解码器
                         /**入参有 5 个，分别解释如下
                          maxFrameLength：框架的最大长度。如果帧的长度大于此值，则将抛出
TooLongFrameException
                          lengthFieldOffset：长度属性的偏移量。即对应的长度属性在整个消息数据中
的位置
                          lengthFieldLength：长度字段的长度。如果长度属性是 int 型，那么这个值就
是 4（long 型就是 8）
                          lengthAdjustment：要添加到长度属性值的补偿值
                          initialBytesToStrip：从解码帧中去除的第一个字节数
                          */
                         pipeline.addLast(new
LengthFieldBasedFrameDecoder(Integer.MAX_VALUE, 0, 4, 0, 4));
                         //自定义协议编码器
                         pipeline.addLast(new LengthFieldPrepender(4));
                         //对象参数类型编码器
                         pipeline.addLast("encoder",new ObjectEncoder());
                         //对象参数类型解码器
                         pipeline.addLast("decoder",new ObjectDecoder(Integer.MAX_VALUE,

    ClassResolvers.cacheDisabled(null)));
                         pipeline.addLast(new RegistryHandler());
                     }
                  })
                  .option(ChannelOption.SO_BACKLOG, 128)
                  .childOption(ChannelOption.SO_KEEPALIVE, true);
```

```
            ChannelFuture future = b.bind(port).sync();
            System.out.println("GP RPC Registry start listen at " + port );
            future.channel().closeFuture().sync();
        } catch (Exception e) {
             bossGroup.shutdownGracefully();
             workerGroup.shutdownGracefully();
        }
    }


    public static void main(String[] args) throws Exception {
        new RpcRegistry(8080).start();
    }
}
```

在 RegistryHandler 中实现注册的具体逻辑，上面的代码主要实现服务注册和服务调用的功能。因为所有模块创建在同一个项目中，所以为了简化，服务端没有采用远程调用，而是直接扫描本地 Class，然后利用反射调用。代码实现如下。

```
package com.gupaoedu.vip.netty.rpc.registry;

import java.io.File;
import java.lang.reflect.Method;
import java.net.URL;
import java.util.ArrayList;
import java.util.List;
import java.util.concurrent.ConcurrentHashMap;

import com.gupaoedu.vip.netty.rpc.protocol.InvokerProtocol;

import io.netty.channel.ChannelHandlerContext;
import io.netty.channel.ChannelInboundHandlerAdapter;

public class RegistryHandler  extends ChannelInboundHandlerAdapter {

    //保存所有可用的服务
    public static ConcurrentHashMap<String, Object> registryMap = new
ConcurrentHashMap<String,Object>();

    //保存所有相关的服务类
    private List<String> classNames = new ArrayList<String>();

    public RegistryHandler(){
    //完成递归扫描
    scannerClass("com.gupaoedu.vip.netty.rpc.provider");
    doRegister();
```

```
    }

    @Override
    public void channelRead(ChannelHandlerContext ctx, Object msg) throws Exception {
     Object result = new Object();
        InvokerProtocol request = (InvokerProtocol)msg;

        //当客户端建立连接时，需要从自定义协议中获取信息，以及具体的服务和实参
        //使用反射调用
        if(registryMap.containsKey(request.getClassName())){
         Object clazz = registryMap.get(request.getClassName());
         Method method = clazz.getClass().getMethod(request.getMethodName(),
request.getParames());
         result = method.invoke(clazz, request.getValues());
        }
        ctx.write(result);
        ctx.flush();
        ctx.close();
    }

    @Override
    public void exceptionCaught(ChannelHandlerContext ctx, Throwable cause) throws Exception
{
         cause.printStackTrace();
         ctx.close();
    }


    /*
     * 递归扫描
     */
     private void scannerClass(String packageName){
        URL url = this.getClass().getClassLoader().getResource(packageName.replaceAll("\\.",
"/"));
         File dir = new File(url.getFile());
         for (File file : dir.listFiles()) {
             //如果是一个文件夹，继续递归
             if(file.isDirectory()){
                 scannerClass(packageName + "." + file.getName());
             }else{
                 classNames.add(packageName + "." + file.getName().replace(".class",
"").trim());
             }
         }
     }
```

```
    /**
     * 完成注册
     */
    private void doRegister(){
        if(classNames.size() == 0){ return; }
        for (String className : classNames) {
            try {
                Class<?> clazz = Class.forName(className);
                Class<?> i = clazz.getInterfaces()[0];
                registryMap.put(i.getName(), clazz.newInstance());
            } catch (Exception e) {
                e.printStackTrace();
            }
        }
    }
}
```

至此，注册中心的基本功能就完成了。

5.3.5　实现 Consumer 远程调用

梳理一下基本的实现思路，主要完成一个这样的功能：API 模块中的接口功能在服务端实现（并没有在客户端实现）。因此，客户端调用 API 中定义的某一个接口方法时，实际上是要发起一次网络请求去调用服务端的某一个服务。而这个网络请求被注册中心接受，由注册中心先确定需要调用的服务的位置，再将请求转发至真实的服务实现，最终调用服务端代码，将返回值通过网络传输给客户端。整个过程对于客户端而言是完全无感知的，就像调用本地方法一样。具体调用过程如下图所示。

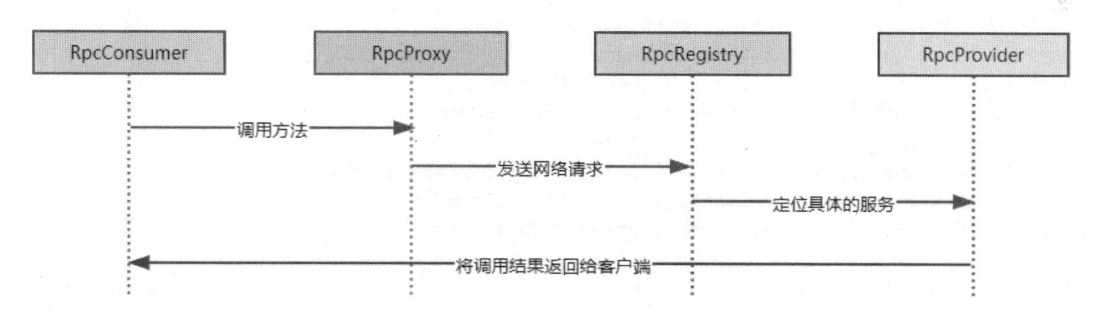

下面来看代码实现，创建 RpcProxy 类。

```
package com.gupaoedu.vip.netty.rpc.consumer.proxy;

import java.lang.reflect.Proxy;
```

```
public class RpcProxy {

    public static <T> T create(Class<?> clazz){
          //clazz 传进来本身就是 interface
        MethodProxy proxy = new MethodProxy(clazz);
        Class<?> [] interfaces = clazz.isInterface() ?
                              new Class[]{clazz} :
                              clazz.getInterfaces();
        T result = (T) Proxy.newProxyInstance(clazz.getClassLoader(),interfaces,proxy);
        return result;
    }

}
```

在 RpcProxy 类的内部实现远程方法调用的代理类，由 Netty 发送网络请求，具体代码如下。

```
package com.gupaoedu.vip.netty.rpc.consumer.proxy;

import java.lang.reflect.InvocationHandler;
import java.lang.reflect.Method;
import java.lang.reflect.Proxy;

import com.gupaoedu.vip.netty.rpc.protocol.InvokerProtocol;

import io.netty.bootstrap.Bootstrap;
import io.netty.channel.ChannelFuture;
import io.netty.channel.ChannelInitializer;
import io.netty.channel.ChannelOption;
import io.netty.channel.ChannelPipeline;
import io.netty.channel.EventLoopGroup;
import io.netty.channel.nio.NioEventLoopGroup;
import io.netty.channel.socket.SocketChannel;
import io.netty.channel.socket.nio.NioSocketChannel;
import io.netty.handler.codec.LengthFieldBasedFrameDecoder;
import io.netty.handler.codec.LengthFieldPrepender;
import io.netty.handler.codec.serialization.ClassResolvers;
import io.netty.handler.codec.serialization.ObjectDecoder;
import io.netty.handler.codec.serialization.ObjectEncoder;

public class RpcProxy {

    public static <T> T create(Class<?> clazz){
        ...
    }

    private static class MethodProxy implements InvocationHandler {
```

```
    private Class<?> clazz;
    public MethodProxy(Class<?> clazz){
        this.clazz = clazz;
    }

    public Object invoke(Object proxy, Method method, Object[] args)  throws Throwable {
        //如果传进来是一个已实现的具体类（本次演示略过此逻辑)
        if (Object.class.equals(method.getDeclaringClass())) {
            try {
                return method.invoke(this, args);
            } catch (Throwable t) {
                t.printStackTrace();
            }
            //如果传进来的是一个接口（核心）
        } else {
            return rpcInvoke(proxy,method, args);
        }
        return null;
    }


    /**
     * 实现接口的核心方法
     * @param method
     * @param args
     * @return
     */
    public Object rpcInvoke(Object proxy,Method method,Object[] args){

        //传输协议封装
        InvokerProtocol msg = new InvokerProtocol();
        msg.setClassName(this.clazz.getName());
        msg.setMethodName(method.getName());
        msg.setValues(args);
        msg.setParames(method.getParameterTypes());

        final RpcProxyHandler consumerHandler = new RpcProxyHandler();
        EventLoopGroup group = new NioEventLoopGroup();
        try {
            Bootstrap b = new Bootstrap();
            b.group(group)
                    .channel(NioSocketChannel.class)
                    .option(ChannelOption.TCP_NODELAY, true)
                    .handler(new ChannelInitializer<SocketChannel>() {
                        @Override
                        public void initChannel(SocketChannel ch) throws Exception {
```

```
    ChannelPipeline pipeline = ch.pipeline();
    //自定义协议解码器
    /** 入参有 5 个，分别解释如下
    maxFrameLength：框架的最大长度。如果帧的长度大于此值，则将抛出 TooLongFrameException
    lengthFieldOffset：长度属性的偏移量。即对应的长度属性在整个消息数据中的位置
    lengthFieldLength：长度属性的长度。如果长度属性是 int 型，那么这个值就是 4（long 型就是 8）
    lengthAdjustment：要添加到长度属性值的补偿值
    initialBytesToStrip：从解码帧中去除的第一个字节数
    */
    pipeline.addLast("frameDecoder",
                         new LengthFieldBasedFrameDecoder(Integer.MAX_VALUE, 0, 4, 0, 4));
    //自定义协议编码器
    pipeline.addLast("frameEncoder",new LengthFieldPrepender(4));
    //对象参数类型编码器
    pipeline.addLast("encoder", new ObjectEncoder());
    //对象参数类型解码器
    pipeline.addLast("decoder",
                new ObjectDecoder(Integer.MAX_VALUE,ClassResolvers.cacheDisabled(null)));
    pipeline.addLast("handler",consumerHandler);

                         }
                     });

                ChannelFuture future = b.connect("localhost", 8080).sync();
                future.channel().writeAndFlush(msg).sync();
                future.channel().closeFuture().sync();
            } catch(Exception e){
                e.printStackTrace();
            }finally {
                group.shutdownGracefully();
            }
            return consumerHandler.getResponse();
        }

    }
}
```

接收网络调用的返回值，代码如下。

```
package com.gupaoedu.vip.netty.rpc.consumer.proxy;

import io.netty.channel.ChannelHandlerContext;
import io.netty.channel.ChannelInboundHandlerAdapter;

public class RpcProxyHandler extends ChannelInboundHandlerAdapter {

    private Object response;
```

```
    public Object getResponse() {
        return response;
    }

    @Override
    public void channelRead(ChannelHandlerContext ctx, Object msg) throws Exception {
        response = msg;
    }

    @Override
    public void exceptionCaught(ChannelHandlerContext ctx, Throwable cause) throws Exception
{
        System.out.println("client exception is general");
    }
}
```

完成客户端调用的代码如下。

```
package com.gupaoedu.vip.netty.rpc.consumer;

import com.gupaoedu.vip.netty.rpc.api.IRpcService;
import com.gupaoedu.vip.netty.rpc.api.IRpcHelloService;
import com.gupaoedu.vip.netty.rpc.consumer.proxy.*;

public class RpcConsumer {

    public static void main(String [] args){
        IRpcHelloService rpcHello = RpcProxy.create(IRpcHelloService.class);

        System.out.println(rpcHello.hello("Tom 老师"));

        IRpcService service = RpcProxy.create(IRpcService.class);

        System.out.println("8 + 2 = " + service.add(8, 2));
        System.out.println("8 - 2 = " + service.sub(8, 2));
        System.out.println("8 * 2 = " + service.mult(8, 2));
        System.out.println("8 / 2 = " + service.div(8, 2));
    }

}
```

5.3.6　Monitor 监控

Dubbo 中的 Monitor 是用 Spring 的 AOP 埋点来实现的，我们没有引入 Spring 框架，在本节中不实现监控的功能。感兴趣的小伙伴可以回顾之前 Spring AOP 的内容自行完善此功能。

5.4 运行效果演示

第一步，启动注册中心，运行结果如下图所示。

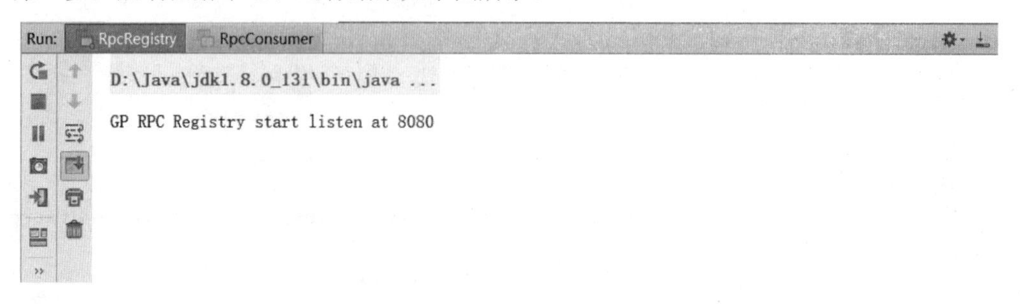

第二步，运行客户端，调用结果如下图所示。

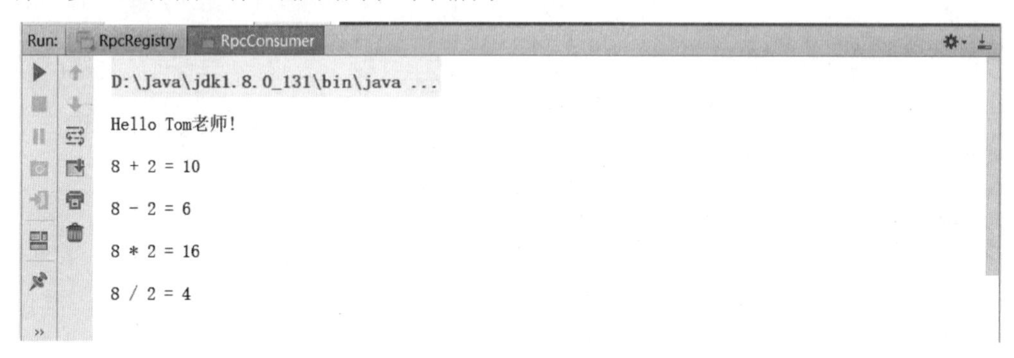

通过以上案例演示，相信小伙伴们对 Netty 的应用已经有了一个比较深刻的印象。在之后的内容中，我们继续深入分析 Netty 的底层原理。本章内容只是对 RPC 的基本实现原理做了一个简单的实现，感兴趣的小伙伴可以在本项目的基础上继续完善 RPC 的其他细节。

3

第 3 篇

Netty 核心篇

第 6 章 Netty 高性能之道

6.1 背景介绍

6.1.1 Netty 惊人的性能数据

早前，我们做过一次测验，使用 Netty（NIO 框架）完成一次 RPC 网络通信，相比于传统的基于 Java 序列化+BIO 的通信框架，其通信性能提升了 8 倍多。然而，笔者对这个数据并不感到惊讶，根据笔者多年的 NIO 编程经验，通过选择合适的 NIO 框架，精心地设计 Reactor 线程模型，达到上述性能指标是完全有可能的。

6.1.2 传统 RPC 调用性能差的“三宗罪”

第一宗罪：网络传输方式存在弊端。传统的 RPC 框架或者基于 RMI 等方式的远程服务（过程）调用都是采用 BIO，当客户端的并发压力或网络时延增大的时候，BIO 会因频繁的“wait”导致 I/O 线程经常出现阻塞的情况，由于线程本身无法高效地工作，I/O 处理能力自然就会下降。下面通过 BIO 通信模型图看一下 BIO 通信的弊端。

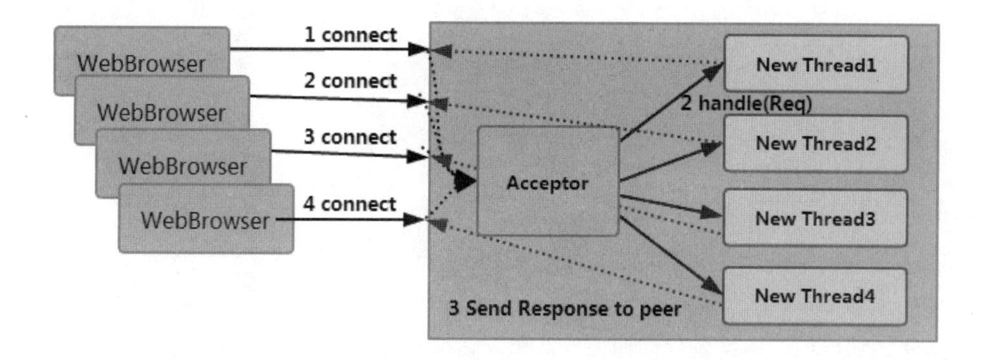

采用 BIO 通信模型的服务端，通常由一个独立的 Acceptor 线程负责监听客户端的连接，接收到客户端连接之后为客户端连接创建一个新的线程处理请求消息，处理完成之后，返回应答消息给客户端，线程销毁，这就是典型的一请求一应答模型。这样的架构设计，最大的问题就是无法进行弹性伸缩。当用户访问量剧增时，并发量自然上升，而服务端的线程个数和并发访问数成线性正比。由于线程是 JVM 非常宝贵的系统资源，所以随着并发量的持续增加、线程数急剧膨胀，系统的性能也急剧下降，可能会发生句柄溢出和线程堆栈溢出等问题，最终可能会导致服务器宕机。

第二宗罪：序列化方式存在弊端。Java 序列化存在如下几个较为典型的问题。

（1）Java 序列化是 Java 内部针对对象（Object）设计的编解码技术，无法跨语言使用。如果在异构系统之间对接，Java 序列化后的字节码流需要能够通过其他语言反序列化成原始对象（即副本），在目前的技术环境下无法支持。

（2）相比于其他开源的序列化框架，Java 序列化后的字节码流占用的空间太大，无论是网络传输还是持久化到磁盘，都会增加资源的消耗。

（3）序列化性能较差，在编解码过程中需要占用更高的 CPU 资源。

第三宗罪：线程模型存在弊端。由于传统的 RPC 框架均采用 BIO 模型，这使得每个 TCP 连接都需要分配 1 个线程，而线程资源是 JVM 非常宝贵的资源，当 I/O 读写阻塞导致线程无法及时释放时，会导致系统性能急剧下降，甚至会导致虚拟机无法创建新的线程。

6.1.3　Netty 高性能的三个主题

Netty 高性能的三个主题如下图所示。

（1）I/O 传输模型：用什么样的通道将数据发送给对方，是 BIO、NIO 还是 AIO，I/O 传输模型在很大程度上决定了框架的性能。

（2）数据协议：采用什么样的通信协议，是 HTTP 还是内部私有协议。协议的选择不同，性能模型也就不同。一般来说内部私有协议比公有协议的性能更高。

（3）线程模型：线程模型涉及如何读取数据包，读取之后的编解码在哪个线程中进行，编解码后的消息如何派发等方面。线程模型设计得不同，对性能也会产生非常大的影响。

6.2 Netty 高性能之核心法宝

6.2.1 异步非阻塞通信

在 I/O 编程过程中，当需要同时处理多个客户端接入请求的时候，我们可以利用多线程或者多路复用 I/O 技术来实现。多路复用 I/O 就是把多个 I/O 的阻塞复用到同一个 Selector 的阻塞上，从而达到系统在单线程的情况下也可以同时处理多个客户端请求的目的。与传统的多进程/线程模型相比，多路复用 I/O 的最大优势是系统开销小，系统不再需要新的进程或线程，也不再需要维护新创建的进程或线程的运行，降低了系统的维护工作量，减轻了系统开销。

从 JDK 1.4 开始，Java API 就提供了对非阻塞 I/O（即 NIO）的支持，从 JDK 1.5_update10 开始，采用 epoll 模型替代传统的 select/poll 模型，极大地提升了 NIO 通信的性能。与 Socket 类和 ServerSocket 类相对应，NIO 也提供了 SocketChannel 和 ServerSocketChannel 两种不同的套接字通道实现。这两种新增的 Channel 都支持阻塞和非阻塞两种 I/O 模式。阻塞 I/O 模式的 API 使用起来非常简单，但在实际应用中，其性能和可靠性都表现得不太好。而非阻塞 I/O 模式恰好相反，API 使用相对复杂，但性能和可靠性表现更优。开发者可以根据具体的业务场景来选择更合适的 I/O 模式。一般来说，低负载、低并发的应用程序可以选择 BIO 以降低编程复杂度。但是对于高负载、高并发的网络应用程序，通常会采用 NIO 的非阻塞模式进行开发。

接下来要介绍的Netty就是一个满足高性能、高并发的网络通信框架。Netty底层是采用Reactor线程模型来设计和实现的，先来看Netty服务端API的通信步骤，其序列图如下图所示。

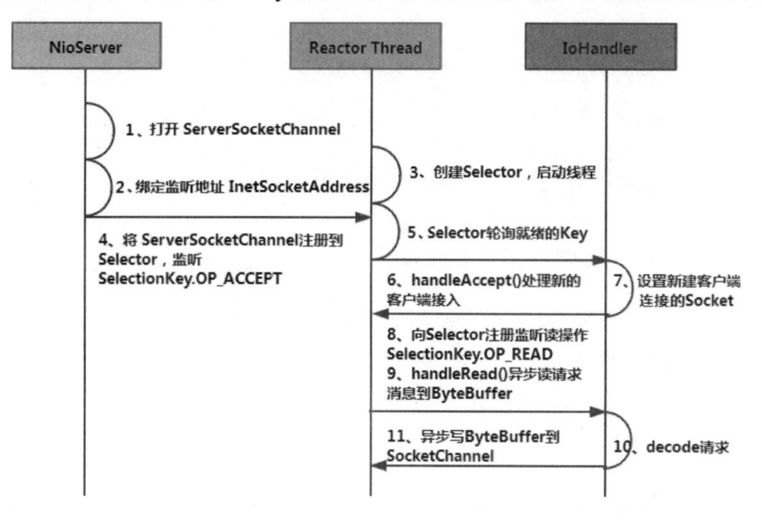

再来看一下Netty客户端API的通信步骤，其序列图如下图所示。

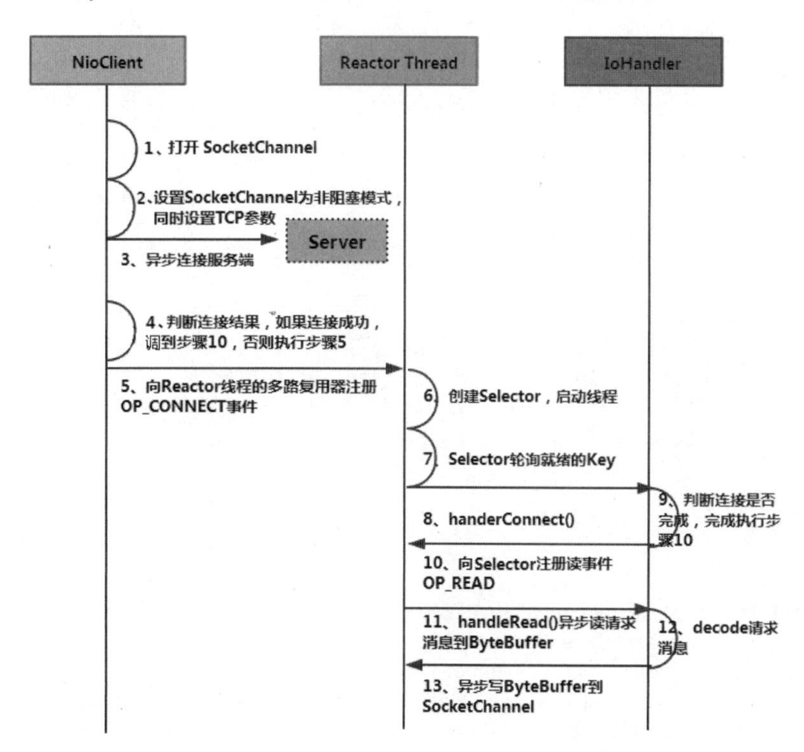

通过上面的序列图，我们大概能够了解到 Netty 的 I/O 线程 NioEventLoop 聚合了 Selector，可以同时并发处理成百上千个客户端 Channel，而且它的读写操作都是非阻塞的，这可以大幅提升 I/O 线程的运行效率，避免由于频繁 I/O 阻塞导致的线程挂起。另外，由于 Netty 采用的是异步通信模式，单个 I/O 线程也可以并发处理多个客户端连接和读写操作，所以从根本上解决了传统 BIO 的单连接单线程模型的弊端，使整个系统的性能、弹性伸缩性能和可靠性都得到了极大的提升。

6.2.2 零拷贝

Netty 的零拷贝主要体现在如下三个方面。

（1）Netty 接收和发送 ByteBuffer 采用 DirectBuffer，使用堆外直接内存进行 Socket 读写，不需要进行字节缓冲区的二次拷贝。如果使用传统的堆存（Heap Buffer）进行 Socket 读写，那么 JVM 会将堆存拷贝一份到直接内存中，然后才写入 Socket。相比于堆外直接内存，消息在发送过程中多了一次缓冲区的内存拷贝。

（2）Netty 提供了组合 Buffer 对象，可以聚合多个 ByteBuffer 对象，用户可以像操作一个 Buffer 那样方便地对组合 Buffer 进行操作，避免了传统的通过内存拷贝的方式将几个小 Buffer 合并成一个大 Buffer 的烦琐操作。

（3）Netty 中文件传输采用了 transferTo()方法，它可以直接将文件缓冲区的数据发送到目标 Channel，避免了传统通过循环 write()方式导致的内存拷贝问题。

下面我们针对上述三种对零拷贝的描述在源码中进行验证，以加深理解。先看第一种 Netty 对于堆外直接内存的使用，AbstractNioByteChannel$NioByteUnsafe 的源码如下。

```
public final void read() {
    final ChannelConfig config = config();
    final ChannelPipeline pipeline = pipeline();
    final ByteBufAllocator allocator = config.getAllocator();
    final RecvByteBufAllocator.Handle allocHandle = recvBufAllocHandle();
    allocHandle.reset(config);

    ByteBuf byteBuf = null;
    boolean close = false;
    try {
        do {
            byteBuf = allocHandle.allocate(allocator);
            allocHandle.lastBytesRead(doReadBytes(byteBuf));
            if (allocHandle.lastBytesRead() <= 0) {
                    // nothing was read. release the buffer.
                    byteBuf.release();
```

```
                    byteBuf = null;
                    close = allocHandle.lastBytesRead() < 0;
                    break;
                }

                allocHandle.incMessagesRead(1);
                readPending = false;
                pipeline.fireChannelRead(byteBuf);
                byteBuf = null;
            } while (allocHandle.continueReading());

                allocHandle.readComplete();
                pipeline.fireChannelReadComplete();

                if (close) {
                    closeOnRead(pipeline);
                }
        } catch (Throwable t) {
                handleReadException(pipeline, byteBuf, t, close, allocHandle);
        } finally {
                //Check if there is a readPending which was not processed yet.
                //This could be for two reasons:
                //* The user called Channel.read() or ChannelHandlerContext.read() in
channelRead(...) method
                //* The user called Channel.read() or ChannelHandlerContext.read() in
channelReadComplete(...) method
                //
                //See https://github.com/netty/netty/issues/2254
                if (!readPending && !config.isAutoRead()) {
                    removeReadOp();
                }
        }
}
```

再找到 do while 中的 allocHandle.allocate() 方法，实际上调用的是 DefaultMaxMessageRecvByteBufAllocator$MaxMessageHandle 的 allocate()方法。

```
public ByteBuf allocate(ByteBufAllocator alloc) {
    return alloc.ioBuffer(guess());
}
```

相当于每循环读取一次消息，就通过 ByteBufferAllocator 的 ioBuffer()方法获取 ByteBuf 对象，继续看它的接口定义。

```
public abstract class ByteBuf implements ReferenceCounted, Comparable<ByteBuf> {
    ...
}
```

当 Socket 进行 I/O 读写的时候，为了避免从堆内存拷贝一份副本到直接内存，Netty 的 ByteBuf 分配器直接创建非堆内存避免缓冲区的二次拷贝，通过零拷贝来提升读写性能。在这里，大家先有个印象，后续章节我们还会详细分析。

下面继续看第二种零拷贝组合 Buffer 的实现类 CompositeByteBuf 类，它将多个 ByteBuf 封装成一个 ByteBuf，对外提供统一封装后的 ByteBuf 接口，它的类定义如下图所示。

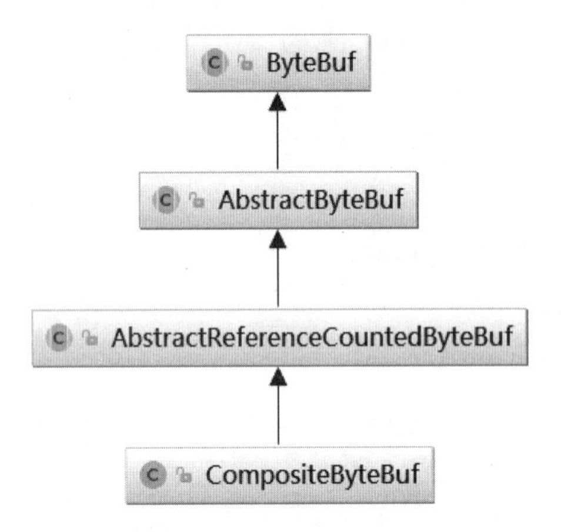

通过继承关系可以看出 CompositeByteBuf 实际就是一个 ByteBuf 的包装器，它将多个 ByteBuf 组合成一个集合，然后对外提供统一的 ByteBuf 接口，相关定义如下。

```
private static final ByteBuffer EMPTY_NIO_BUFFER = Unpooled.EMPTY_BUFFER.nioBuffer();
private static final Iterator<ByteBuf> EMPTY_ITERATOR =
Collections.<ByteBuf>emptyList().iterator();

private final ByteBufAllocator alloc;
private final boolean direct;
private final List<Component> components;
private final int maxNumComponents;

private boolean freed;
```

添加 ByteBuf 不需要做内存拷贝，相关代码如下。

```
private int addComponent0(boolean increaseWriterIndex, int cIndex, ByteBuf buffer) {
    assert buffer != null;
    boolean wasAdded = false;
    try {
        checkComponentIndex(cIndex);

        int readableBytes = buffer.readableBytes();
```

```
        @SuppressWarnings("deprecation")
        Component c = new Component(buffer.order(ByteOrder.BIG_ENDIAN).slice());
        if (cIndex == components.size()) {
            wasAdded = components.add(c);
            if (cIndex == 0) {
                c.endOffset = readableBytes;
            } else {
                Component prev = components.get(cIndex - 1);
                c.offset = prev.endOffset;
                c.endOffset = c.offset + readableBytes;
            }
        } else {
            components.add(cIndex, c);
            wasAdded = true;
            if (readableBytes != 0) {
                updateComponentOffsets(cIndex);
            }
        }
        if (increaseWriterIndex) {
            writerIndex(writerIndex() + buffer.readableBytes());
        }
        return cIndex;
    } finally {
        if (!wasAdded) {
            buffer.release();
        }
    }
}
```

最后，看一下第三种文件传输的零拷贝，transferTo()方法的应用如下。

```
public long transferTo(WritableByteChannel target, long position) throws IOException {
    long count = this.count - position;
    if (count < 0 || position < 0) {
        throw new IllegalArgumentException(
                "position out of range: " + position +
                " (expected: 0 - " + (this.count - 1) + ')');
    }
    if (count == 0) {
        return 0L;
    }
    if (refCnt() == 0) {
        throw new IllegalReferenceCountException(0);
    }
    //Call open to make sure fc is initialized. This is a no-oop if we called it before.
    open();
```

```
    long written = file.transferTo(this.position + position, count, target);
    if (written > 0) {
        transferred += written;
    }
    return written;
}
```

Netty 文件传输 DefaultFileRegion 通过 transferTo()方法将文件发送到目标 Channel 中，下面重点看 FileChannel 的 transferTo()方法，它的 API DOC 说明如下。

```
    /**
     * Transfers bytes from this channel's file to the given writable byte
     * channel.
     *
     * <p> This method is potentially much more efficient than a simple loop
     * that reads from this channel and writes to the target channel.  Many
     * operating systems can transfer bytes directly from the filesystem cache
     * to the target channel without actually copying them.  </p>
     */
    public abstract long transferTo(long position, long count,WritableByteChannel target)
throws IOException;
```

API DOC 说明翻译如下。

> 将文件 Channel 的数据写入指定的 Channel。
>
> 这个方法可能比简单地将数据从一个 Channel 循环读到另一个 Channel 更有效。
>
> 许多操作系统可以直接从文件系统缓存传输字节到目标通道，而不实际拷贝它们。

API DOC 的意思是，我们调用 FileChannel 的 transferTo()方法就实现了零拷贝（想实现零拷贝并不止这一种方法，有更优雅的方法，这里只是作为一个演示）。当然也要看操作系统支不支持底层零拷贝。因为这部分工作其实是由操作系统来完成的。

对于很多操作系统，它接收文件缓冲区的内容直接发送给目标 Channel，而不需要从内核再拷贝到应用程序内存，这种更加高效的传输实现了文件传输的零拷贝。

6.2.3 内存池

随着 JVM 和 JIT（Just-In-Time）即时编译技术的发展，对象的分配和回收已然是一个非常轻量级的工作。但是对于缓冲区来说还有些特殊，尤其是对于堆外直接内存的分配和回收，是一种耗时的操作。为了尽量重复利用缓冲区内存，Netty 设计了一套基于内存池的缓冲区重用机制。下面一起看一下 Netty ByteBuf 的实现，如下图所示。

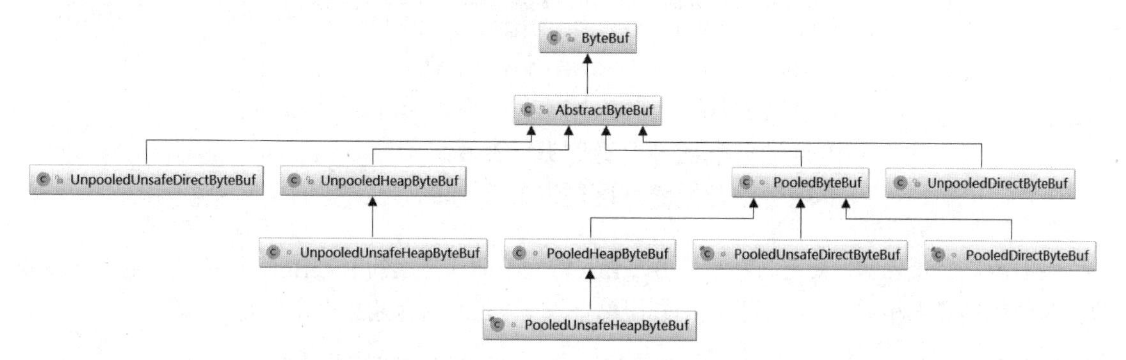

Netty 提供了多种内存管理策略，通过在启动辅助类中配置相关参数，可以实现差异化的个性定制。

下面通过性能测试来看一下基于内存池循环利用的 ByteBuf 和普通 ByteBuf 的性能差异。

编写如下代码，采用内存池分配器创建直接缓冲区。

```
final byte[] CONTENT = new byte[1024];
int loop = 1800000;
long startTime = System.currentTimeMillis();
ByteBuf poolBuffer = null;
for (int i = 0; i < loop; i++) {
        poolBuffer = PooledByteBufAllocator.DEFAULT.directBuffer(1024);
        poolBuffer.writeBytes(CONTENT);
        poolBuffer.release();
}
long endTime = System.currentTimeMillis();
System.out.println("内存池分配缓冲区耗时" + (endTime - startTime) + "ms.");
```

再编写如下代码，采用非堆内存分配器创建直接缓冲区。

```
long startTime2 = System.currentTimeMillis();
ByteBuf buffer = null;
for (int i = 0; i < loop; i++) {
        buffer = Unpooled.directBuffer(1024);
        buffer.writeBytes(CONTENT);
}
endTime = System.currentTimeMillis();
System.out.println("非内存池分配缓冲区耗时" + (endTime - startTime2) + "ms.");
```

各执行 180 万次，其性能对比结果如下图所示。

```
Console ☒  Problems  Progress  Search
<terminated> ChatServer (1) [Java Application] /Library/Java/JavaVi
log4j:WARN No appenders could be
log4j:WARN Please initialize the
log4j:WARN See http://logging.apa
内存池分配缓冲区耗时1079ms.
非内存池分配缓冲区耗时1313ms.
```

性能测试经验表明，采用内存池的 ByteBuf 相比于非内存池的 ByteBuf，性能明显提升（性能数据与使用场景强相关）。下面我们一起简单分析下 Netty 内存池的内存分配。

```
public ByteBuf directBuffer(int initialCapacity, int maxCapacity) {
    if (initialCapacity == 0 && maxCapacity == 0) {
        return emptyBuf;
    }
    validate(initialCapacity, maxCapacity);
    return newDirectBuffer(initialCapacity, maxCapacity);
}
```

继续看 newDirectBuffer()方法。我们发现它是一个抽象方法，由 AbstractByteBufAllocator 的子类负责具体实现，其类关系如下图所示。

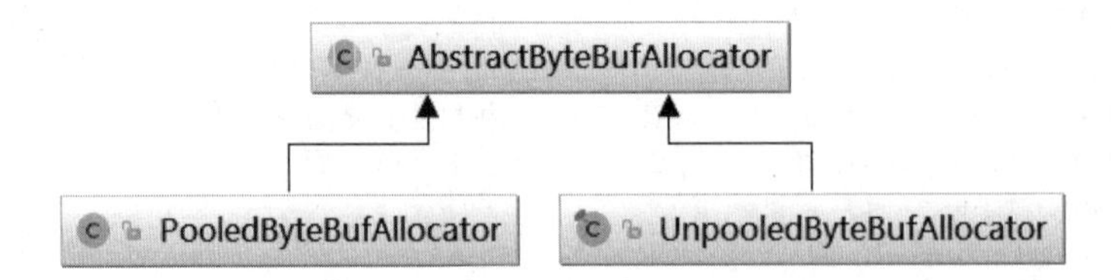

继续进入 PooledByteBufAllocator 的 newDirectBuffer()方法，从 Cache 中获取内存区域 PoolArena，调用它的 allocate()方法进行内存分配。

```
protected ByteBuf newDirectBuffer(int initialCapacity, int maxCapacity) {
    PoolThreadCache cache = threadCache.get();
    PoolArena<ByteBuffer> directArena = cache.directArena;

    ByteBuf buf;
    if (directArena != null) {
        buf = directArena.allocate(cache, initialCapacity, maxCapacity);
    } else {
        if (PlatformDependent.hasUnsafe()) {
            buf = UnsafeByteBufUtil.newUnsafeDirectByteBuf(this, initialCapacity,
maxCapacity);
        } else {
```

```
            buf = new UnpooledDirectByteBuf(this, initialCapacity, maxCapacity);
        }
    }

    return toLeakAwareBuffer(buf);
}
```

PoolArena 的 allocate()方法如下。

```
PooledByteBuf<T> allocate(PoolThreadCache cache, int reqCapacity, int maxCapacity) {
    PooledByteBuf<T> buf = newByteBuf(maxCapacity);
    allocate(cache, buf, reqCapacity);
    return buf;
}
```

下面重点分析 newByteBuf()方法的实现，它同样是一个抽象方法，由子类 DirectArena 和 HeapArena 来实现不同类型的缓冲区分配，如下图所示。

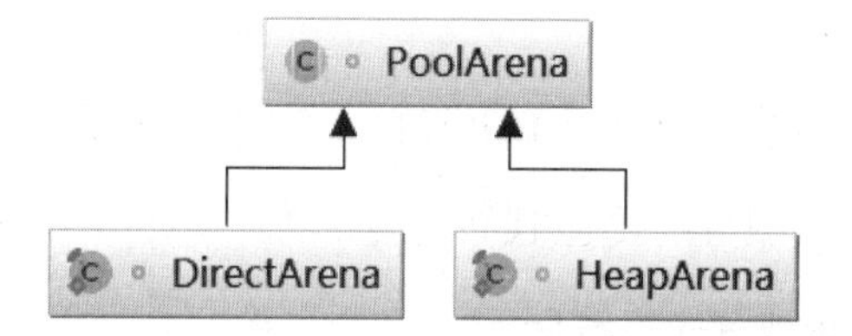

由于测试用例使用的是堆外内存，因此重点分析 DirectArena 的实现。

如果没有开启和使用 JDK 内置的 Unsafe，则执行如下代码。

```
protected PooledByteBuf<ByteBuffer> newByteBuf(int maxCapacity) {
    if (HAS_UNSAFE) {
        return PooledUnsafeDirectByteBuf.newInstance(maxCapacity);
    } else {
        return PooledDirectByteBuf.newInstance(maxCapacity);
    }
}
```

执行 PooledDirectByteBuf 的 newInstance()方法，代码如下。

```
static PooledDirectByteBuf newInstance(int maxCapacity) {
    PooledDirectByteBuf buf = RECYCLER.get();
    buf.reuse(maxCapacity);
    return buf;
}
```

从以上代码中可以看出，通过调用 RECYCLER 的 get()方法获取循环使用的 ByteBuf 对象，判断如果是非内存池实现，则直接创建一个新的 ByteBuf 对象。从缓冲区中获取 ByteBuf 之后，调用 AbstractReferenceCountedByteBuf 的 setRefCnt()方法设置引用计数器，用于对象的引用计数和内存回收（这和 JVM 的垃圾回收机制非常类似）。这里，我们先做简单了解，后面的章节还会深入介绍。

6.2.4 高效的 Reactor 线程模型

常用的 Reactor 线程模型有三种，分别如下。

（1）Reactor 单线程模型。

（2）Reactor 多线程模型。

（3）主从 Reactor 多线程模型。

Reactor 单线程模型，指的是所有的 I/O 操作都在同一个 NIO 线程中完成，NIO 线程的职责如下。

（1）作为 NIO 服务端，接收客户端的 TCP 连接。

（2）作为 NIO 客户端，向服务端发起 TCP 连接。

（3）读取通信对端的请求或者应答消息。

（4）向通信对端发送消息请求或者应答消息。

Reactor 单线程模型的工作方式如下图所示。

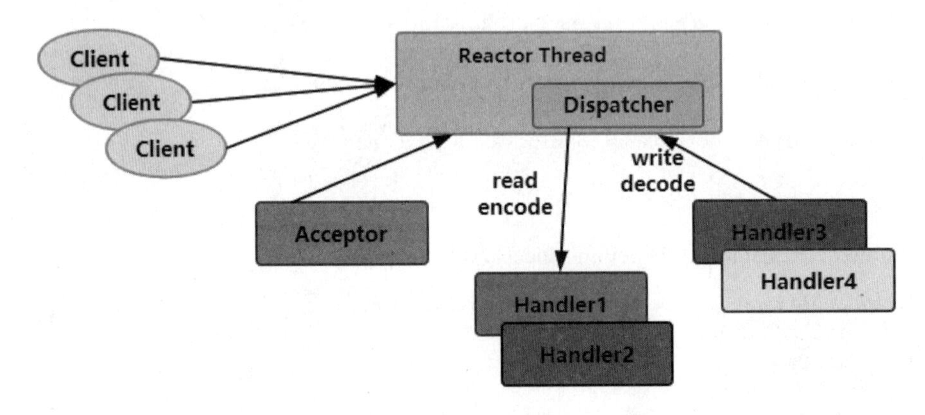

由于 Reactor 模式使用的是 NIO，所有的 I/O 操作都不会阻塞，理论上一个线程可以独立处理所有 I/O 相关的操作。从架构层面看，一个 NIO 线程确实可以完成其承担的职责。从上图中看到，Acceptor 负责接收客户端的 TCP 连接请求消息，链路建立成功之后，通过 Dispatcher 将对应的

ByteBuffer 派发到指定的 Handler 上进行消息解码，用户 Handler 通过 NIO 线程将消息发送给客户端。

对于并发量较小的业务场景，可以使用单线程模型。但单线程模型不适用于高负载、高并发的场景，主要原因如下。

（1）一个 NIO 线程如果同时处理成百上千的链路，则机器在性能上无法满足，即便是 NIO 线程的 CPU 负载达到 100%，也无法满足海量消息的编码、解码、读取和发送。

（2）如果 NIO 线程负载过重，那么处理速度将变慢，从而导致大量客户端连接超时，超时之后往往会进行重发，这更加重了 NIO 线程的负载，最终会导致大量消息积压和处理超时，NIO 线程就会成为系统的性能瓶颈。

（3）一旦 NIO 线程发生意外或者进入死循环状态，就会导致整个系统通信模块不可用，从而不能接收和处理外部消息，造成节点故障。

Reactor 多线程模型就是为了解决以上这些问题而被设计出来的，下面介绍 Reactor 多线程模型。

Rector 多线程模型与单线程模型最大的区别就是设计了一个 NIO 线程池处理 I/O 操作，它的原理如下图所示。

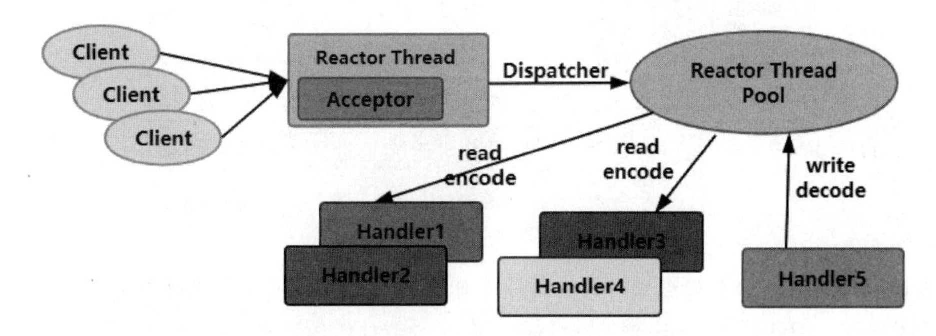

从上图可以看出，Reactor 多线程模型有以下特点。

（1）有一个专门的 NIO 线程 Acceptor 用于监听服务端、接收客户端的 TCP 连接请求。

（2）网络 I/O 的读、写等操作只由一个 NIO 线程池负责，可以采用标准的 JDK 线程池来实现，它包含一个任务队列和多个可用的线程，由这些 NIO 线程负责消息的读取、解码、编码和发送。

（3）一个 NIO 线程可以同时处理多条请求链路，但是一条链路只对应一个 NIO 线程，防止发生并发串行。

在绝大多数场景下，Reactor 多线程模型都可以满足性能需求。但是，在极特殊的应用场景中，一个 NIO 线程负责监听和处理所有的客户端连接也可能会存在性能问题。例如百万客户端并发连接，或者服务端需要对客户端的握手消息进行安全认证，认证本身消耗性能很大。在这类场景下，单个 Acceptor 线程可能会存在性能不足的问题，为了解决性能问题，就出现了主从 Reactor 多线程模型。

主从 Reactor 多线程模型的特点是：服务端用于接收客户端连接的不再是单个 NIO 线程，而是分配了一个独立的 NIO 线程池。Acceptor 接收到客户端 TCP 连接请求并处理完成后（可能包含接入认证等），将新创建的 SocketChannel 注册到 I/O 线程池（Sub Reactor 子线程池）的某个 I/O 线程上，由它负责 SocketChannel 的读写和编解码工作。Acceptor 线程池仅仅用于客户端的登录、握手和安全认证，一旦链路建立成功，就将链路注册到后端 Sub Reactor 子线程池的 I/O 线程上，再由 I/O 线程负责后续的 I/O 操作。其线程模型工作原理如下图所示。

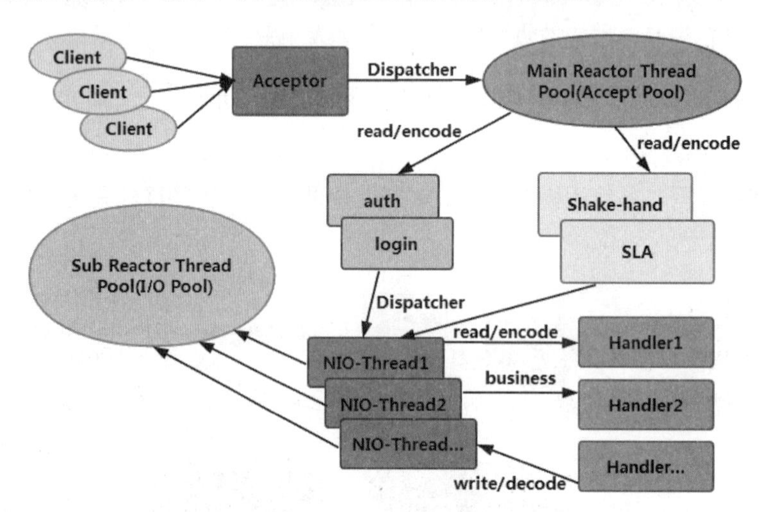

利用主从 Reactor 多线程模型可以解决一个服务端监听线程无法有效处理所有客户端连接的性能不足的问题。因此，在 Netty 的官方 Demo 中，推荐使用该线程模型。

事实上，Netty 的线程模型并非固定不变，通过在启动辅助类中创建不同的 EventLoopGroup 实例并通过适当的参数配置，就可以自由选择上述三种 Reactor 线程模型。正是因为 Netty 对 Reactor 线程模型的支持提供了灵活的定制能力，所以可以满足不同业务场景下的性能需求。

6.2.5 无锁化的串行设计理念

在大多数应用场景下，并行多线程处理可以提升系统的并发性能。但是，如果对共享资源的并发访问处理不当，就会造成严重的锁竞争，最终导致系统性能的下降。为了尽可能避免锁竞争带来的性能损耗，可以通过串行化设计来避免多线程竞争和同步锁，即消息的处理尽可能在同一个线程内完成，不进行线程切换。

为了尽可能提升性能，Netty 采用了无锁化串行设计，在 I/O 线程内部进行串行操作，避免多线程竞争导致的性能下降。表面上看，似乎串行设计的 CPU 利用率不高，并发程度不够。但是，通过调整 NIO 线程池的线程参数，可以同时启动多个串行的线程并行运行，这种局部无锁化的串行线程设计相比一个队列-多个工作线程的模型性能更优。

Netty 的串行设计工作原理如下图所示。

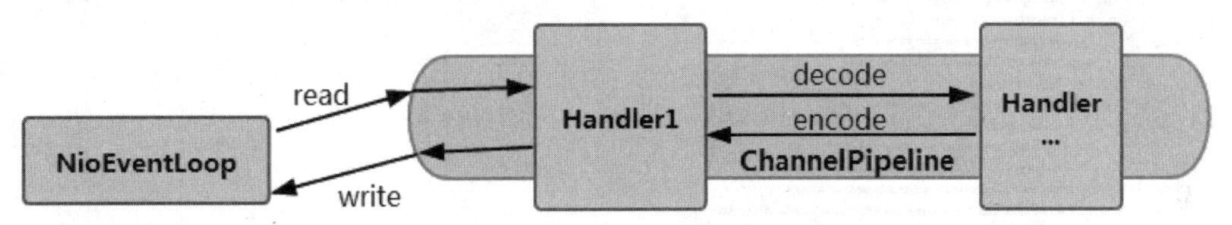

Netty 的 NioEventLoop 读取消息之后，直接调用 ChannelPipeline 的 fireChannelRead(Object msg)，只要用户不主动切换线程，NioEventLoop 就会调用用户的 Handler，期间不进行线程切换，这种串行处理方式避免了多线程操作导致的锁竞争，从性能角度看是最优的。

6.2.6　高效的并发编程

Netty 的高效并发编程主要体现在如下几点。

（1）volatile 关键字的大量且正确的使用。

（2）CAS 和原子类的广泛使用。

（3）线程安全容器的使用。

（4）通过读写锁提升并发性能。

6.2.7　对高性能的序列化框架的支持

影响序列化性能的关键因素总结如下。

（1）序列化后的码流大小（网络带宽的占用）。

（2）序列化/反序列化的性能（CPU 资源占用）。

（3）是否支持跨语言（异构系统的对接和开发语言切换）。

Netty 默认提供了对 Google Protobuf 的支持，用户也可以通过扩展 Netty 的编解码接口接入其他高性能的序列化框架进行编解码，例如 Thrift 的压缩二进制编解码框架。下面我们简单了解一下当前市面上比较流行的序列化和反序列化框架，序列化后的字节码流大小对照如下图所示。

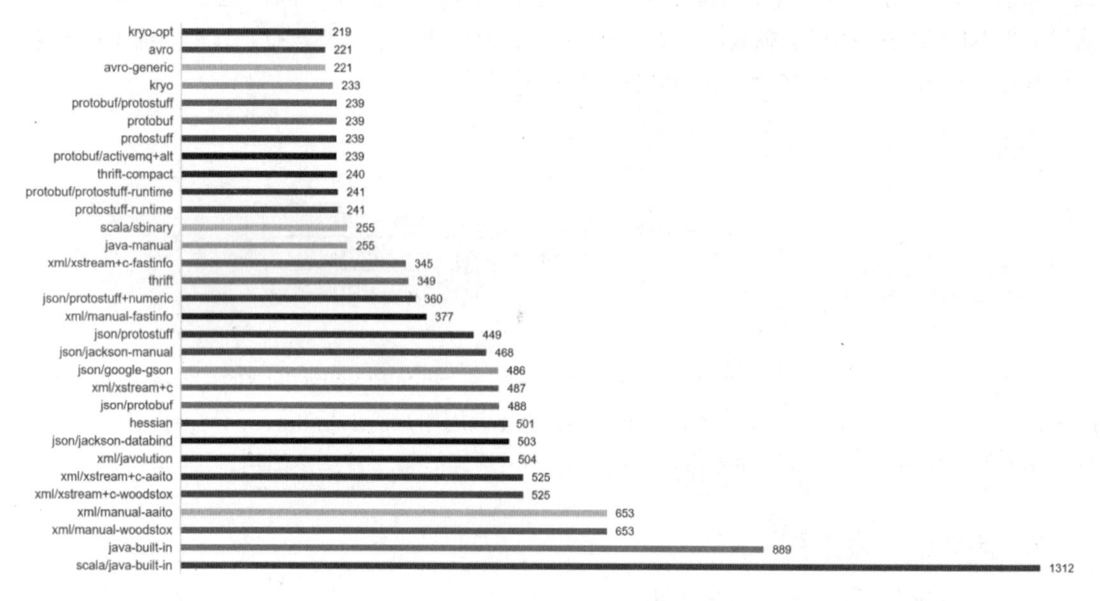

从上图可以看出，Protobuf 序列化后的码流只有 Java 序列化的 1/4 左右。正是由于 Java 原生序列化性能表现太差，才催生出了各种高性能的开源序列化技术和框架（性能差只是其中的一个原因，还有跨语言、IDL 定义等其他因素）。

6.2.8 灵活的 TCP 参数配置能力

合理设置 TCP 参数在某些场景下对性能的提升具有显著的效果，例如 SO_RCVBUF 和 SO_SNDBUF。如果设置不当，对性能的影响也是非常大的。下面我们总结一下对性能影响比较大的几个配置项。

（1）SO_RCVBUF 和 SO_SNDBUF：通常建议值为 128KB 或者 256KB。

（2）SO_TCPNODELAY：Nagle 算法通过将缓冲区内的小封包自动相连，组成较大的封包，阻止大量小封包的发送阻塞网络，从而提高网络应用效率。但是对于延时敏感的应用场景需要关闭该优化算法。

解释： Nagle 算法是以其发明人 John Nagle 的名字命名的，它用于将小的碎片数据连接成更大的报文来最小化所发送的报文的数量。如果需要发送一些较小的报文，则需要禁用该算法。Netty 默认禁用该算法，从而使报文传输延时最小化。

（3）软中断：如果 Linux 内核版本支持 RPS（2.6.35 版本以上），开启 RPS 后可以实现软中断，提升网络吞吐量。RPS 会根据数据包的源地址、目的地址，以及源端口和目的端口进行计算得出一个 Hash 值，然后根据这个 Hash 值来选择软中断 CPU 的运行。从上层来看，也就是说将每个连接和 CPU 绑定，并通过这个 Hash 值在多个 CPU 上均衡软中断，提升网络并行处理性能。

Netty 在启动辅助类中可以灵活地配置 TCP 参数，满足不同的用户场景。相关配置接口定义如下表所示。

参　数	作用范围	解　释
ALLOCATOR	Netty全局参数	ByteBuf的分配器，默认值为ByteBufAllocator.DEFAULT，4.0版本为UnpooledByteBufAllocator，4.1版本为PooledByteBufAllocator，分别对应的字符串值为“unpooled”和“pooled”
ALLOW_HALF_CLOSURE	Netty全局参数	一个连接的远端关闭时本地端是否关闭，默认值为false。值为false时，连接自动关闭
AUTO_READ	Netty全局参数	自动读取，默认值为true。Netty只在必要的时候才设置关心相应的I/O事件。对于读操作，需要调用channel.read()设置关心的I/O事件为OP_READ，这样若有数据到达才能读取以供用户处理
CONNECT_TIMEOUT_MILLIS	Netty全局参数	连接超时毫秒数，默认值为30 000ms，即30s
MESSAGE_SIZE_ESTIMATOR	Netty全局参数	消息大小估算器，默认值为DefaultMessageSizeEstimator.DEFAULT。估算ByteBuf、ByteBufHolder和FileRegion的大小，其中ByteBuf和ByteBufHolder为实际大小，FileRegion估算值为0。该值估算的字节数在计算水位时使用，FileRegion为0可知FileRegion不影响高低水位
RCVBUF_ALLOCATOR	Netty全局参数	用于Channel分配接收Buffer的分配器，默认值为AdaptiveRecvByteBufAllocator.DEFAULT，是一个自适应的接收缓冲区分配器，能根据接收的数据自动调节大小。可选值为FixedRecvByteBufAllocator，固定大小的接收缓冲区分配器
SINGLE_EVENTEXECUTOR_PER_GROUP	Netty全局参数	单线程执行ChannelPipeline中的事件，默认值为true。该值控制执行ChannelPipeline中执行ChannelHandler的线程。如果为true，整个pipeline由一个线程执行，这样不需要进行线程切换以及线程同步，是Netty 4的推荐做法；如果为false，ChannelHandler中的处理过程会由Group中的不同线程执行

续表

参　数	作用范围	解　释
WRITE_BUFFER_WATER_MARK	Netty全局参数	设置某个连接上可以暂存的最大最小Buffer，若该连接等待发送的数据量大于设置的值，则isWritable()会返回不可写。这样，客户端可以不再发送，防止这个量不断地积压，最终可能让客户端挂掉
WRITE_SPIN_COUNT	Netty全局参数	一个Loop写操作执行的最大次数，默认值为16。也就是说，对于大数据量的写操作至多进行16次，如果16次仍没有全部写完数据，那么此时会提交一个新的写任务给EventLoop，任务将在下次调度继续执行。这样，其他写请求才能被响应，不会因为单个大数据量写请求而耽误
IP_MULTICAST_ADDR	IP参数	对应IP参数IP_MULTICAST_IF，设置对应地址的网卡为多播模式
IP_MULTICAST_IF	IP参数	对应IP参数IP_MULTICAST_IF2，同上，但支持IPv6
IP_MULTICAST_LOOP_DISABLED	IP参数	对应IP参数IP_MULTICAST_LOOP，设置本地回环接口的多播功能。由于IP__MULTICAST_LOOP返回true表示关闭，所以Netty加上后缀_DISABLED防止歧义
IP_MULTICAST_TTL	IP参数	多播数据报的Time-to-Live，即存活跳数
IP_TOS	IP参数	设置IP头部的Type-of-Service属性，用于描述IP包的优先级和QoS选项
SO_BACKLOG	Socket参数	服务端接收连接的队列长度，如果队列已满，客户端连接将被拒绝。默认值，Windows为200，其他为128
SO_BROADCAST	Socket参数	设置广播模式
SO_KEEPALIVE	Socket参数	连接保活，默认值为false。启用该功能时，TCP会主动探测空闲连接的有效性。可以将此功能视为TCP的心跳机制，需要注意的是：默认的心跳间隔是7 200s，即2h。Netty默认关闭该功能
SO_LINGER	Socket参数	关闭Socket的延迟时间，默认值为–1，表示禁用该功能。–1表示socket.close()方法立即返回，但操作系统底层会将发送缓冲区的数据全部发送到对端。0表示socket.close()方法立即返回，操作系统放弃发送缓冲区的数据直接向对端发送RST包，对端收到复位错误。非0整数值表示调用socket.close()方法的线程被阻塞直到延迟时间到或缓冲区的数据发送完毕，若超时，则对端会收到复位错误
SO_RCVBUF	Socket参数	TCP数据接收缓冲区大小。该缓冲区即TCP接收滑动窗口，Linux操作系统可使用命令cat /proc/sys/net/ipv4/tcp_rmem查询其大小。一般情况下，该值可由用户在任意时刻设置，但当设置值超过64KB时，需要在连接到远端之前设置

续表

参　　数	作用范围	解　　释
SO_REUSEADDR	Socket参数	地址复用，默认值为false。有四种情况可以使用：（1）当有一个有相同本地地址和端口的Socket1处于TIME_WAIT状态时，你希望启动的程序的Socket2要占用该地址和端口，比如重启服务且保持先前端口。（2）有多块网卡或用IP Alias技术的机器在同一端口启动多个进程，但每个进程绑定的本地IP地址不能相同。（3）单个进程绑定相同的端口到多个Socket上，但每个Socket绑定的IP地址不同。（4）完全相同的地址和端口的重复绑定。但这只用于UDP的多播，不用于TCP
SO_SNDBUF	Socket参数	TCP数据发送缓冲区大小。该缓冲区即TCP发送滑动窗口，Linux操作系统可使用命令cat /proc/sys/net/ipv4/tcp_smem查询其大小
SO_TIMEOUT	Socket参数	用于设置接收数据的等待的超时时间，单位为ms，默认值为0，表示无限等待
TCP_NODELAY	TCP参数	立即发送数据，默认值为ture（Netty默认为true而操作系统默认为false）。该值设置Nagle算法的启用

此表还有不详尽之处，目前大概了解，用作以后备查。

7

第 7 章 揭开 Bootstrap 的神秘面纱

7.1 客户端 Bootstrap

7.1.1 Channel 简介

在 Netty 中，Channel 相当于一个 Socket 的抽象，它为用户提供了关于 Socket 状态（是连接还是断开）及对 Socket 的读、写等操作。每当 Netty 建立了一个连接，都创建一个与其对应的 Channel 实例。

除了 TCP，Netty 还支持很多其他的连接协议，并且每种协议还有 NIO（非阻塞 I/O）和 OIO（Old-I/O，即传统的阻塞 I/O）版本的区别。不同协议不同阻塞类型的连接都有不同的 Channel 类型与之对应，下表对一些常用的 Channel 做了简单介绍。

类　名	解　释
NioSocketChannel	异步非阻塞的客户端TCP Socket连接
NioServerSocketChannel	异步非阻塞的服务端TCP Socket连接
NioDatagramChannel	异步非阻塞的UDP连接
NioSctpChannel	异步的客户端SCTP（Stream Control Transmission Protocol，流控制传输协议）连接

续表

类　名	解　释
NioSctpServerChannel	异步的SCTP服务端连接
OioSocketChannel	同步阻塞的客户端TCP Socket连接
OioServerSocketChannel	同步阻塞的服务端TCP Socket连接
OioDatagramChannel	同步阻塞的UDP连接
OioSctpChannel	同步的SCTP服务端连接
OioSctpServerChannel	同步的客户端TCP Socket连接

我们来看一下 Channel 的总体类图，如下图所示。

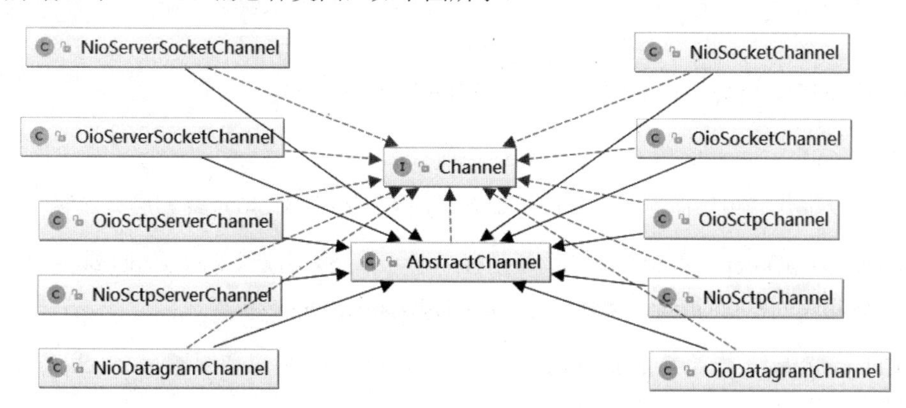

7.1.2　NioSocketChannel 的创建

Bootstrap 是 Netty 提供的一个便利的工厂类，可以通过它来完成客户端或服务端的 Netty 初始化。

先来看一个例子，从客户端和服务端分别分析 Netty 的程序是如何启动的。首先，从客户端的代码片段开始。

```
public class ChatClient {
    public ChatClient connect(int port,String host,final String nickName){
        EventLoopGroup group = new NioEventLoopGroup();
        try {
            Bootstrap bootstrap = new Bootstrap();
            bootstrap.group(group)
            .channel(NioSocketChannel.class)
            .option(ChannelOption.SO_KEEPALIVE, true)
            .handler(new ChannelInitializer<SocketChannel>() {
                @Override
                protected void initChannel(SocketChannel ch) throws Exception {
                    ...
```

```
                }
            });
            //发起同步连接操作
            ChannelFuture channelFuture = bootstrap.connect(host, port).sync();
            channelFuture.channel().closeFuture().sync();
        } catch (InterruptedException e) {
            e.printStackTrace();
        }finally{
            //关闭，释放线程资源
            group.shutdownGracefully();
        }
        return this;
    }

    public static void main(String[] args) {
        new ChatClient().connect(8080, "localhost","Tom 老师");
    }
}
```

上面的客户端代码虽然简单，但是展示了 Netty 客户端初始化时所需的所有内容。

（1）EventLoopGroup：不论是服务端还是客户端，都必须指定 EventLoopGroup。在这个例子中，指定了 NioEventLoopGroup，表示一个 NIO 的 EventLoopGroup。

（2）ChannelType：指定 Channel 的类型。因为是客户端，所以使用了 NioSocketChannel。

（3）Handler：设置处理数据的 Handler。

下面继续看一下客户端启动 Bootstrap 后都做了哪些工作？我们看一下 NioSocketChannel 的类层次结构，如下图所示。

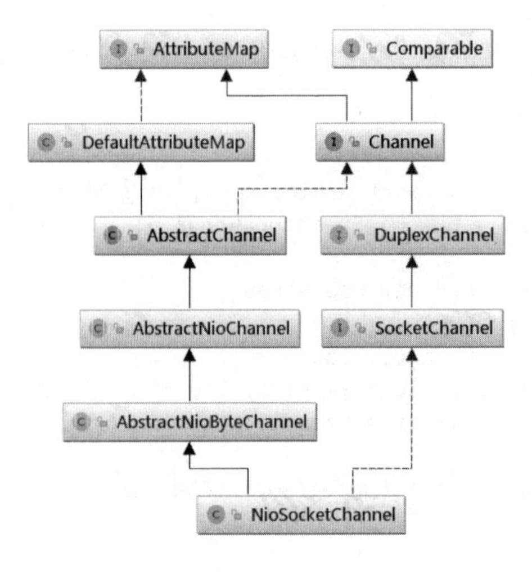

回到我们在客户端连接代码的初始化 Bootstrap 中，该方法调用了一个 channel()方法，传入的参数是 NioSocketChannel.class，在这个方法中其实就是初始化了一个 ReflectiveChannelFactory 的对象，代码实现如下。

```
public B channel(Class<? extends C> channelClass) {
    if (channelClass == null) {
            throw new NullPointerException("channelClass");
    }
    return channelFactory(new ReflectiveChannelFactory<C>(channelClass));
}
```

而 ReflectiveChannelFactory 实现了 ChannelFactory 接口，它提供了唯一的方法，即 newChannel()方法。顾名思义，ChannelFactory 就是创建 Channel 的工厂类。进入 ReflectiveChannelFactory 的 newChannel()方法，我们看其实现代码。

```
public T newChannel() {
    //删除了 try...catch 块
    return clazz.newInstance();
}
```

根据上面代码的提示，我们可以得出下面的结论。

（1）Bootstrap 中的 ChannelFactory 实现类是 ReflectiveChannelFactory。

（2）通过 channel()方法创建的 Channel 具体类型是 NioSocketChannel。

Channel 的实例化过程其实就是调用 ChannelFactory 的 newChannel()方法，而实例化的 Channel 具体类型又和初始化 Bootstrap 时传入的 channel()方法的参数相关。因此对于客户端的 Bootstrap 而言，创建的 Channel 实例就是 NioSocketChannel。

7.1.3　客户端 Channel 的初始化

7.1.2 节中我们已经知道了如何设置一个 Channel 的类型，并且了解到 Channel 是通过 ChannelFactory 的 newChannel()方法来实例化的，那么 ChannelFactory 的 newChannel()方法在哪里调用呢？继续跟踪，我们发现其调用链如下图所示。

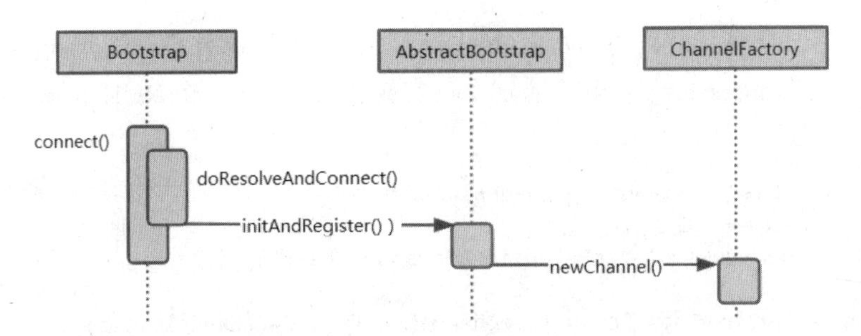

在 AbstractBootstrap 的 initAndRegister()方法中，调用 ChannelFactory 的 newChannel()方法来创建一个 NioSocketChannel 的实例，代码如下。

```
final ChannelFuture initAndRegister() {
    //去掉非关键代码
    Channel channel = channelFactory.newChannel();
    init(channel);

    ChannelFuture regFuture = config().group().register(channel);
    //去掉非关键代码
    return regFuture;
}
```

在 newChannel()方法中，利用反射机制调用类对象的 newInstance()方法来创建一个新的 Channel 实例，相当于调用 NioSocketChannel 的默认构造器。NioSocketChannel 的默认构造器代码如下。

```
public NioSocketChannel() {
    this(DEFAULT_SELECTOR_PROVIDER);
}
```

这里的代码比较关键，我们看到，在这个构造器中首先会调用 newSocket()方法来打开一个新的 Java NIO 的 SocketChannel 对象。

```
private static SocketChannel newSocket(SelectorProvider provider) {
        //删除了 try...catch 块
            return provider.openSocketChannel();
}
```

然后会调用父类，即 AbstractNioByteChannel 的构造器。

```
AbstractNioByteChannel(Channel parent, SelectableChannel ch)
```

同时，传入参数，parent 的值默认为 null，ch 为之前调用 newSocket()方法创建的 Java NIO 的 SocketChannel 对象，因此新创建的 NioSocketChannel 对象中的 parent 暂时是 null。具体代码如下。

```
protected AbstractNioByteChannel(Channel parent, SelectableChannel ch) {
    super(parent, ch, SelectionKey.OP_READ);
}
```

接着会调用父类 AbstractNioChannel 的构造器，并传入实际参数 readInterestOp = SelectionKey.OP_READ。

```
protected AbstractNioChannel(Channel parent, SelectableChannel ch, int readInterestOp) {
       super(parent);
       this.ch = ch;
       this.readInterestOp = readInterestOp;
    //省略...catch 块
       //设置 Java NIO SocketChannel 为非阻塞的
       ch.configureBlocking(false);
}
```

最后会调用父类 AbstractChannel 的构造器。

```
protected AbstractChannel(Channel parent) {
       this.parent = parent;
       id = newId();
       unsafe = newUnsafe();
       pipeline = newChannelPipeline();
}
```

至此，NioSocketChannel 就完成了初始化，我们可以稍微总结一下 NioSocketChannel 初始化所做的工作内容。

（1）调用 NioSocketChannel.newSocket(DEFAULT_SELECTOR_PROVIDER)打开一个新的 Java NioSocketChannel。

（2）初始化 AbstractChannel(Channel parent)对象并给属性赋值，具体赋值的属性如下。

- id：每个 Channel 都会被分配一个唯一的 id。
- parent：属性值默认为 null。
- unsafe：通过调用 newUnsafe()方法实例化一个 Unsafe 对象，它的类型是 AbstractNioByteChannel.NioByteUnsafe 内部类。
- pipeline：是通过调用 new DefaultChannelPipeline(this)新创建的实例。

（3）AbstractNIOChannel 中被赋值的属性如下。

- ch：被赋值为 Java 原生 SocketChannel，即 NioSocketChannel 的 newSocket()方法返回的 Java NIO SocketChannel。

- readInterestOp：被赋值为 SelectionKey.OP_READ。
- ch：被配置为非阻塞，即调用 ch.configureBlocking(false)方法。

（4）NioSocketChannel 中被赋值的属性：config = new NioSocketChannelConfig(this, socket.socket())。

7.1.4 Unsafe 属性的初始化

上节我们简单地提到了，在实例化 NioSocketChannel 的过程中，会在父类 AbstractChannel 的构造方法中调用 newUnsafe()方法来获取一个 Unsafe 实例。那么 Unsafe 是怎么初始化的呢?它的作用是什么?

在实例化 NioSocketChannel 的过程中，Unsafe 就特别关键。Unsafe 其实是对 Java 底层 Socket 操作的封装，因此，它实际上是沟通 Netty 上层和 Java 底层的重要桥梁。下面我们看一下 Unsafe 接口所提供的方法。

```
interface Unsafe {
       RecvByteBufAllocator.Handle recvBufAllocHandle();
       SocketAddress localAddress();
       SocketAddress remoteAddress();
       void register(EventLoop eventLoop, ChannelPromise promise);
       void bind(SocketAddress localAddress, ChannelPromise promise);
       void connect(SocketAddress remoteAddress, SocketAddress localAddress, ChannelPromise
promise);
       void disconnect(ChannelPromise promise);
       void close(ChannelPromise promise);
       void closeForcibly();
       void deregister(ChannelPromise promise);
       void beginRead();
       void write(Object msg, ChannelPromise promise);
       void flush();

       ChannelPromise voidPromise();
       ChannelOutboundBuffer outboundBuffer();
   }
}
```

从上述代码中可以看出，这些方法其实都是与 Java 底层的相关 Socket 的操作相对应的。

继续回到 AbstractChannel 的构造方法中，在这里调用了 newUnsafe()方法获取一个新的 Unsafe 对象，而 newUnsafe()方法在 NioSocketChannel 中被重写了，代码如下。

```
protected AbstractNioUnsafe newUnsafe() {
    return new NioSocketChannelUnsafe();
}
```

NioSocketChannel 的 newUnsafe()方法会返回一个 NioSocketChannelUnsafe 实例。从这里我们就可以确定，在实例化的 NioSocketChannel 中的 Unsafe 属性其实是一个 NioSocketChannelUnsafe 的实例。

7.1.5　ChannelPipeline 的初始化

7.1.3 节中我们分析了 NioSocketChannel 的大体初始化过程，但是漏掉了一个关键的部分，即 ChannelPipeline 的初始化。在 Pipeline 的注释说明中写道“Each channel has its own pipeline and it is created automatically when a new channel is created”。我们知道，在实例化一个 Channel 时，必然都要实例化一个 ChannelPipeline。而我们确实在 AbstractChannel 的构造器中看到了 Pipeline 属性被初始化为 DefaultChannelPipeline 的实例。DefaultChannelPipeline 构造器的代码如下。

```
protected DefaultChannelPipeline(Channel channel) {
    this.channel = ObjectUtil.checkNotNull(channel, "channel");
    succeededFuture = new SucceededChannelFuture(channel, null);
    voidPromise =  new VoidChannelPromise(channel, true);

    tail = new TailContext(this);
    head = new HeadContext(this);

    head.next = tail;
    tail.prev = head;
}
```

DefaultChannelPipeline 的构造器需要传入一个 Channel，而这个 Channel 其实就是我们实例化的 NioSocketChannel 对象，DefaultChannelPipeline 会将这个 NioSocketChannel 对象保存在 Channel 属性中。DefaultChannelPipeline 中还有两个特殊的属性，即 Head 和 Tail，这两个属性是双向链表的头和尾。其实在 DefaultChannelPipeline 中维护了一个以 AbstractChannelHandlerContext 为节点元素的双向链表，这个链表是 Netty 实现 Pipeline 机制的关键。关于 DefaultChannelPipeline 中的双向链表及其所起的作用，本节我们暂不讲解，后续再做深入分析。先看 HeadContext 的继承层次结构，如下图所示。

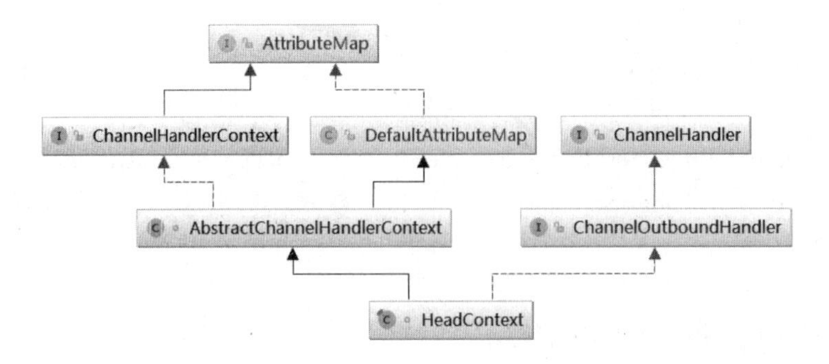

TailContext 的继承层次结构如下图所示。

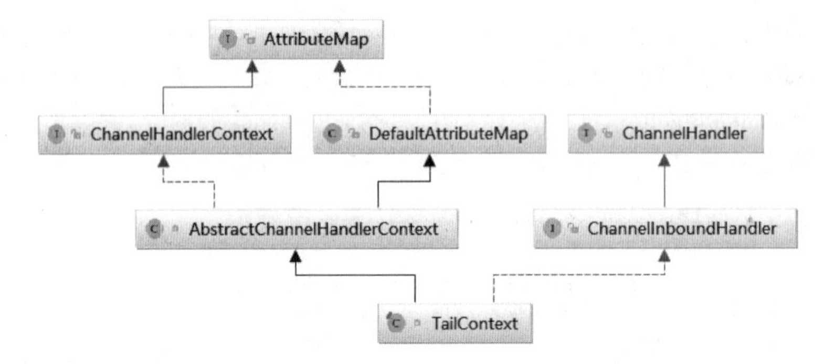

我们可以看到，链表中 Head 是一个 ChannelOutboundHandler，而 Tail 则是一个 ChannelInboundHandler。接着看 HeadContext 的构造器的代码。

```
HeadContext(DefaultChannelPipeline pipeline) {
    super(pipeline, null, HEAD_NAME, false, true);
    unsafe = pipeline.channel().unsafe();
    setAddComplete();
}
```

它调用了父类 AbstractChannelHandlerContext 的构造器，并传入参数 inbound = false，outbound = true。而 TailContext 的构造器与 HeadContext 的相反，它调用了父类 AbstractChannelHandlerContext 的构造器，并传入参数 inbound = true，outbound = false，即 Head 是一个 OutBoundHandler，而 Tail 是一个 InBoundHandler。关于这一特征，大家要特别注意。后续分析到 Netty 的 Pipeline 时，会反复用到 inbound 和 outbound 这两个属性。

7.1.6 EventLoop 的初始化

回到最开始的 ChatClient 用户代码中，我们在一开始就实例化了一个 NioEventLoopGroup 对象，因此就从它的构造器中追踪 EventLoop 的初始化过程。首先来看 NioEventLoopGroup 的类继承层次结构，如下图所示。

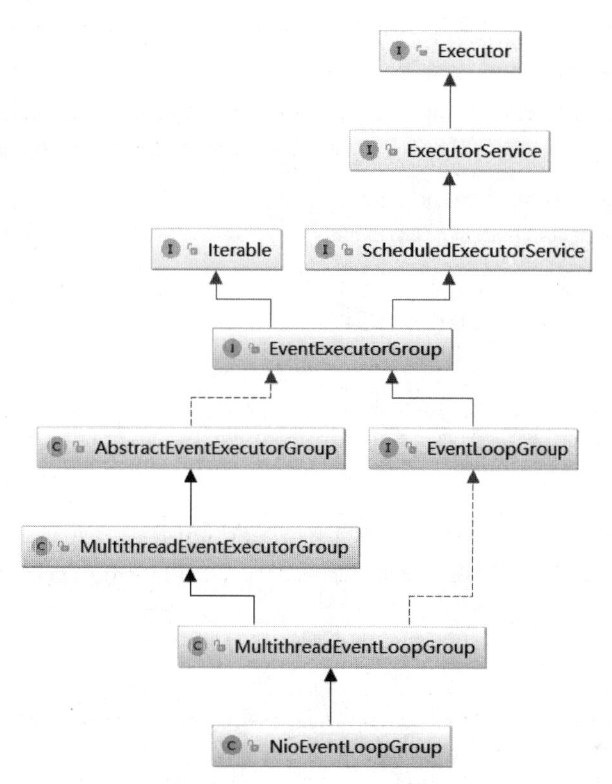

NioEventLoop 有几个重载的构造器，不过内容都没有太大的区别，最终都调用父类 MultithreadEventLoopGroup 的构造器，代码如下。

```
protected MultithreadEventLoopGroup(int nThreads, Executor executor, Object... args) {
    super(nThreads == 0 ? DEFAULT_EVENT_LOOP_THREADS : nThreads, executor, args);
}
```

其中有意思的地方是，如果我们传入的线程数 nThreads 是 0，那么 Netty 会设置默认的线程数 DEFAULT_EVENT_LOOP_THREADS，而这个默认的线程数是怎么确定的呢?

其实很简单，在静态代码块中，首先确定 DEFAULT_EVENT_LOOP_THREADS 的值。

```
static {
    DEFAULT_EVENT_LOOP_THREADS = Math.max(1,

    SystemPropertyUtil.getInt("io.netty.eventLoopThreads",Runtime.getRuntime().availablePr
ocessors() * 2));
}
```

Netty 首先从系统属性中获取“io.netty.eventLoopThreads”的值，如果我们没有设置，就返回默认值，即 CPU 核数×2。回到 MultithreadEventLoopGroup 构造器中会继续调用父类 MultithreadEventExecutorGroup 的构造器。

```
protected MultithreadEventExecutorGroup(int nThreads, Executor executor,
                                        EventExecutorChooserFactory chooserFactory,
Object... args) {
      //去掉了参数检查、异常处理等代码
       children = new EventExecutor[nThreads];

       for (int i = 0; i < nThreads; i ++) {
       //去掉了 try...catch...finally 代码块
            children[i] = newChild(executor, args);
       }
       chooser = chooserFactory.newChooser(children);
       //去掉了包装处理的代码
}
```

继续跟踪 newChooser()方法看其实现逻辑，具体代码如下。

```
public final class DefaultEventExecutorChooserFactory implements EventExecutorChooserFactory
{

    //去掉了定义全局变量的代码

    public EventExecutorChooser newChooser(EventExecutor[] executors) {
        if (isPowerOfTwo(executors.length)) {
            return new PowerOfTwoEventExecutorChooser(executors);
        } else {
            return new GenericEventExecutorChooser(executors);
        }
    }

    private static boolean isPowerOfTwo(int val) {
        return (val & -val) == val;
    }

    private static final class PowerOfTwoEventExecutorChooser implements EventExecutorChooser
{
        private final AtomicInteger idx = new AtomicInteger();
        private final EventExecutor[] executors;

        PowerOfTwoEventExecutorChooser(EventExecutor[] executors) {
            this.executors = executors;
        }
```

```
        public EventExecutor next() {
            return executors[idx.getAndIncrement() & executors.length - 1];
        }
    }

    private static final class GenericEventExecutorChooser implements EventExecutorChooser {
        private final AtomicInteger idx = new AtomicInteger();
        private final EventExecutor[] executors;

        GenericEventExecutorChooser(EventExecutor[] executors) {
            this.executors = executors;
        }

        public EventExecutor next() {
            return executors[Math.abs(idx.getAndIncrement() % executors.length)];
        }
    }
}
```

上面代码主要表达的意思是：如果 nThreads 是 2 的平方，则使用 PowerOfTwoEventExecutorChooser，否则使用 GenericEventExecutorChooser。这两个 Chooser 都重写 next()方法。next()方法的主要功能就是将数组索引循环位移，如下图所示。

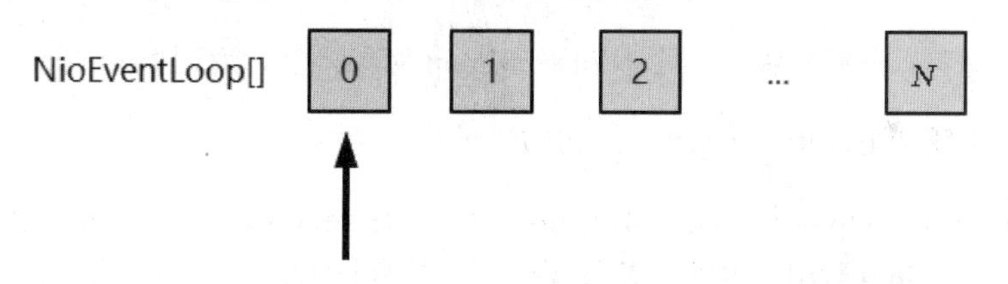

当索引移动到最后一个位置时，再调用 next()方法就会将索引位置重新指向 0，如下图所示。

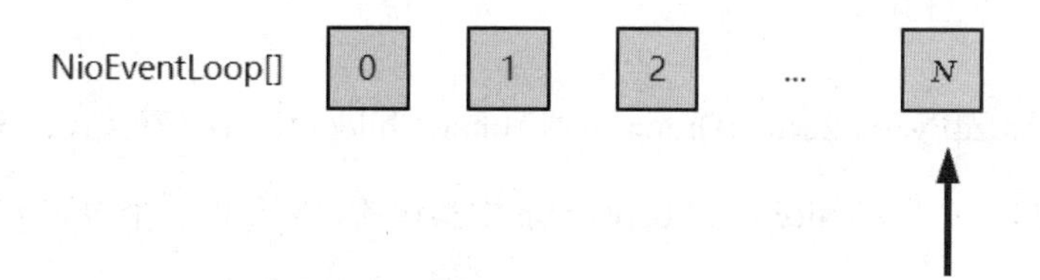

这个运算逻辑其实很简单，就是每次让索引自增后与数组长度取模：idx.getAndIncrement() % executors.length。但是就连一个非常简单的数组索引运算，Netty 都帮我们做了优化。因为在计算机底层，&比%运算效率更高。

分析到这里，我们已经非常清楚 MultithreadEventExecutorGroup 中的处理逻辑，简单总结如下。

（1）创建一个大小为 nThreads 的 SingleThreadEventExecutor 数组。

（2）根据 nThreads 的大小，创建不同的 Chooser，即如果 nThreads 是 2 的平方，则使用 PowerOfTwoEventExecutorChooser，反之使用 GenericEventExecutorChooser。不论使用哪个 Chooser，它们的功能都是一样的，即从 children 数组中选出一个合适的 EventExecutor 实例。

（3）调用 newChild()方法初始化 children 数组。

根据上面的代码，我们也能知道 MultithreadEventExecutorGroup 内部维护了一个 EventExecutor 数组，而 Netty 的 EventLoopGroup 的实现机制其实就建立在 MultithreadEventExecutorGroup 之上。每当 Netty 需要一个 EventLoop 时，都会调用 next()方法获取一个可用的 EventLoop。

上面代码的最后一部分是 newChild()方法，这是一个抽象方法，它的任务是实例化 EventLoop 对象。我们跟踪一下它的代码，可以发现，这个方法在 NioEventLoopGroup 类中有实现，其内容如下。

```
protected EventLoop newChild(Executor executor, Object... args) throws Exception {
    return new NioEventLoop(this, executor, (SelectorProvider) args[0],
                                    ((SelectStrategyFactory)
args[1]).newSelectStrategy(), (RejectedExecutionHandler) args[2]);
}
```

其实逻辑很简单，就是实例化一个 NioEventLoop 对象，然后返回 NioEventLoop 对象。

最后总结一下整个 EventLoopGroup 的初始化过程。

（1）EventLoopGroup（其实是 MultithreadEventExecutorGroup）内部维护一个类型为 EventExecutor 的 children 数组，其大小是 nThreads，这样就构成了一个线程池。

（2）我们在实例化 NioEventLoopGroup 时，如果指定线程池大小，则 nThreads 就是指定的值，反之是 CPU 核数×2。

（3）在 MultithreadEventExecutorGroup 中调用 newChild()象方法来初始化 children 数组。

（4）newChild()方法是在 NioEventLoopGroup 中实现的，它返回一个 NioEventLoop 实例。

（5）初始化 NioEventLoop 对象并给属性赋值，具体赋值的属性如下。

- provider：就是在 NioEventLoopGroup 构造器中，调用 SelectorProvider.provider()方法获取的 SelectorProvider 对象。
- selector：就是在 NioEventLoop 构造器中，调用 provider.openSelector()方法获取的 Selector 对象。

7.1.7　将 Channel 注册到 Selector

在 7.1.3 节的分析中，我们提到 Channel 会在 Bootstrap 的 initAndRegister()中进行初始化，但是这个方法还会将初始化好的 Channe 注册到 NioEventLoop 的 Selector 中。接下来我们分析一下 Channel 注册的过程。

先回顾一下 AbstractBootstrap 的 initAndRegister()方法，代码如下。

```
final ChannelFuture initAndRegister() {
    //删除了非关键代码
    Channelchannel = channelFactory.newChannel();
    init(channel);

    ChannelFuture regFuture = config().group().register(channel);

    return regFuture;
}
```

当 Channel 初始化后，紧接着会调用 group().register()方法来向 Selector 注册 Channel。继续跟踪的话，会发现其调用链如下图所示。

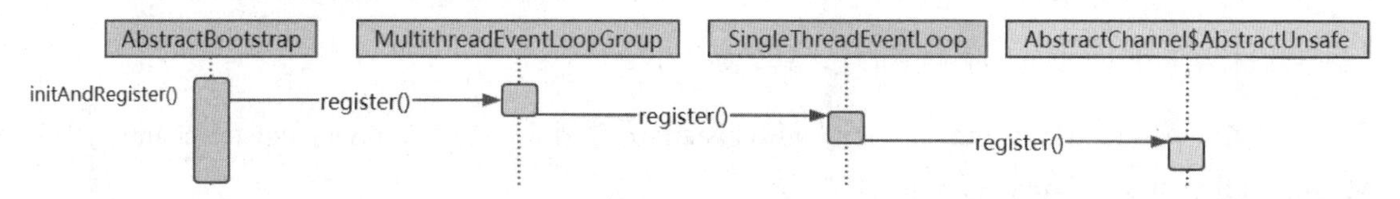

通过跟踪调用链，我们最终发现在 AbstractBootstrap 的 initAndRegister()方法中调用的是 Unsafe 的 register()方法，接下来看一下 AbstractChannel$AbstractUnsafe.register()方法的具体实现代码。

```
public final void register(EventLoop eventLoop, final ChannelPromise promise) {
        //省略了条件判断和错误处理的代码
        AbstractChannel.this.eventLoop = eventLoop;
        register0(promise);
}
```

首先，将 EventLoop 赋值给 Channel 的 eventLoop 属性，我们知道 EventLoop 对象其实是通过 MultithreadEventLoopGroup 的 next()方法获取的，根据前面的分析，可以确定 next()方法返回的 eventLoop 对象是 NioEventLoop 实例。register()方法接着调用了 register0()方法，代码如下。

```
private void register0(ChannelPromise promise) {
        //省略了非关键代码
            boolean firstRegistration = neverRegistered;
            doRegister();
            neverRegistered = false;
            registered = true;
```

```
            pipeline.invokeHandlerAddedIfNeeded();

            safeSetSuccess(promise);
            pipeline.fireChannelRegistered();

            if (isActive()) {
                if (firstRegistration) {
                    pipeline.fireChannelActive();
                }
            }
}
```

register0()方法又调用了 AbstractNioChannel 的 doRegister()方法，代码如下。

```
protected void doRegister() throws Exception {
    //省略了错误处理的代码
    selectionKey = javaChannel().register(eventLoop().selector, 0, this);
}
```

看到 javaChannel()这个方法，我们在前面已经知道了，它返回的是一个 Java NIO 的 SocketChannel 对象，这里我们将 SocketChannel 注册到与 eventLoop 关联的 Selector 上。

我们总结一下 Channel 的注册过程，具体如下。

（1）在 AbstractBootstrap 的 initAndRegister()方法中，通过 group().register(channel)调用 MultithreadEventLoopGroup 的 register()方法。

（2）在 MultithreadEventLoopGroup 的 register()方法中，调用 next()方法获取一个可用的 SingleThreadEventLoop，然后调用它的 register()方法。

（3）在 SingleThreadEventLoop 的 register()方法中，调用 channel.unsafe().register(this, promise)方法获取 Channel 的 unsafe()底层操作对象，然后调用 Unsafe 的 register()方法。

（4）在 AbstractUnsafe 的 register()方法中，调用 register0()方法注册 Channel 对象。

（5）在 AbstractUnsafe 的 register0()方法中，调用 AbstractNioChannel 的 doRegister()方法。

（6）AbstractNioChannel 的 doRegister()方法通过 javaChannel().register(eventLoop().selector, 0, this)将 Channel 对应的 Java NIO 的 SocketChannel 注册到一个 eventLoop 的 Selector 中，并且将当前 Channel 作为 Attachment 与 SocketChannel 关联。

总的来说，Channel 的注册过程所做的工作就是将 Channel 与对应的 EventLoop 进行关联。因此，在 Netty 中，每个 Channel 都会关联一个特定的 EventLoop，并且这个 Channel 中的所有 I/O

操作都是在这个 EventLoop 中执行的；当关联好 Channel 和 EventLoop 后，会继续调用底层 Java NIO 的 SocketChannel 对象的 register()方法，将底层 Java NIO 的 SocketChannel 注册到指定的 Selector 中。通过这两步，就完成了 Netty 对 Channel 的注册过程。

7.1.8 Handler 的添加过程

Netty 有一个强大和灵活之处就是基于 Pipeline 的自定义 Handler 机制。基于此，我们可以像添加插件一样自由组合各种各样的 Handler 来完成业务逻辑。例如我们需要处理 HTTP 数据，那么就可以在 Pipeline 前添加一个针对 HTTP 编解码的 Handler，然后添加我们自己的业务逻辑的 Handler，这样网络上的数据流就像通过一个管道一样，从不同的 Handler 中流过并进行编解码，最终到达我们自定义的 Handler 中。

说到这里，有些小伙伴肯定会好奇，既然这个 Pipeline 机制这么强大，那么它是怎么实现的呢?在此我们不进行详细讲解，而是从简单的内容入手，先体验一下自定义的 Handler 是如何及何时添加到 ChannelPipeline 中的。我们看一下用户代码片段。

```
        //此处省略 N 句代码
        .handler(new ChannelInitializer<SocketChannel>() {
                @Override
                protected void initChannel(SocketChannel ch) throws Exception {
                    ChannelPipeline pipeline = ch.pipeline();
                    pipeline.addLast(new StringDecoder());
                    pipeline.addLast(new StringEncoder());
                    pipeline.addLast(new ChatClientHandler(nickName));
                }
    });
```

这个代码片段就实现了 Handler 的添加功能。我们看到，Bootstrap 的 handler()方法接收一个 ChannelHandler，而我们传入的参数是一个派生于抽象类 ChannelInitializer 的匿名类，它也实现了 ChannelHandler 接口。我们来看 ChannelInitializer 类中到底有什么玄机，代码如下。

```
public abstract class ChannelInitializer<C extends Channel> extends
ChannelInboundHandlerAdapter {
    private static final InternalLogger logger =
InternalLoggerFactory.getInstance(ChannelInitializer.class);
    private final ConcurrentMap<ChannelHandlerContext, Boolean> initMap =
PlatformDependent.newConcurrentHashMap();

    protected abstract void initChannel(C ch) throws Exception;

    @Override
    @SuppressWarnings("unchecked")
```

```
    public final void channelRegistered(ChannelHandlerContext ctx) throws Exception {
        if (initChannel(ctx)) {
            ctx.pipeline().fireChannelRegistered();
        } else {
            ctx.fireChannelRegistered();
        }
    }
    //这个方法在 channelRegistered 中被调用
    private boolean initChannel(ChannelHandlerContext ctx) throws Exception {
        initChannel((C) ctx.channel());
        remove(ctx);
        return false;
    }
    //省略...
}
```

ChannelInitializer 是一个抽象类，它有一个抽象的 initChannel()方法，我们看到的匿名类正是实现了这个方法，并在这个方法中添加了自定义的 Handler。那么 initChannel()方法是在哪里被调用的呢?其实是在 ChannelInitializer 的 channelRegistered()方法中。

接下来关注一下 channelRegistered()方法。我们从上面的代码中可以看到，在 channelRegistered()方法中，会调用 initChannel()方法，将自定义的 Handler 添加到 ChannelPipeline 中，然后调用 ctx.pipeline().remove(this)方法将自己从 ChannelPipeline 中删除。

一开始，ChannelPipeline 中只有三个 Handler，分别是 Head、Tail 和我们添加的 ChannelInitializer，如下图所示。

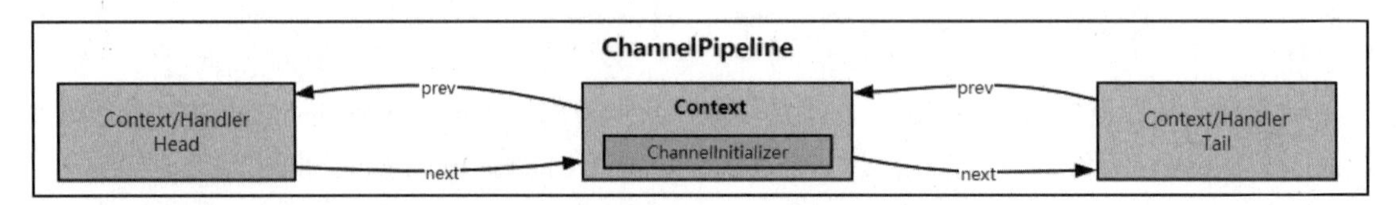

接着调用 initChannel()方法，添加自定义的 Handler，如下图所示。

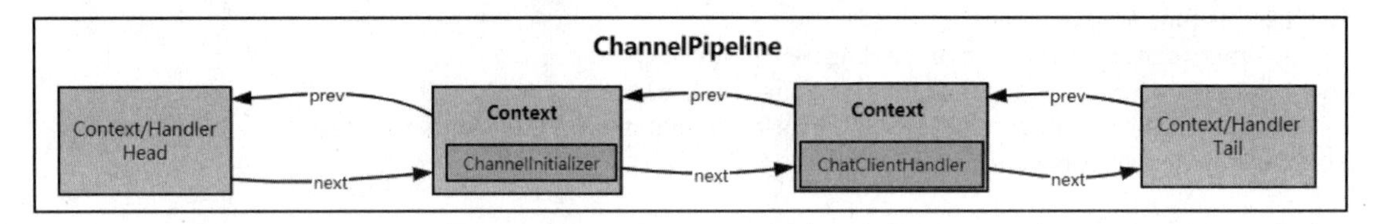

最后将 ChannelInitializer 删除，如下图所示。

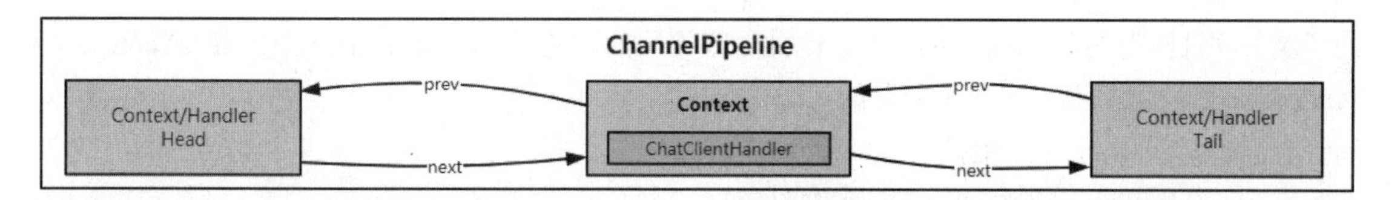

分析到这里，我们已经简单了解了自定义的 Handler 是如何添加到 ChannelPipeline 中的，之后的章节我们再进行深入的探讨。

7.1.9　客户端发起连接请求

经过上面的各种分析后，我们已经大致了解 Netty 客户端初始化时所做的工作，那么接下来就直奔主题，分析一下客户端是如何发起 TCP 连接的。

客户端通过调用 Bootstrap 的 connect()方法进行连接。在 connect()方法中进行一些参数检查，并调用 doConnect()方法，其代码实现如下。

```
private static void doConnect(
        final SocketAddress remoteAddress,
        final SocketAddress localAddress,
        final ChannelPromise connectPromise) {
        final Channel channel = connectPromise.channel();
        channel.eventLoop().execute(new Runnable() {
            @Override
            public void run() {
                if (localAddress == null) {
                    channel.connect(remoteAddress, connectPromise);
                } else {
                    channel.connect(remoteAddress, localAddress, connectPromise);
                }
                connectPromise.addListener(ChannelFutureListener.CLOSE_ON_FAILURE);
            }
       });
}
```

在 doConnect()方法中，eventLoop 线程会调用 Channel 的 connect()方法，而这个 Channel 的具体类型实际就是 NioSocketChannel，前面已经分析过。继续跟踪 channel.connect()方法，我们发现它调用的是 DefaultChannelPipeline 的 connect()方法，Pipeline 的 connect()方法的代码如下。

```
public final ChannelFuture connect(SocketAddress remoteAddress, ChannelPromise promise) {
      return tail.connect(remoteAddress, promise);
}
```

我们已经分析过，Tail 是一个 TailContext 的实例，而 TailContext 又是 AbstractChannelHandlerContext

的子类，并且没有实现 connect()方法，因此这里调用的其实是 AbstractChannelHandlerContext 的 connect()方法，我们看一下这个方法的实现代码。

```
public ChannelFuture connect(
            final SocketAddress remoteAddress,
            final SocketAddress localAddress, final ChannelPromise promise) {
        //删除参数检查的代码
         final AbstractChannelHandlerContext next = findContextOutbound();
         EventExecutor executor = next.executor();
         if (executor.inEventLoop()) {
            next.invokeConnect(remoteAddress, localAddress, promise);
         } else {
            safeExecute(executor, new Runnable() {
                @Override
                public void run() {
                    next.invokeConnect(remoteAddress, localAddress, promise);
                }
            }, promise, null);
         }
         return promise;
}
```

上面代码片段中有一句非常关键的代码，即 final AbstractChannelHandlerContext next = findContextOutbound()，这里调用 findContextOutbound()方法，从 DefaultChannelPipeline 内的双向链表的 Tail 开始，不断向前找到第一个 Outbound 为 true 的 AbstractChannelHandlerContext，然后调用它的 invokeConnect()方法，代码如下。

```
private void invokeConnect(SocketAddress remoteAddress, SocketAddress localAddress,
ChannelPromise promise) {
 //忽略 try...catch 块
 ((ChannelOutboundHandler) handler()).connect(this, remoteAddress, localAddress, promise);

}
```

在 7.1.5 节，我们提到，在 DefaultChannelPipeline 的构造器中，实例化了两个对象：Head 和 Tail，并形成了双向链表的头和尾。Head 是 HeadContext 的实例，它实现了 ChannelOutboundHandler 接口，并且它的 Outbound 设置为 true。因此在 findContextOutbound()方法中，找到的 AbstractChannelHandlerContext 对象其实就是 Head，进而在 invokeConnect()方法中，我们向上转换为 ChannelOutboundHandler 就没问题了。而又因为 HeadContext 重写了 connect()方法，所以实际上调用的是 HeadContext 的 connect()方法。接着跟踪 HeadContext 的 connect()方法。

```
public void connect(
            ChannelHandlerContext ctx,
            SocketAddress remoteAddress, SocketAddress localAddress,
```

```
            ChannelPromise promise) throws Exception {
        unsafe.connect(remoteAddress, localAddress, promise);
}
```

这个 connect()方法很简单，只是调用了 Unsafe 的 connect()方法。回顾一下 HeadContext 的构造器，我们发现这个 Unsafe 其实就是 pipeline.channel().unsafe()返回的 Channel 的 Unsafe 属性。到这里为止，我们应该已经知道，其实是 AbstractNioByteChannel.NioByteUnsafe 内部类转了一大圈。最后，我们找到创建 Socket 连接的关键代码继续跟踪，其实调用的就是 AbstractNioUnsafe 的 connect()方法。

```
public final void connect(
            final SocketAddress remoteAddress, final SocketAddress localAddress, final
ChannelPromise promise) {

            //省去前面的判断
            boolean wasActive = isActive();
            if (doConnect(remoteAddress, localAddress)) {
                fulfillConnectPromise(promise, wasActive);
            } else {
                //此处省略 N 行代码
            }
}
```

在这个 connect()方法中，又调用了 doConnect()方法。注意：这个方法并不是 AbstractNioUnsafe 的方法，而是 AbstractNioChannel 的抽象方法。doConnect()方法是在 NioSocketChannel 中实现的，因此进入 NioSocketChannel 的 doConnect()方法，代码如下。

```
protected boolean doConnect(SocketAddress remoteAddress, SocketAddress localAddress) throws
Exception {
        if (localAddress != null) {
            doBind0(localAddress);
        }

        boolean success = false;
        try {
            boolean connected = javaChannel().connect(remoteAddress);
            if (!connected) {
                selectionKey().interestOps(SelectionKey.OP_CONNECT);
            }
            success = true;
            return connected;
        } finally {
            if (!success) {
                doClose();
            }
```

```
        }
}
```

我们终于看到最关键的部分了，庆祝一下！上面代码的功能是，首先获取 Java NIO 的 SocketChannel，然后获取 NioSocketChannel 的 newSocket()方法返回的 SocketChannel 对象；再调用 SocketChannel 的 connect()方法完成 Java NIO 底层的 Socket 连接。总结一下，客户端 Bootstrap 发起连接请求的流程可以用如下时序图直观地展示。

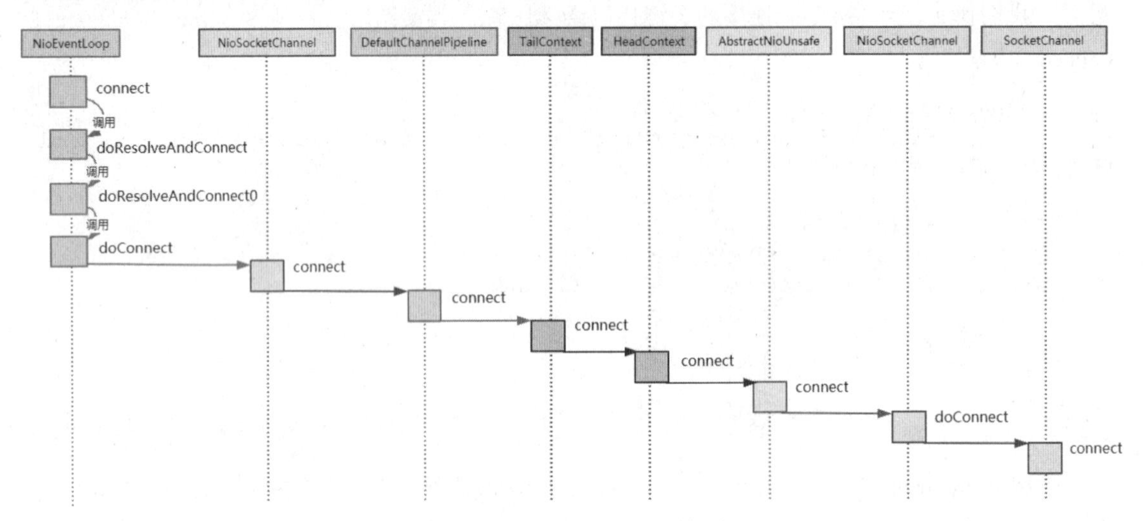

7.2 服务端 ServerBootstrap

在分析客户端的代码时，我们已经对 Bootstrap 启动 Netty 有了一个大致的认识，接下来在分析服务端时，就会相对简单一些了。下面来看服务端的启动代码。

```
public class ChatServer {
    public void start(int port) throws Exception{
        EventLoopGroup bossGroup = new NioEventLoopGroup();
        EventLoopGroup workerGroup = new NioEventLoopGroup();
        try {
            ServerBootstrap b = new ServerBootstrap();
            b.group(bossGroup, workerGroup)
             .channel(NioServerSocketChannel.class)
             .childHandler(new ChannelInitializer<SocketChannel>(){
                  @Override
                  public void initChannel(SocketChannel ch) throws Exception {
                      ...
```

```
                    }
                })
                .option(ChannelOption.SO_BACKLOG, 128)
                .childOption(ChannelOption.SO_KEEPALIVE, true);

            System.out.println("服务已启动,监听端口" + port + "");

            //绑定端口，开始接收连接请求
            ChannelFuture f = b.bind(port).sync();

            //等待服务器 Socket 关闭
            f.channel().closeFuture().sync();

        } finally {
            workerGroup.shutdownGracefully();
            bossGroup.shutdownGracefully();

            System.out.println("服务已关闭");
        }
    }


    public static void main(String[] args) {
        try {
            new ChatServer().start(8080);
        } catch (Exception e) {
            e.printStackTrace();
        }
    }
}
```

服务端基本写法与客户端的代码相比，没有很大的差别，基本上也是进行如下几个部分的初始化。

（1）EventLoopGroup：无论是服务端还是客户端，都必须指定 EventLoopGroup。在上面的代码中，指定了 NioEventLoopGroup，表示一个 NIO 的 EventLoopGroup，不过服务端需要指定两个 EventLoopGroup，一个是 bossGroup，用于处理客户端的连接请求；另一个是 workerGroup，用于处理与各个客户端连接的 I/O 操作。

（2）指定 Channel 的类型。这里是服务端，所以使用了 NioServerSocketChannel。

（3）配置自定义的业务处理器 Handler。

7.2.1 NioServerSocketChannel 的创建

我们在分析客户端 Channel 的初始化过程时已经提到，Channel 是对 Java 底层 Socket 连接的抽象，并且知道客户端 Channel 的具体类型是 NioSocketChannel，由此可知，服务端 Channel 的类型就是 NioServerSocketChannel。

我们按照分析客户端的流程，对服务端的代码也同样分析一遍，这样会方便我们对比服务端和客户端有哪些不同。通过 7.1 节的分析，我们知道，在客户端中，Channel 类型的指定是在初始化时通过 Bootstrap 的 channel()方法设置的，服务端也是同样的方式。

再看服务端代码，我们调用 ServerBootstarap 的 channel(NioServerSocketChannel.class)方法，传入的参数是 NioServerSocketChannel.class 对象。如此，按照与客户端代码同样的流程，我们可以确定 NioServerSocketChannel 的实例化也是通过 ReflectiveChannelFactory 工厂类来完成的，而 ReflectiveChannelFactory 中的 clazz 属性被赋值为 NioServerSocketChannel.class，因此当调用 ReflectiveChannelFactory 的 newChannel()方法时，就能获取一个 NioServerSocketChannel 的实例。newChannel()方法的代码如下。

```
public T newChannel() {
     //删除了 try 代码块
     return clazz.newInstance();
}
```

下面总结一下。

（1）ServerBootstrap 中的 ChannelFactory 的实现类是 ReflectiveChannelFactory 类。

（2）创建的 Channel 具体类型是 NioServerSocketChannel。

Channel 的实例化过程，其实就是调用 ChannelFactory 的 newChannel()方法，而实例化的 Channel 具体类型就是初始化 ServerBootstrap 时传给 channel()方法的实参。因此，上面代码案例中的服务端 ServerBootstrap 创建的 Channel 实例就是 NioServerSocketChannel 的实例。

7.2.2 服务端 Channel 的初始化

我们来分析 NioServerSocketChannel 的实例化过程，先看一下 NioServerSocketChannel 的类层次结构，如下图所示。

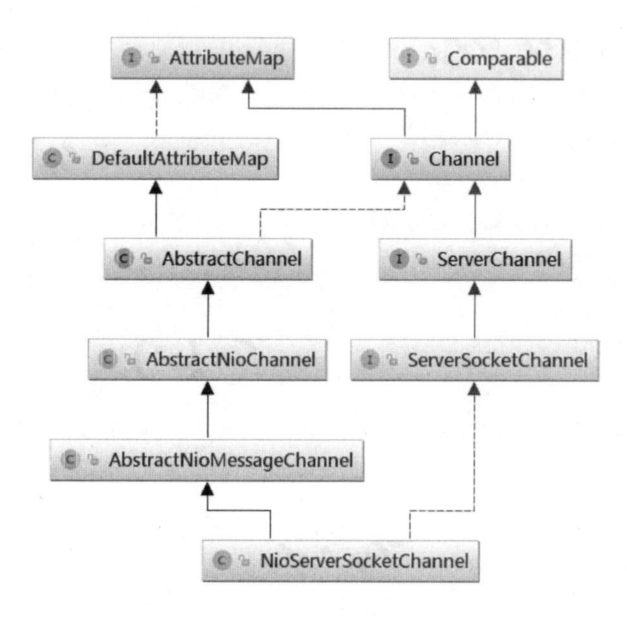

首先，我们来看一下 NioServerSocketChannel 的默认构造器。与 NioSocketChannel 类似，构造器都是调用 newSocket()方法来打开一个 Java NIO Socket。不过需要注意的是，客户端的 newSocket()方法调用的是 openSocketChannel，而服务端的 newSocket()方法调用的是 openServerSocketChannel。顾名思义，一个是客户端的 Java SocketChannel，另一个是服务端的 Java ServerSocketChannel，代码如下。

```
private static ServerSocketChannel newSocket(SelectorProvider provider) {
     return provider.openServerSocketChannel();
}
public NioServerSocketChannel() {
     this(newSocket(DEFAULT_SELECTOR_PROVIDER));
}
```

然后，调用重载构造方法，代码如下。

```
public NioServerSocketChannel(ServerSocketChannel channel) {
       super(null, channel, SelectionKey.OP_ACCEPT);
       config = new NioServerSocketChannelConfig(this, javaChannel().socket());
}
```

在这个构造方法中，调用父类构造方法时传入的参数是 SelectionKey.OP_ACCEPT。作为对比，我们回顾一下，在客户端的 Channel 初始化时，传入的参数是 SelectionKey.OP_READ。在服务启动后需要监听客户端的连接请求，因此在这里我们设置 SelectionKey.OP_ACCEPT，也就是通知 Selector 我们对客户端的连接请求感兴趣。

接下来，和客户端对比分析一下，逐级调用父类的构造器，首先调用 NioServerSocketChannel 的构造器，其次调用 AbstractNioMessageChannel 的构造器，然后调用 AbstractNioChannel 的构造器，最后调用 AbstractChannel 的构造器。同样地，在 AbstractChannel 中实例化一个 Unsafe 和 Pipeline，代码如下。

```
protected AbstractChannel(Channel parent) {
        this.parent = parent;
        id = newId();
        unsafe = newUnsafe();
        pipeline = newChannelPipeline();
}
```

不过，这里需要注意的是，客户端的 Unsafe 是 AbstractNioByteChannel.NioByteUnsafe 的实例，而服务端的 Unsafe 是 AbstractNioMessageChannel.AbstractNioUnsafe 的实例。AbstractNioMessageChannel 重写了 newUnsafe()方法，代码如下。

```
protected AbstractNioUnsafe newUnsafe() {
        return new NioMessageUnsafe();
}
```

最后，总结一下在 NioServerSocketChannel 实例化过程中执行的逻辑。

（1）调用 NioServerSocketChannel.newSocket(DEFAULT_SELECTOR_PROVIDER)方法创建一个新的 Java NIO 原生的 ServerSocketChannel 对象。

（2）实例化 AbstractChannel 对象并给属性赋值，具体赋值的属性如下。

- parent：设置为默认值 null。
- unsafe：通过调用 newUnsafe()方法，实例化一个 Unsafe 对象，其类型是 AbstractNioMessageChannel.AbstractNioUnsafe 内部类。
- pipeline：赋值的是 DefaultChannelPipeline 的实例。

（3）实例化 AbstractNioChannel 对象并给属性赋值，具体赋值的属性如下。

- ch：被赋值为 Java NIO 原生的 ServerSocketChannel 对象，通过调用 NioServerSocketChannel 的 newSocket()方法获取。
- readInterestOp：被赋值为默认值 SelectionKey.OP_ACCEPT。
- Ch：被设置为非阻塞，也就是调用 ch.configureBlocking(false)方法。

（4）给 NioServerSocketChannel 对象的 config 属性赋值为 new NioServerSocketChannelConfig (this, javaChannel().socket())。

7.2.3 服务端 ChannelPipeline 的初始化

服务端 ChannelPipeline 的初始化和客户端一致，也可以参考 7.1.5 节的内容，不再单独分析。

7.2.4 将服务端 Channel 注册到 Selector

服务端 Channel 的注册过程和客户端一致，也可以参考 7.1.7 节的内容，不再单独分析。

7.2.5 bossGroup 与 workerGroup

在客户端初始化的时候，我们初始化了一个 EventLoopGroup 对象，而在服务端的初始化时，我们设置了两个 EventLoopGroup：一个是 bossGroup，另一个是 workerGroup。那么这两个 EventLoopGroup 都是干什么用的呢?接下来我们详细探究一下。其实，bossGroup 只用于服务端的 accept，也就是用于处理客户端新连接接入的请求。我们可以把 Netty 比作一个餐馆，bossGroup 就像一个大堂经理，当客户来到餐馆吃饭时，大堂经理就会引导顾客就座，为顾客端茶送水等。而 workerGroup 就是实际干活的厨师，它们负责客户端连接通道的 I/O 操作：当大堂经理接待顾客后，顾客可以稍做休息，而此时后厨里的厨师们（workerGroup）就开始忙碌地准备饭菜了。bossGroup 与 workerGroup 的关系如下图所示，前面的章节我们也分析过，这里再巩固一下。

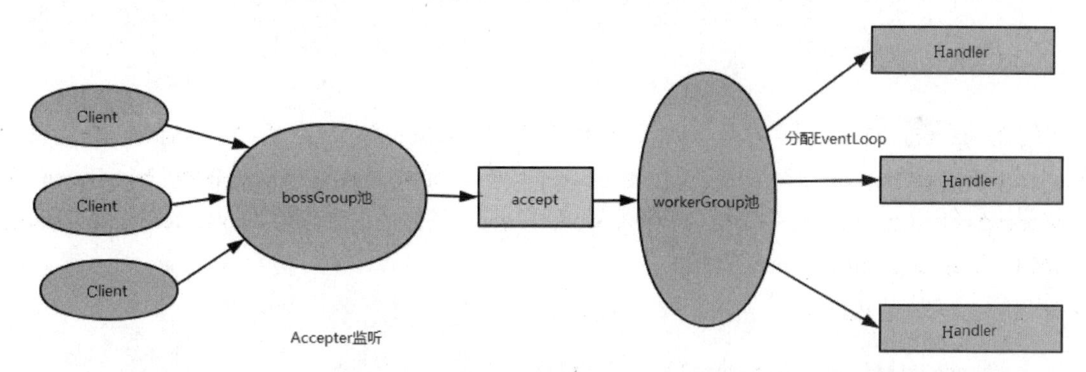

首先，服务端的 bossGroup 不断地监听是否有客户端的连接，当发现有一个新的客户端连接到来时，bossGroup 就会为此连接初始化各项资源；然后，从 workerGroup 中选出一个 EventLoop 绑定到此客户端连接中；接下来，服务端与客户端的交互过程将全部在此分配的 EventLoop 中完成。下面我们结合代码进行分析。

在 ServerBootstrap 初始化时调用了 b.group(bossGroup, workerGroup)，并设置了两个 EventLoopGroup，代码如下。

```
public ServerBootstrap group(EventLoopGroup parentGroup, EventLoopGroup childGroup) {
        super.group(parentGroup);
       //此处省略一部分代码
        this.childGroup = childGroup;
        return this;
}
```

显然，这个方法初始化了两个属性。一个是 group = parentGroup，它是在 super.group(parentGroup)中完成初始化的；另一个是 childGroup = childGroup。接着从应用程序的启动代码看，调用了 b.bind()方法来监听一个本地端口。bind()方法会触发如下调用链。

```
AbstractBootstrap.bind() -> AbstractBootstrap.doBind() ->
AbstractBootstrap.initAndRegister()
```

代码看到这里，我们发现对于 AbstractBootstrap 的 initAndRegister()方法，我们已经很熟悉，在分析客户端程序时和它打过很多交道，现在再来回顾一下这个方法。

```
final ChannelFuture initAndRegister() {
        //省略异常判断
        Channel channel = channelFactory.newChannel();
        init(channel);
        //省略非关键代码
        ChannelFuture regFuture = config().group().register(channel);

        return regFuture;
}
```

这里 group()方法返回的是上面我们提到的 bossGroup，而这里的 Channel 其实就是 NioServerSocketChannel 的实例，因此可以猜测 group().register(channel)将 bossGroup 和 NioServerSocketChannel 关联起来了。那么 workerGroup 具体是在哪里与 NioServerSocketChannel 关联的呢？继续看 init(channel)方法。

```
void init(Channel channel) throws Exception {
      //省略参数判断
       ChannelPipeline p = channel.pipeline();

       final EventLoopGroup currentChildGroup = childGroup;
       final ChannelHandler currentChildHandler = childHandler;
       final Entry<ChannelOption<?>, Object>[] currentChildOptions;
       final Entry<AttributeKey<?>, Object>[] currentChildAttrs;

      //省略非关键代码

       p.addLast(new ChannelInitializer<Channel>() {
           @Override
           public void initChannel(Channel ch) throws Exception {
```

```
                final ChannelPipeline pipeline = ch.pipeline();
                ChannelHandler handler = config.handler();
                if (handler != null) {
                    pipeline.addLast(handler);
                }
                ch.eventLoop().execute(new Runnable() {
                    @Override
                    public void run() {
                        pipeline.addLast(new ServerBootstrapAcceptor(
                                currentChildGroup, currentChildHandler, currentChildOptions,
currentChildAttrs));
                    }
                });
            }
        });
}
```

实际上，init()方法在 ServerBootstrap 中被重写了，从上面的代码片段中我们看到，它为 Pipeline 添加了一个 ChannelInitializer，而这个 ChannelInitializer 中添加了一个非常关键的 ServerBootstrapAcceptor 的 Handler。关于 Handler 的添加与初始化的过程，我们留到之后的章节再详细分析。现在，先来关注 ServerBootstrapAcceptor 类。在 ServerBootstrapAcceptor 中重写了 channelRead()方法，其主要代码如下。

```
public void channelRead(ChannelHandlerContext ctx, Object msg) {
            final Channel child = (Channel) msg;
            child.pipeline().addLast(childHandler);
            //省略非关键代码
                childGroup.register(child).addListener(...);
}
```

ServerBootstrapAcceptor 中的 childGroup 是构造此对象时传入的 currentChildGroup，也就是 workerGroup 对象。而这里的 Channel 是 NioSocketChannel 的实例，因此 childGroup 的 register()方法就是将 workerGroup 中的某个 EventLoop 和 NioSocketChannel 进行关联。那么，ServerBootstrapAcceptor 的 channelRead()方法是在哪里被调用的呢?其实当一个 Client 连接到 Server 时，Java 底层 NIO 的 ServerSocketChannel 就会有一个 SelectionKey.OP_ACCEPT 的事件就绪，接着就会调用 NioServerSocketChannel 的 doReadMessages()方法，代码如下。

```
protected int doReadMessages(List<Object> buf) throws Exception {
        SocketChannel ch = javaChannel().accept();
        //省略异常处理
        buf.add(new NioSocketChannel(this, ch));
        return 1;
        //省略错误处理
}
```

在 doReadMessages()方法中，通过调用 javaChannel().accept()方法获取客户端新连接的 SocketChannel 对象，紧接着实例化一个 NioSocketChannel，并且传入 NioServerSocketChannel 对象(即 this)。由此可知，我们创建的 NioSocketChannel 的父类 Channel 就是 NioServerSocketChannel 实例。接下来利用 Netty 的 ChannelPipeline 机制，将读取事件逐级发送到各个 Handler 中，于是就会触发 ServerBootstrapAcceptor 的 channelRead()方法。

7.2.6 服务端 Selector 事件轮询

回到服务端 ServerBootStrap 的启动代码，它是从 bind()方法开始的。ServerBootStrapt 的 bind()方法实际上就是其父类 AbstractBootstrap 的 bind()方法，来看代码。

```
    private static void doBind0(
            final ChannelFuture regFuture, final Channel channel,
            final SocketAddress localAddress, final ChannelPromise promise) {

        channel.eventLoop().execute(new Runnable() {
            @Override
            public void run() {
                if (regFuture.isSuccess()) {
                    channel.bind(localAddress,
promise).addListener(ChannelFutureListener.CLOSE_ON_FAILURE);
                } else {
                    promise.setFailure(regFuture.cause());
                }
            }
        });
    }
```

在 doBind0()方法中，调用 EventLoop 的 execute()方法，代码如下。

```
public void execute(Runnable task) {
        //省略了空判断

        boolean inEventLoop = inEventLoop();
        if (inEventLoop) {
            addTask(task);
        } else {
            startThread();
            addTask(task);
            //省略删除任务的逻辑
        }

        //省略判断逻辑
    }
```

execute()方法主要就是创建线程，将线程添加到 EventLoop 的无锁化串行任务队列。重点关注 startThread()方法，代码如下。

```
            private void startThread() {
        if (state == ST_NOT_STARTED) {
            if (STATE_UPDATER.compareAndSet(this, ST_NOT_STARTED, ST_STARTED)) {
                doStartThread();
            }
        }
    }

    private void doStartThread() {
                    //省略部分代码
                SingleThreadEventExecutor.this.run();
               //省略部分代码
    }
```

我们发现 startThread()方法最终调用的是 SingleThreadEventExecutor.this.run()方法，这个 this 就是 NioEventLoop 对象，代码如下。

```
protected void run() {
        for (;;) {
                                    switch (selectStrategy.calculateStrategy(selectNowSupplier,
hasTasks())) {
                    case SelectStrategy.CONTINUE:
                        continue;
                    case SelectStrategy.SELECT:
                        select(wakenUp.getAndSet(false));
                                                //省略 select 的唤醒逻辑
                    default:
                }

               cancelledKeys = 0;
                needsToSelectAgain = false;
                final int ioRatio = this.ioRatio;
                if (ioRatio == 100) {
                        processSelectedKeys();
                    //省略异常处理
                } else {
                    final long ioStartTime = System.nanoTime();
                            processSelectedKeys();
                    //省略异常处理
                }
        }
    }
```

终于看到似曾相识的代码。上面代码主要就是用一个死循环不断地轮询 SelectionKey 的。select()方法主要用来解决 JDK 空轮询 Bug，而 processSelectedKeys()就是针对不同的轮询事件进行处理。如果客户端有数据写入，最终也会调用 AbstractNioMessageChannel 的 doReadMessages()方法。下面我们来总结一下 Selector 的轮询流程。

（1）Selector 事件轮询是从 EventLoop 的 execute()方法开始的。

（2）在 EventLoop 的 execute()方法中，会为每一个任务都创建一个独立的线程，并保存到无锁化串行任务队列。

（3）线程任务队列的每个任务实际调用的是 NioEventLoop 的 run()方法。

（4）在 run()方法中调用 processSelectedKeys()处理轮询事件。

7.2.7 Netty 解决 JDK 空轮询 Bug

大家应该早就听说过臭名昭著的 Java NIO epoll 的 Bug，它会导致 Selector 空轮询，最终导致 CPU 使用率达到 100%。官方声称 JDK 1.6 的 update18 修复了该问题，但是直到 JDK 1.7 该问题仍旧存在，只不过该 Bug 发生概率降低了一些而已，并没有被根本解决。出现此 Bug 是因为当 Selector 轮询结果为空时，没有进行 wakeup 或对新消息及时进行处理，导致发生了空轮询，CPU 使用率达到了 100%。我们来看一下这个问题在 issue 中的原始描述。

> This is an issue with poll (and epoll) on Linux. If a file descriptor for a connected socket is polled with a request event mask of 0, and if the connection is abruptly terminated (RST) then the poll wakes up with the POLLHUP (and maybe POLLERR) bit set in the returned event set. The implication of this behaviour is that Selector will wakeup and as the interest set for the SocketChannel is 0 it means there aren't any selected events and the select method returns 0.

具体解释为：在部分 Linux Kernel 2.6 中，poll 和 epoll 对于突然中断的 Socket 连接会对返回的 EventSet 事件集合置为 POLLHUP，也可能是 POLLERR，EventSet 事件集合发生了变化，这就可能导致 Selector 会被唤醒。

这是与操作系统机制有关系的，JDK 虽然仅仅是一个兼容各个操作系统平台的软件，但遗憾的是在 JDK 5 和 JDK 6 最初的版本中，这个问题并没有得到解决，而将这个“帽子”抛给了操作系统方，这就是这个 Bug 一直到 2013 年才最终修复的原因。

在 Netty 中最终的解决办法是：创建一个新的 Selector，将可用事件重新注册到新的 Selector 中来终止空轮询。我们来回顾一下事件轮询的关键代码。

```
protected void run() {
        for (;;) {
             switch (selectStrategy.calculateStrategy(selectNowSupplier, hasTasks())) {
                   case SelectStrategy.CONTINUE:
                       continue;
                   case SelectStrategy.SELECT:
                       select(wakenUp.getAndSet(false));
                        //省略 select 的唤醒逻辑
                   default:
               }

              //事件轮询处理逻辑
        }
}
```

前面我们提到 select()方法解决了 JDK 空轮询的 Bug，那么它到底是如何解决的呢？下面我们来一探究竟，先来看一下 select()方法的源码。

```
public final class NioEventLoop extends SingleThreadEventLoop {

    ...

    int selectorAutoRebuildThreshold = SystemPropertyUtil.getInt("io.netty.
selectorAutoRebuildThreshold", 512);
//省略判断代码
       SELECTOR_AUTO_REBUILD_THRESHOLD = selectorAutoRebuildThreshold;
...
private void select(boolean oldWakenUp) throws IOException {
      Selector selector = this.selector;
       long currentTimeNanos = System.nanoTime();
       for (;;) {
            //省略非关键代码
            long timeoutMillis = (selectDeadLineNanos - currentTimeNanos + 500000L) / 1000000L;
            int selectedKeys = selector.select(timeoutMillis);
            selectCnt ++;

            //省略非关键代码

           long time = System.nanoTime();
           if (time - TimeUnit.MILLISECONDS.toNanos(timeoutMillis) >= currentTimeNanos) {
               // timeoutMillis elapsed without anything selected.
               selectCnt = 1;
           } else if (SELECTOR_AUTO_REBUILD_THRESHOLD > 0 &&
                   selectCnt >= SELECTOR_AUTO_REBUILD_THRESHOLD) {
                 //日志打印代码

                 rebuildSelector();
```

```
                    selector = this.selector;

                    // Select again to populate selectedKeys.
                    selector.selectNow();
                    selectCnt = 1;
                    break;
                }

                currentTimeNanos = time;
            }

        //省略非关键代码
        }
}
   ...
}
```

从上面的代码中可以看出，Selector 每一次轮询都计数 selectCnt++，开始轮询会将系统时间戳赋值给 timeoutMillis，轮询完成后再将系统时间戳赋值给 time，这两个时间会有一个时间差，而这个时间差就是每次轮询所消耗的时间。从上面的逻辑可以看出，如果每次轮询消耗的时间为 0 s，且重复次数超过 512 次，则调用 rebuildSelector()方法，即重构 Selector，具体实现代码如下。

```
public void rebuildSelector() {
        //省略判断语句
        rebuildSelector0();
    }

    private void rebuildSelector0() {
        final Selector oldSelector = selector;
        final SelectorTuple newSelectorTuple;

            newSelectorTuple = openSelector();
       //省略非关键代码

        // Register all channels to the new Selector.
        int nChannels = 0;
        for (SelectionKey key: oldSelector.keys()) {
           //省略非关键代码和异常处理
                key.cancel();
                SelectionKey newKey =
key.channel().register(newSelectorTuple.unwrappedSelector, interestOps, a);

        }

        //省略非关键代码
    }
```

实际上，在 rebuildSelector()方法中，主要做了以下三件事情。

（1）创建一个新的 Selector。

（2）将原来 Selector 中注册的事件全部取消。

（3）将可用事件重新注册到新的 Selector，并激活。就这样，Netty 完美解决了 JDK 的空轮询 Bug。看到这里，是不是感觉没那么神秘了？

7.2.8 Netty 对 Selector 中 KeySet 的优化

分析完 Netty 对 JDK 空轮询 Bug 的解决方案，接下来看一个很有意思的细节。Netty 对 Selector 中存储 SelectionKey 的 HashSet 也做了优化。在前面的分析中，Netty 对 Selector 有重构，创建一个新的 Selector 就会调用 openSelector()方法，来看代码。

```
private void rebuildSelector0() {
    final Selector oldSelector = selector;
    final SelectorTuple newSelectorTuple;
        newSelectorTuple = openSelector();
   //省略非关键代码
}
```

我们再来看一下 openSelector()方法的代码。

```
private SelectorTuple openSelector() {
        final Selector unwrappedSelector;
        //省略异常处理代码
        unwrappedSelector = provider.openSelector();
       //省略非关键代码

      final SelectedSelectionKeySet selectedKeySet = new SelectedSelectionKeySet();

      Object maybeSelectorImplClass = AccessController.doPrivileged(
                                                  new PrivilegedAction<Object>() {
        @Override
        public Object run() {
                return Class.forName(
                        "sun.nio.ch.SelectorImpl",
                        false,
                        PlatformDependent.getSystemClassLoader());
            //省略异常处理代码
      }
   });

   //省略非关键代码
```

```
    final Class<?> selectorImplClass = (Class<?>) maybeSelectorImplClass;

    Object maybeException = AccessController.doPrivileged(new PrivilegedAction<Object>() {
          @Override
          public Object run() {

              Field selectedKeysField = selectorImplClass.getDeclaredField("selectedKeys");
              Field publicSelectedKeysField =
                              selectorImplClass.getDeclaredField("publicSelectedKeys");

               //省略非关键代码

               selectedKeysField.set(unwrappedSelector, selectedKeySet);
               publicSelectedKeysField.set(unwrappedSelector, selectedKeySet);
               return null;
            //省略异常处理代码
          }
       });

       //省略非关键代码
   }
```

上面代码的主要功能就是利用反射机制，获取 JDK 底层的 Selector 的 Class 对象，用反射方法从 Class 对象中获得两个属性：selectedKeys 和 publicSelectedKeys，这两个属性就是用来存储已注册事件的。然后，将这两个对象重新赋值为 Netty 创建的 SelectedSelectionKeySet，是不是有种"偷梁换柱"的感觉？

我们先看 selectedKeys 和 publicSelectedKeys 到底是什么类型，打开 SelectorImpl 的源码，看其构造方法的代码。

```
public abstract class SelectorImpl extends AbstractSelector {
   protected Set<SelectionKey> selectedKeys = new HashSet();
   protected HashSet<SelectionKey> keys = new HashSet();
   private Set<SelectionKey> publicKeys;
   private Set<SelectionKey> publicSelectedKeys;

   protected SelectorImpl(SelectorProvider var1) {
      //省略非关键代码
         this.publicKeys = this.keys;
         this.publicSelectedKeys = this.selectedKeys;
          //省略非关键代码
}
   ...
}
```

我们发现 selectedKeys 和 publicSelectedKeys 就是 HashSet。我们再来看 Netty 创建的 SelectedSelectionKeySet 对象的源代码。

```
final class SelectedSelectionKeySet extends AbstractSet<SelectionKey> {

    SelectionKey[] keys;
    int size;

    SelectedSelectionKeySet() {
        keys = new SelectionKey[1024];
    }

    @Override
    public boolean add(SelectionKey o) {
        if (o == null) {
            return false;
        }

        keys[size++] = o;
        if (size == keys.length) {
            increaseCapacity();
        }

        return true;
    }

    @Override
    public int size() {
        return size;
    }

    @Override
    public boolean remove(Object o) {
        return false;
    }

    @Override
    public boolean contains(Object o) {
        return false;
    }

    @Override
    public Iterator<SelectionKey> iterator() {
        throw new UnsupportedOperationException();
    }
```

```
    void reset() {
        reset(0);
    }

    void reset(int start) {
        Arrays.fill(keys, start, size, null);
        size = 0;
    }

    private void increaseCapacity() {
        SelectionKey[] newKeys = new SelectionKey[keys.length << 1];
        System.arraycopy(keys, 0, newKeys, 0, size);
        keys = newKeys;
    }
}
```

源码篇幅不长，但很精辟。SelectedSelectionKeySet 同样继承了 AbstractSet，因此赋值给 selectedKeys 和 publicSelectedKeys 不存在类型强制转换的问题。细心的小伙伴应该已经发现在 SelectedSelectionKeySet 中禁用了 remove()方法、contains()方法和 iterator()方法，只保留了 add()方法，而且底层存储结构用的是数组 SelectionKey[] keys。那么，Netty 为什么要这样设计呢？主要目的还是简化我们在轮询事件时的操作，不需要每次轮询都移除 Key。

7.2.9 Handler 的添加过程

服务端 Handler 的添加过程和客户端的有点区别，跟 EventLoopGroup 一样，服务端的 Handler 也有两个：一个是通过 handler()方法设置的 Handler，另一个是通过 childHandler()方法设置的 childHandler。通过前面的 bossGroup 和 workerGroup 的分析，其实我们可以在这里大胆地猜测：Handler 与 accept 过程有关，即 Handler 负责处理客户端新连接接入的请求；而 childHandler 就是负责和客户端连接的 I/O 交互。那么实际上是不是这样的呢？我们继续用代码来验证。在前面章节我们已经了解到 ServerBootstrap 重写了 init()方法，在这个方法中也添加了 Handler，代码如下。

```
void init(Channel channel) throws Exception {
       //省去逻辑判断
        ChannelPipeline p = channel.pipeline();

        final EventLoopGroup currentChildGroup = childGroup;
        final ChannelHandler currentChildHandler = childHandler;
        final Entry<ChannelOption<?>, Object>[] currentChildOptions;
        final Entry<AttributeKey<?>, Object>[] currentChildAttrs;

        p.addLast(new ChannelInitializer<Channel>() {
            @Override
```

```
            public void initChannel(Channel ch) throws Exception {
                final ChannelPipeline pipeline = ch.pipeline();
                ChannelHandler handler = config.handler();
                if (handler != null) {
                    pipeline.addLast(handler);
                }

                ch.eventLoop().execute(new Runnable() {
                    @Override
                    public void run() {
                        pipeline.addLast(new ServerBootstrapAcceptor(
                                currentChildGroup, currentChildHandler, currentChildOptions,
currentChildAttrs));
                    }
                });
            }
        });
}
```

在上面代码的 initChannel()方法中，首先通过 handler()方法获取一个 Handler，如果获取的 Handler 不为空，则添加到 Pipeline 中，然后添加一个 ServerBootstrapAcceptor 的实例。这里的 handler()方法返回的是哪个对象呢?其实它返回的是 Handler 属性，而这个属性就是我们在服务端的启动代码中设置的。

```
b.group(bossGroup, workerGroup)
```

这个时候，Pipeline 中的 Handler 情况如下图所示。

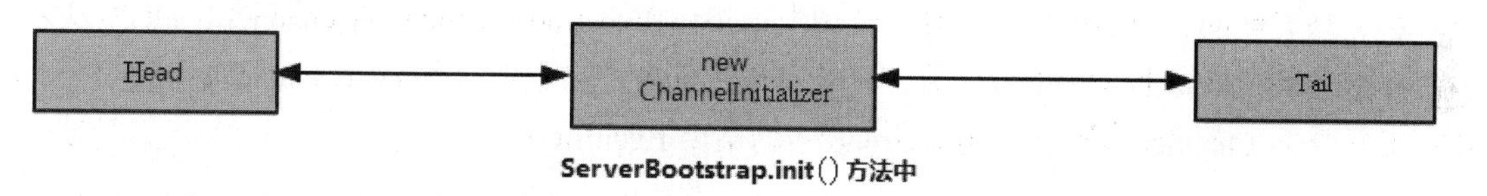

ServerBootstrap.init()方法中

根据对原来客户端代码的分析，将 Channel 绑定到 EventLoop（这里是指 NioServerSocketChannel 绑定到 bossGroup）后，会在 Pipeline 中触发 fireChannelRegistered 事件，接着会触发对 ChannelInitializer 的 initChannel()方法的调用。因此在绑定完成后，此时的 Pipeline 的内容如下图所示。

channelRegistered后的Pipeline内容

在我们分析 bossGroup 和 workerGroup 时，已经知道 ServerBootstrapAcceptor 的 channelRead() 方法会为新建的 Channel 设置 Handler 并注册到一个 EventLoop 中。

```
public void channelRead(ChannelHandlerContext ctx, Object msg) {
        final Channel child = (Channel) msg;
        child.pipeline().addLast(childHandler);
        //省去非关键代码
        childGroup.register(child).addListener(...);

}
```

而这里的 childHandler 就是我们在服务端启动代码中设置的 Handler。

```
...
.childHandler(new ChannelInitializer<SocketChannel>(){
                @Override
                public void initChannel(SocketChannel ch) throws Exception {
                    ...
                }
            })
```

后续的步骤我们基本上已经清楚了，在客户端连接 Channel 注册后，就会触发 ChannelInitializer 的 initChannel()方法的调用。最后我们总结一下服务端 Handler 与 childHandler 的区别与联系。

（1）在服务端 NioServerSocketChannel 对象的 Pipeline 中添加了 Handler 对象和 ServerBootstrapAcceptor 对象。

（2）当有新的客户端连接请求时，会调用 ServerBootstrapAcceptor 的 channelRead()方法创建此连接对应的 NioSocketChannel 对象，并将 childHandler 添加到 NioSocketChannel 对应的 Pipeline 中，而且将此 Channel 绑定到 workerGroup 中的某个 EventLoop 中。

（3）Handler 对象只在 accept()阻塞阶段起作用，它主要处理客户端发送过来的连接请求。

（4）childHandler 在客户端连接建立以后起作用，它负责客户端连接的 I/O 交互。

最后来看一张图，加深理解。下图描述了服务端从启动初始化到有新连接接入的变化过程。

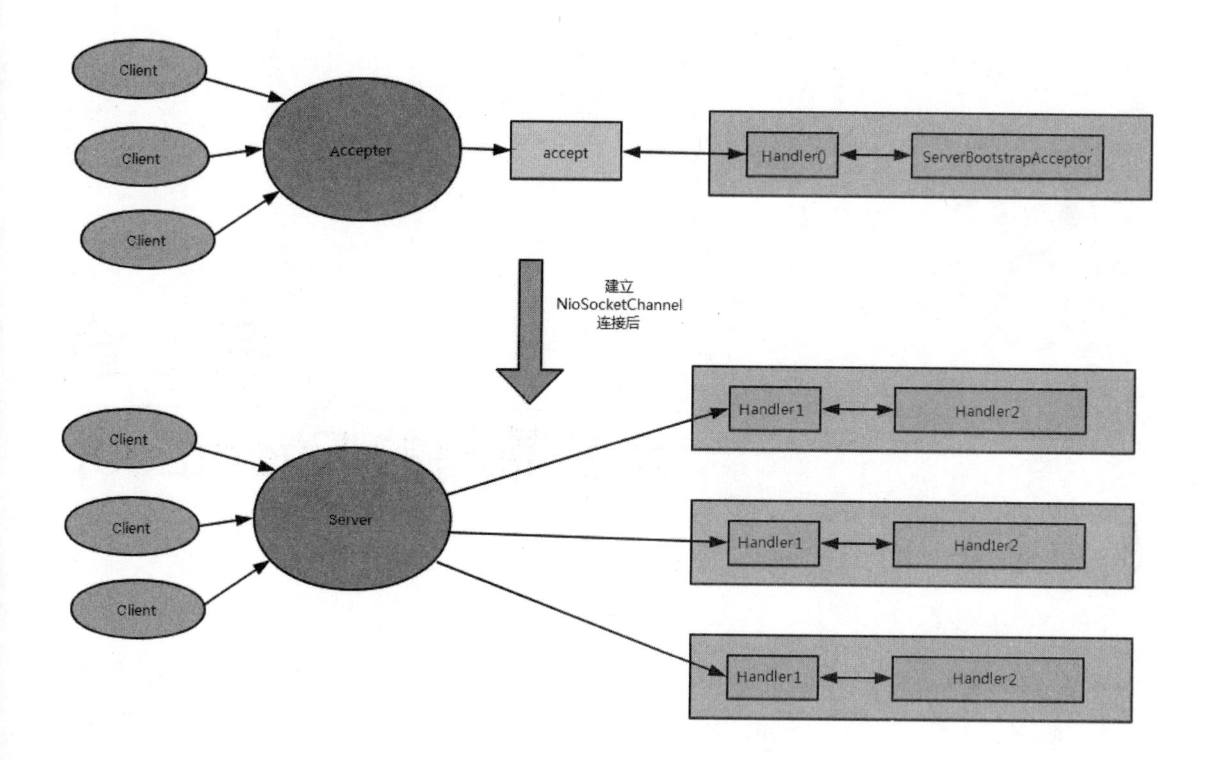
Client
Client
Client
Accepter
accept
Handler()
ServerBootstrapAcceptor
建立
NioSocketChannel
连接后
Client
Client
Client
Server
Handler1
Handler2
Handler1
Handler2
Handler1
Handler2

8

第 8 章 大名鼎鼎的 EventLoop

8.1 EventLoopGroup 与 Reactor

上一章中我们已经知道，一个 Netty 程序启动时，至少要指定一个 EventLoopGroup，在使用 NIO 的过程中通常是指 NioEventLoopGroup。那么，这个 NioEventLoopGroup 在 Netty 中到底扮演着什么角色呢？我们知道 Netty 是 Reactor 模型的一个实现，因此就从 Reactor 的线程模型开始分析。

8.1.1 再谈 Reactor 线程模型

第 6 章中我们对 Reactor 的三种线程模型——单线程模型、多线程模型、主从多线程模型做了介绍，这里我们具体分析 Reactor 在 Netty 中的应用。先来看单线程模型，如下图所示。

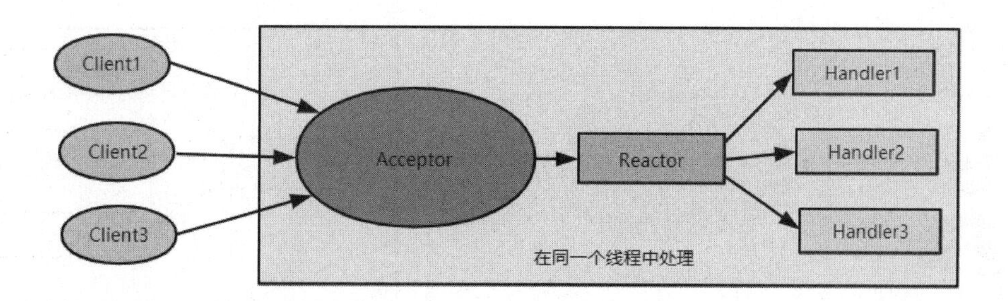

所谓单线程，即 Acceptor 的处理和 Handler 的处理都在同一个线程中。这个模型的弊端显而易见：当其中某个 Handler 阻塞时，会导致其他所有的 Client 的 Handler 都无法执行，并且更严重的是，Handler 的阻塞也会导致整个服务不能接收新的 Client 请求（因为 Acceptor 也被阻塞了）。因为有这么多的缺陷，所以单线程 Reactor 模型在 Netty 中的应用场景比较少。

Netty 中 Reactor 多线程模型的应用如下图所示。

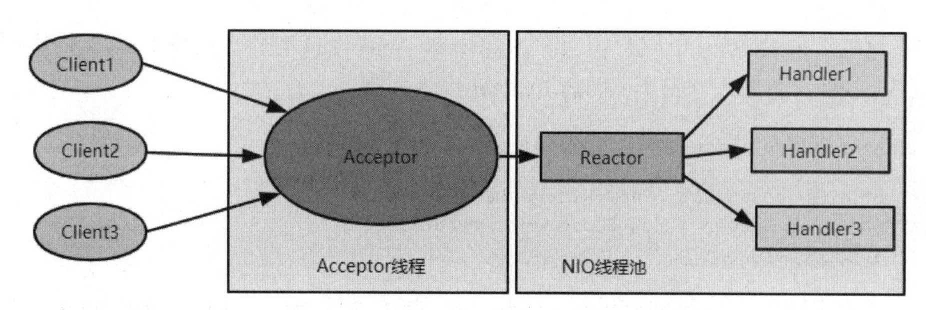

（1）设计一个专门的线程 Acceptor，用于监听客户端的 TCP 连接请求。

（2）客户端连接的 I/O 操作都由一个特定的 NIO 线程池负责。每个客户端连接都与一个特定的 NIO 线程绑定，因此在这个客户端连接中的所有 I/O 操作都是在同一个线程中完成的。

（3）客户端连接有很多，但是 NIO 线程数是比较少的，因此一个 NIO 线程可以同时绑定到多个客户端连接中。

我们再来看主从 Reactor 多线程模型在 Netty 中的应用。一般情况下，Reactor 的多线程模型已经适用于大部分业务场景。但如果服务端需要同时处理大量的客户端连接请求，或者需要在客户端连接时增加一些诸如权限的校验等操作，那么单个 Acceptor 就很有可能处理不过来，将会造成大量的客户端连接超时。主从 Reactor 多线程模型将服务端接收客户端的连接请求专门设计为一个独立的线程池，其线程模型如下图所示。

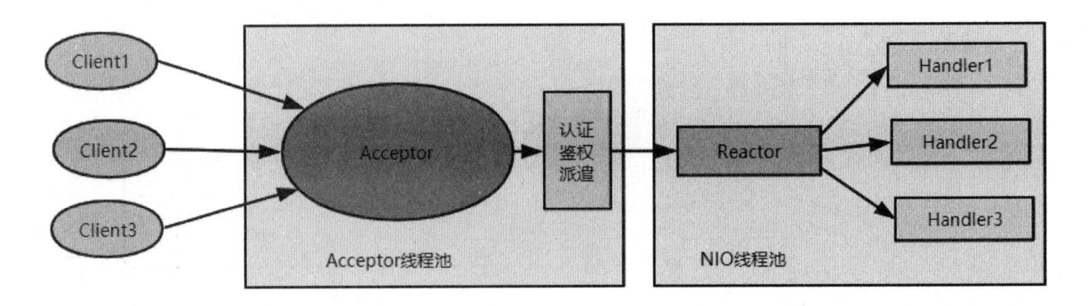

从上图可以看出，主从 Reactor 多线程模型和 Reactor 多线程模型很类似，只是在主从 Reactor 多线程模型的 Acceptor 线程池中获取线程，通过认证鉴权后进行派遣，再分配给 Reactor 线程池来处理客户端请求。

8.1.2 EventLoopGroup 与 Reactor 关联

我们了解的三种 Reactor 的线程模型和 NioEventLoopGroup 有什么关系呢?其实，不同的设置 NioEventLoopGroup 的方式对应了不同的 Reactor 线程模型。

（1）单线程模型在 Netty 中的应用代码如下。

```
EventLoopGroup bossGroup = new NioEventLoopGroup(1);
ServerBootstrap server = new ServerBootstrap();
server.group(bossGroup);
```

在此，我们实例化了一个 NioEventLoopGroup，接着调用 server.group(bossGroup)设置服务端的 EventLoopGroup。有人可能会有疑惑：在启动服务端的 Netty 程序时，需要设置 bossGroup 和 workerGroup，为何这里只设置一个 bossGroup？其实原因很简单，ServerBootstrap 重写了 group 方法，代码如下。

```
public ServerBootstrap group(EventLoopGroup group) {
      return group(group, group);
}
```

因此当传入一个 group 时，bossGroup 和 workerGroup 就是同一个 NioEventLoopGroup，并且这个 NioEventLoopGroup 线程池数量只设置了一个线程，也就是说 Netty 中的 Acceptor 和后续的所有客户端连接的 I/O 操作都是在一个线程中处理的。那么对应到 Reactor 的线程模型中，我们这样设置 NioEventLoopGroup 时，就相当于 Reactor 的单线程模型。

（2）多线程模型在 Netty 中的应用代码如下。

```
EventLoopGroup bossGroup = new NioEventLoopGroup(128);
ServerBootstrap server = new ServerBootstrap();
server.group(bossGroup);
```

从上面代码中可以看出，我们只需要将 bossGroup 的参数设置为大于 1 的数，其实就是 Reactor 多线程模型。

（3）主从 Reactor 多线程模型在 Netty 中的应用代码如下。

```
EventLoopGroup bossGroup = new NioEventLoopGroup();
EventLoopGroup workerGroup = new NioEventLoopGroup();
ServerBootstrap b = new ServerBootstrap();
b.group(bossGroup, workerGroup);
```

bossGroup 为主线程，而 workerGroup 中的线程数是 CPU 核数×2，因此对应到 Reactor 线程模型中，我们知道，这样设置的 NioEventLoopGroup 其实就是主从 Reactor 多线程模型。

8.1.3 EventLoopGroup 的实例化

首先，我们来看 EventLoopGroup 的类结构图，如下图所示。

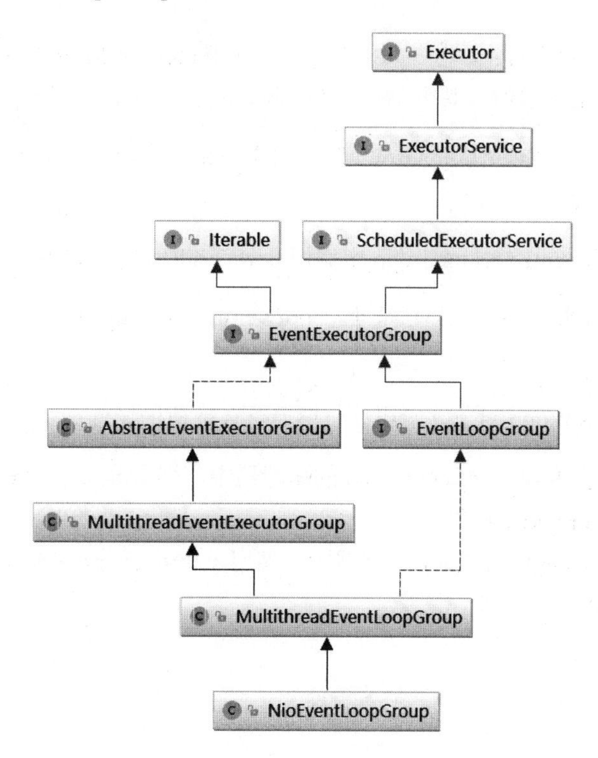

然后，我们通过时序图来了解一下 NioEventLoopGroup 初始化的基本过程，如下图所示。

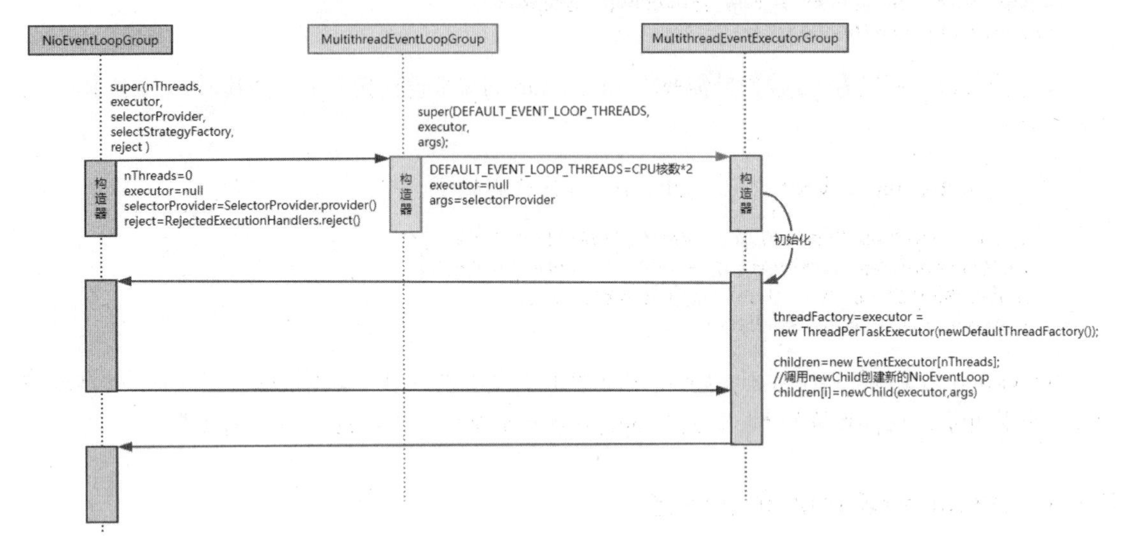

基本步骤如下。

（1）EventLoopGroup（其实是 MultithreadEventExecutorGroup）内部维护一个属性为 EventExecutor 的 children 的数组，其大小是 nThreads，这样就初始化了一个线程池。

（2）我们在实例化 NioEventLoopGroup 时，如果指定线程池大小，则 nThreads 就是指定的值，否则是 CPU 核数×2。

（3）在 MultithreadEventExecutorGroup 中调用 newChild()抽象方法来初始化 children 数组。

（4）newChild() 抽象方法实际上是在 NioEventLoopGroup 中实现的，由它返回一个 NioEventLoop 实例。

（5）初始化 NioEventLoop 对象并给属性赋值，具体赋值的属性如下。

- provider：就是在 NioEventLoopGroup 的构造器中，调用 SelectorProvider 的 provider()方法获取的 SelectorProvider 对象。
- selector：就是在 NioEventLoop 构造器中，调用 selector = provider.openSelector()方法获取的 Selector 对象。

8.2　任务执行者 EventLoop

NioEventLoop 继承自 SingleThreadEventLoop，而 SingleThreadEventLoop 又继承自 SingleThreadEventExecutor。而 SingleThreadEventExecutor 是 Netty 对本地线程的抽象，它内部有一个 Thread 属性，实际上就是存储了一个本地 Java 线程。因此我们可以简单地认为，一个 NioEventLoop 对象其实就是和一个特定的线程进行绑定，并且在 NioEventLoop 生命周期内，其绑定的线程都不会再改变。NioEventLoop 的类层次结构图如下图所示。

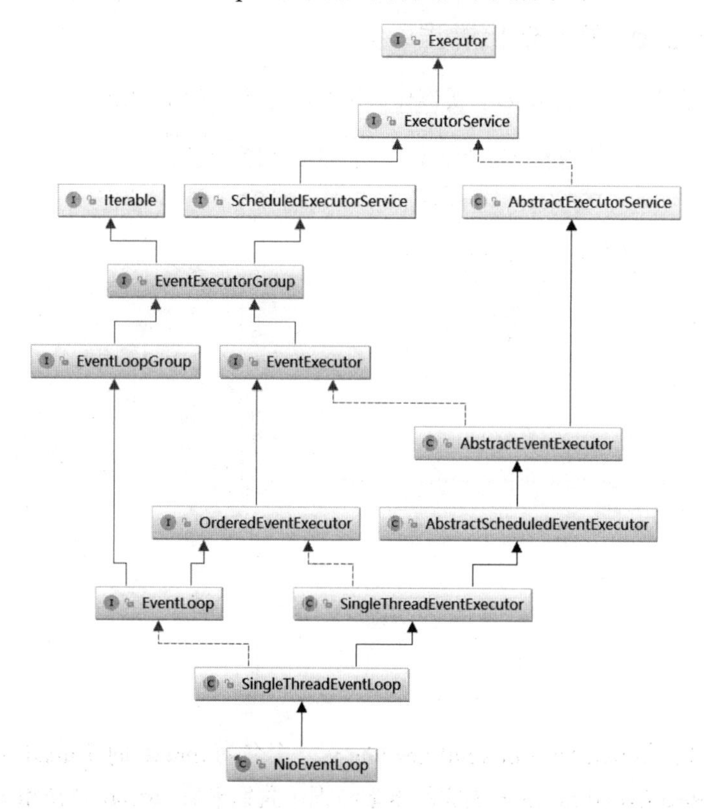

从上图可以看出，NioEventLoop 的类层次结构图比较复杂，不过我们只需要关注重点即可。首先来看 NioEventLoop 的继承关系：NioEventLoop 继承 SingleThreadEventLoop，SingleThreadEventLoop 继承 SingleThreadEventExecutor，SingleThreadEventExecutor 继承 AbstractScheduledEventExecutor。

在 AbstractScheduledEventExecutor 中，Netty 实现了 NioEventLoop 的 Schedule 功能，即我们可以通过调用一个 NioEventLoop 实例的 schedule 方法来运行一些定时任务。而在 SingleThreadEventLoop

中，又实现了任务队列的功能。通过它，我们可以调用一个 NioEventLoop 实例的 execute()方法向任务队列中添加一个 Task，并由 NioEventLoop 进行调度执行。

通常来说，NioEventLoop 负责执行两个任务：第一个任务是作为 I/O 线程，执行与 Channel 相关的 I/O 操作，包括调用 Selector 等待就绪的 I/O 事件、读写数据与数据处理等；而第二个任务是作为任务队列，执行 taskQueue 中的任务，例如用户调用 eventLoop.schedule 提交的定时任务也是由这个线程执行的。

8.2.1 NioEventLoop 的实例化过程

同样，先简单了解一下 EventLoop 实例化的运行时序图，如下图所示。

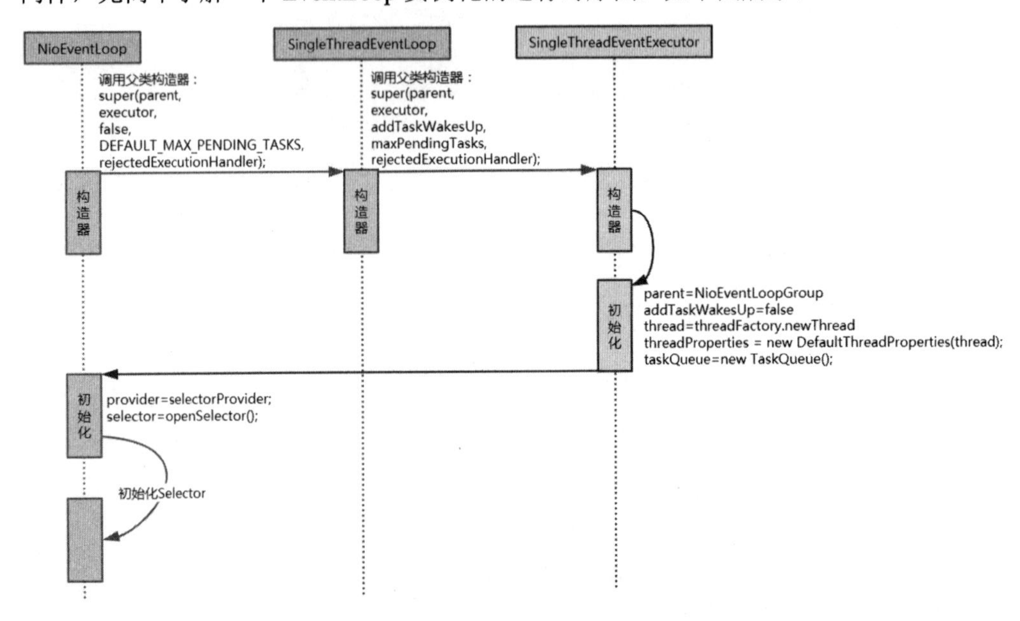

从上图可以看到，SingleThreadEventExecutor 有一个名为 thread 的 Thread 类型属性，这个属性就是与 SingleThreadEventExecutor 关联的本地线程。我们来看 thread 是在哪里被赋值的，代码如下。

```
private void doStartThread() {
    assert thread == null;
    executor.execute(new Runnable() {
        @Override
        public void run() {
            thread = Thread.currentThread();
            boolean success = false;
```

```
            updateLastExecutionTime();
            try {
                SingleThreadEventExecutor.this.run();
                success = true;
            } catch (Throwable t) {
                logger.warn("Unexpected exception from an event executor: ", t);
            } finally {
                //此处省略清理代码              }
        }
    });
}
```

在 7.2.6 节我们分析过，SingleThreadEventExecutor 启动时会调用 doStartThread()方法，然后调用 executor.execute()方法，将当前线程赋值给 thread。在这个线程中所做的事情主要就是调用 SingleThreadEventExecutor.this.run()方法，因为 NioEventLoop 实现了这个方法，所以根据多态性，其实调用的是 NioEventLoop.run()方法。

8.2.2 EventLoop 与 Channel 的关联

在 Netty 中，每个 Channel 都有且仅有一个 EventLoop 与之关联，它们的关联过程如下图所示。

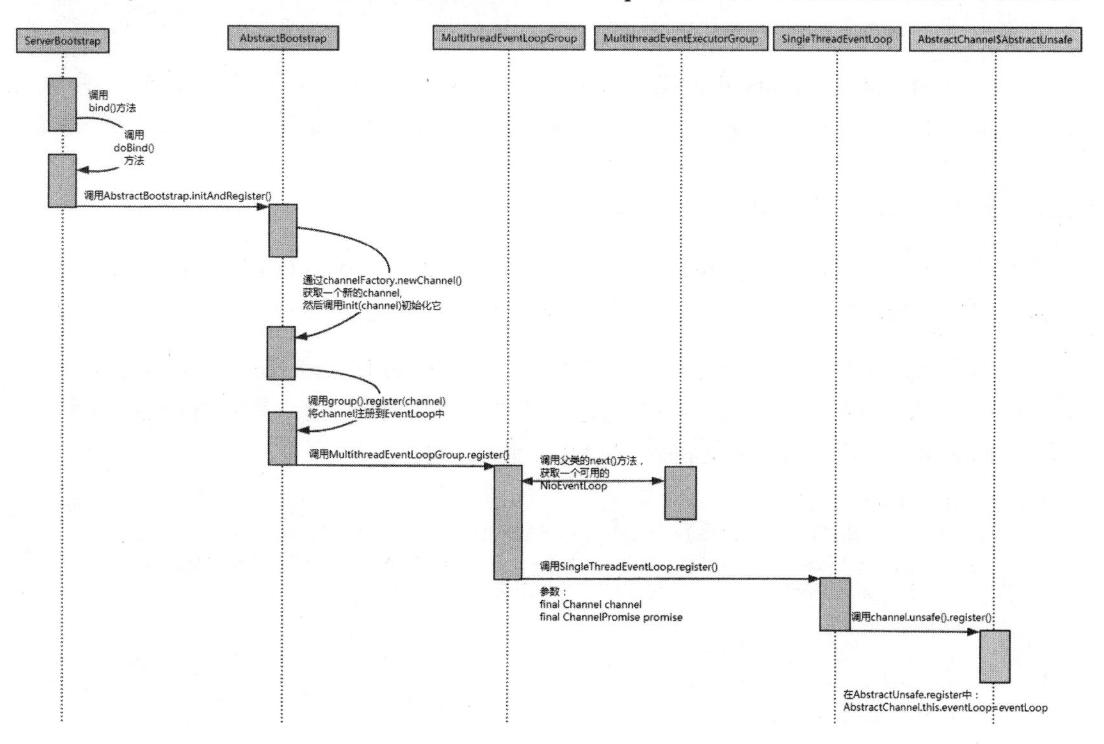

从上图中我们可以看到，当调用 AbstractChannel$AbstractUnsafe.register()方法后，就完成了 Channel 和 EventLoop 的关联。register()方法的具体实现代码如下。

```
public final void register(EventLoop eventLoop, final ChannelPromise promise) {
    //删除条件检查

    AbstractChannel.this.eventLoop = eventLoop;

    if (eventLoop.inEventLoop()) {
        register0(promise);
    } else {
        try {
            eventLoop.execute(new Runnable() {
                @Override
                public void run() {
                    register0(promise);
                }
            });
        } catch (Throwable t) {
            //删除 catch 块内容
        }
    }
}
```

在 AbstractChannel$AbstractUnsafe.register() 方法中，会将一个 EventLoop 赋值给 AbstractChannel 内部的 eventLoop 属性，这句代码就是完成 EventLoop 与 Channel 的关联过程的。

8.2.3 EventLoop 的启动

我们已经知道 NioEventLoop 本身就是一个 SingleThreadEventExecutor，因此 NioEventLoop 的启动，其实就是 NioEventLoop 所绑定的本地 Java 线程的启动。

按照这个思路，我们只需要找到在哪里调用了 SingleThreadEventExecutor 中 thread 属性的 start()方法就可以知道在哪里启动的这个线程了。第 7 章中我们分析过，其实 thread.start()被封装在 SingleThreadEventExecutor.startThread()方法中，来看代码。

```
private void startThread() {
    if (STATE_UPDATER.get(this) == ST_NOT_STARTED) {
        if (STATE_UPDATER.compareAndSet(this, ST_NOT_STARTED, ST_STARTED)) {
            doStartThread();
        }
    }
}
```

STATE_UPDATER 是 SingleThreadEventExecutor 内部维护的一个属性，它的作用是标识当前的 Thread 的状态。在初始的时候，STATE_UPDATER == ST_NOT_STARTED，因此第一次调用 startThread()方法时，就会进入 if 语句内，进而调用 thread.start()方法。而这个关键的 startThread()方法又是在哪里调用的呢?用方法调用关系反向查找功能，我们发现，startThread()方法是在 SingleThreadEventExecutor 的 execute()方法中调用的，代码如下。

```
public void execute(Runnable task) {
    if (task == null) {
        throw new NullPointerException("task");
    }

    boolean inEventLoop = inEventLoop();
    if (inEventLoop) {
        addTask(task);
    } else {
        startThread();  //调用 startThread()方法，启动 EventLoop 线程
        addTask(task);
        if (isShutdown() && removeTask(task)) {
            reject();
        }
    }

    if (!addTaskWakesUp && wakesUpForTask(task)) {
        wakeup(inEventLoop);
    }
}
```

既然如此，我们现在只需要找到第一次调用 SingleThreadEventExecutor 的 execute()方法的位置即可。细心的小伙伴可能已经想到，我们在提到注册 Channel 的过程中，会在 AbstractChannel$AbstractUnsafe 的 register()方法中调用 eventLoop.execute()方法，在 EventLoop 中进行 Channel 注册代码的执行，AbstractChannel$AbstractUnsafe 的 register()方法的部分代码如下。

```
public final void register(EventLoop eventLoop, final ChannelPromise promise) {
        //删除判断
        AbstractChannel.this.eventLoop = eventLoop;

        if (eventLoop.inEventLoop()) {
            register0(promise);
        } else {
            try {
                eventLoop.execute(new Runnable() {
                    @Override
                    public void run() {
                        register0(promise);
```

```
                    }
                });
            } catch (Throwable t) {
                //删除异常处理代码
            }
        }
    }
```

很显然，从 Bootstrap 的 bind()方法一路跟踪到 AbstractChannel$AbstractUnsafe 的 register()方法，整个代码都是在主线程中运行的，因此上面的 eventLoop.inEventLoop()返回值为 false，于是进入 else 分支，在这个分支中调用 eventLoop.execute()方法，而 NioEventLoop 没有实现 execute()方法，因此调用的是 SingleThreadEventExecutor 的 execute()方法，代码如下。

```
public void execute(Runnable task) {
    //条件判断
    boolean inEventLoop = inEventLoop();
    if (inEventLoop) {
        addTask(task);
    } else {
        startThread();
        addTask(task);
        if (isShutdown() && removeTask(task)) {
            reject();
        }
    }
    if (!addTaskWakesUp && wakesUpForTask(task)) {
        wakeup(inEventLoop);
    }
}
```

我们已经分析过，inEventLoop == false，因此执行到 else 分支就调用 startThread()方法来启动 SingleThreadEventExecutor 内部关联的 Java 本地线程。

用一句话总结一下：当 EventLoop 的 execute()方法第一次被调用时，会触发 startThread()方法的调用，进而启动 EventLoop 所对应的 Java 本地线程。

我们将上节中的时序图补全以后，就得到了 EventLoop 启动过程完整的时序图，如下图所示。

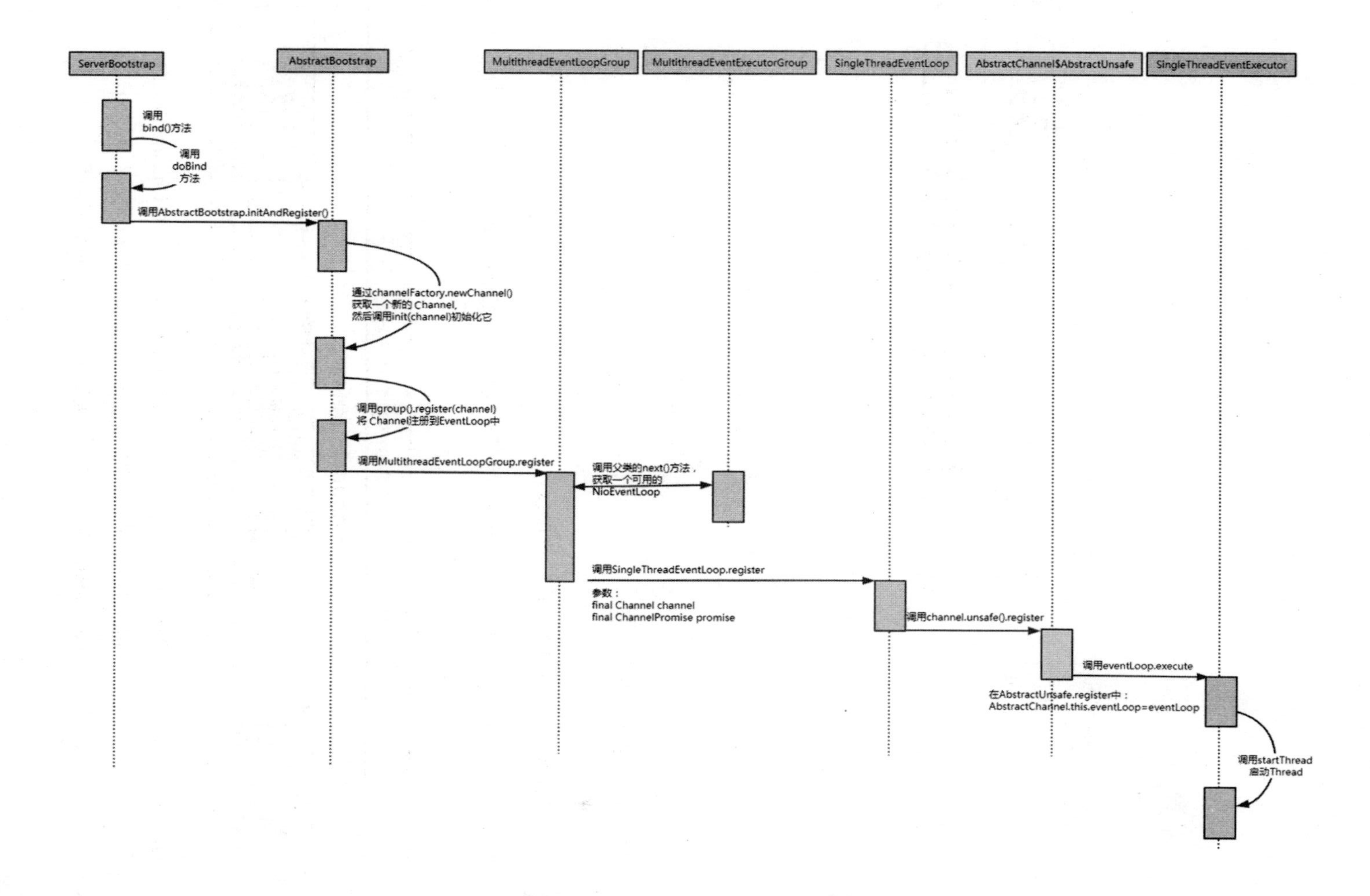
ServerBootstrap
AbstractBootstrap
MultithreadEventLoopGroup
MultithreadEventExecutorGroup
SingleThreadEventLoop
AbstractChannel$AbstractUnsafe
SingleThreadEventExecutor
调用
bind()方法
调用
doBind
方法
调用AbstractBootstrap.initAndRegister()
通过channelFactory.newChannel()
获取一个新的 Channel,
然后调用init(channel)初始化它
调用group().register(channel)
将 Channel注册到EventLoop中
调用MultithreadEventLoopGroup.register
调用父类的next()方法，
获取一个可用的
NioEventLoop
调用SingleThreadEventLoop.register
参数：
final Channel channel
final ChannelPromise promise
调用channel.unsafe().register
调用eventLoop.execute
在AbstractUnsafe.register中：
AbstractChannel.this.eventLoop=eventLoop
调用startThread
启动Thread

9

第 9 章 Netty 大动脉 Pipeline

9.1 Pipeline 设计原理

9.1.1 Channel 与 ChannelPipeline

大家已经知道，在 Netty 中每个 Channel 都有且仅有一个 ChannelPipeline 与之对应，它们的组成关系如下图所示。

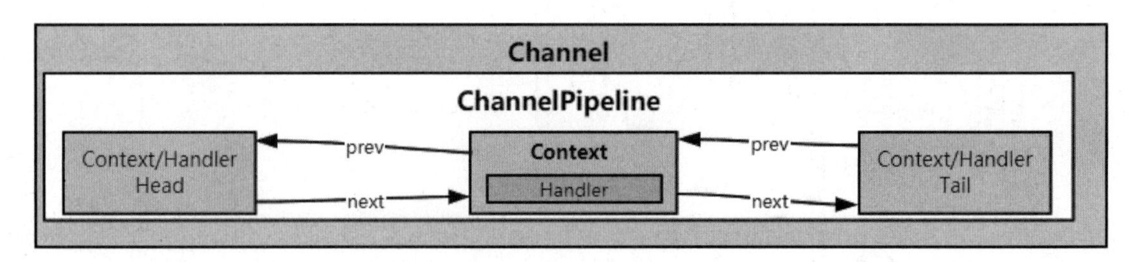

通过上图可以看到，一个 Channel 包含了一个 ChannelPipeline，而 ChannelPipeline 中又维护了一个由 ChannelHandlerContext 组成的双向链表。这个链表的头是 HeadContext，链表的尾是 TailContext，并且每个 ChannelHandlerContext 中又关联着一个 ChannelHandler。

上图给了我们一个对 ChannelPipeline 的直观认识，但是实际上 Netty 实现的 Channel 是否真的这样呢？我们继续通过代码进行分析。我们已经知道了一个 Channel 初始化的基本过程，下面再回顾一下。AbstractChannel 构造器的代码如下。

```
    protected AbstractChannel(Channel parent) {
        this.parent = parent;
        id = newId();
        unsafe = newUnsafe();
        pipeline = newChannelPipeline();
    }
```

AbstractChannel 有一个 pipeline 属性，在构造器中会把它初始化为 DefaultChannelPipeline 的实例。这里的代码就印证了这一点：每个 Channel 都有一个 ChannelPipeline。我们来跟踪一下 DefaultChannelPipeline 的初始化过程，DefaultChannelPipeline 构造器的代码如下。

```
protected DefaultChannelPipeline(Channel channel) {
        this.channel = ObjectUtil.checkNotNull(channel, "channel");
        succeededFuture = new SucceededChannelFuture(channel, null);
        voidPromise =  new VoidChannelPromise(channel, true);

        tail = new TailContext(this);
        head = new HeadContext(this);

        head.next = tail;
        tail.prev = head;
}
```

在 DefaultChannelPipeline 构造器中，首先将与之关联的 Channel 保存到属性 channel 中。然后实例化两个 ChannelHandlerContext：一个是 HeadContext 实例 Head，另一个是 TailContext 实例 Tail。接着将 Head 和 Tail 互相指向，构成一个双向链表。

特别注意的是：在开始的示意图中，Head 和 Tail 并没有包含 ChannelHandler，这是因为 HeadContext 和 TailContext 继承于 AbstractChannelHandlerContext 的同时，也实现了 ChannelHandler 接口，所以它们有 Context 和 Handler 的双重属性。

9.1.2 再谈 ChannelPipeline 的初始化

在第 7 章中我们已经对 ChannelPipeline 的初始化有了一个大致的了解，不过当时没有重点关注 ChannelPipeline，因此没有深入分析它的初始化过程。下面就来看一下 ChannelPipeline 的初始化具体都做了哪些工作。先回顾一下，在实例化一个 Channel 时，会伴随着一个 ChannelPipeline 的实例化，并且此 Channel 会与这个 ChannelPipeline 相互关联，这一点可以通过 NioSocketChannel

的父类 AbstractChannel 的构造器予以佐证，代码如下。

```
protected AbstractChannel(Channel parent) {
        this.parent = parent;
        id = newId();
        unsafe = newUnsafe();
        pipeline = newChannelPipeline();
}
```

当实例化一个 NioSocketChannel 时，其 pipeline 属性就是我们新创建的 DefaultChannelPipeline 对象，再来回顾一下 DefaultChannelPipeline 的构造方法，代码如下。

```
protected DefaultChannelPipeline(Channel channel) {
        this.channel = ObjectUtil.checkNotNull(channel, "channel");
        succeededFuture = new SucceededChannelFuture(channel, null);
        voidPromise =  new VoidChannelPromise(channel, true);

        tail = new TailContext(this);
        head = new HeadContext(this);

        head.next = tail;
        tail.prev = head;
}
```

上面代码中的 Head 实现了 ChannelInboundHandler 接口，而 Tail 实现了 ChannelOutboundHandler 接口，因此可以说 Head 和 Tail 既是 ChannelHandler，又是 ChannelHandlerContext。

9.1.3 ChannelInitializer 的添加

我们在第 7 章已经分析过 Channel 的组成，我们了解到，最开始的时候 ChannelPipeline 中含有两个 ChannelHandlerContext（同时也是 ChannelHandler），但是此时的 Pipeline 并不能实现特定的功能，因为还没有添加自定义的 ChannelHandler。通常来说，在初始化 Bootstrap 时，会添加自定义的 ChannelHandler，下面就以具体的客户端启动代码片段来举例。

```
            Bootstrap bootstrap = new Bootstrap();
            bootstrap.group(group)
            .channel(NioSocketChannel.class)
            .option(ChannelOption.SO_KEEPALIVE, true)
            .handler(new ChannelInitializer<SocketChannel>() {
                @Override
                protected void initChannel(SocketChannel ch) throws Exception {
                    ChannelPipeline pipeline = ch.pipeline();
                    pipeline.addLast(new ChatClientHandler(nickName));
```

```
            }
        });
```

上面代码的初始化过程，相信大家都不陌生。在调用 Handler 时，传入 ChannelInitializer 对象，它提供了一个 initChannel()方法来初始化 ChannelHandler。那么这个初始化过程是怎样的呢？下面我们来进行分析。

通过代码跟踪，我们发现 ChannelInitializer 是在 Bootstrap 的 init()方法中添加到 ChannelPipeline 中的，代码如下。

```
void init(Channel channel) throws Exception {
       ChannelPipeline p = channel.pipeline();
       p.addLast(config.handler());
      //略去 N 句代码
}
```

由上面的代码可见，将 handler()方法返回的 ChannelHandler 添加到 Pipeline 中，而 handler()方法返回的其实就是我们在初始化 Bootstrap 时通过 handler()方法设置的 ChannelInitializer 实例，因此这里就将 ChannelInitializer 插到了 Pipeline 的末端。此时 Pipeline 的结构如下图所示。

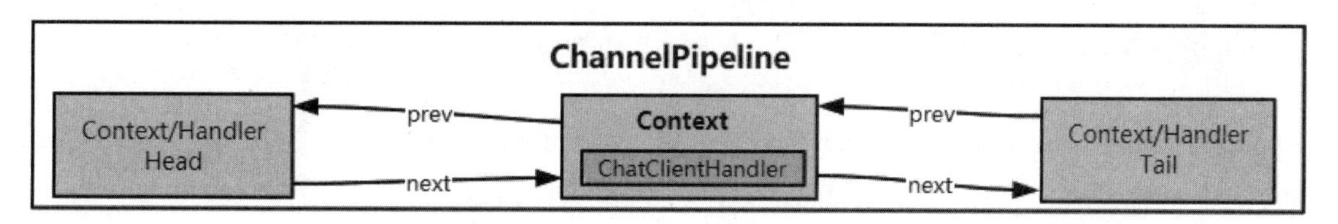

这时候，可能就有小伙伴疑惑了，明明插入的是一个 ChannelInitializer 实例，为什么在 ChannelPipeline 的双向链表中的元素却是一个 ChannelHandlerContext 呢?我们继续去代码中寻找答案。

在上面提到的 Bootstrap 的 init()方法中会调用 p.addLast()方法，将 ChannelInitializer 插入链表的末端，代码如下。

```
    public final ChannelPipeline addLast(EventExecutorGroup group, String name, ChannelHandler
handler) {
        final AbstractChannelHandlerContext newCtx;
        synchronized (this) {
            checkMultiplicity(handler);
            newCtx = newContext(group, filterName(name, handler), handler);
            addLast0(newCtx);
            //略去 N 句代码
        return this;
    }
```

```
    private AbstractChannelHandlerContext newContext(EventExecutorGroup group, String name,
ChannelHandler handler) {
        return new DefaultChannelHandlerContext(this, childExecutor(group), name, handler);
    }
```

addLast()方法有很多重载的方法，我们只需关注这个比较重要的方法就可以。上面的 addLast()方法中，首先检查 ChannelHandler 的名字是否重复，如果不重复，则调用 newContex()方法为这个 Handler 创建一个对应的 DefaultChannelHandlerContext 实例，并与之关联起来（Context 中有一个 Handler 属性保存着对应的 Handler 实例）。

为了添加一个 Handler 到 Pipeline 中，必须把此 Handler 包装成 ChannelHandlerContext。因此在上面的代码中，我们新实例化了一个 newCtx 对象，并将 Handler 作为参数传递到构造方法中。那么我们来看一下实例化的 DefaultChannelHandlerContext 到底有什么玄机。首先看它的构造器。

```
DefaultChannelHandlerContext(DefaultChannelPipeline pipeline, EventExecutor executor, String
name, ChannelHandler handler) {
        super(pipeline, executor, name, isInbound(handler), isOutbound(handler));
        if (handler == null) {
            throw new NullPointerException("handler");
        }
        this.handler = handler;
}
```

在 DefaultChannelHandlerContext 的构造器中，调用了两个很有意思的方法：isInbound()方法与 isOutbound()方法，这两个方法是做什么的呢？来看代码。

```
    private static boolean isInbound(ChannelHandler handler) {
        return handler instanceof ChannelInboundHandler;
    }

    private static boolean isOutbound(ChannelHandler handler) {
        return handler instanceof ChannelOutboundHandler;
    }
```

从上面代码中可以看到，当一个 Handler 实现了 ChannelInboundHandler 接口，则 isInbound 返回 true；类似地，当一个 Handler 实现了 ChannelOutboundHandler 接口，则 isOutbound 也返回 true。而这两个 boolean 变量会传递到父类 AbstractChannelHandlerContext 中，并初始化父类的两个属性：inbound 与 outbound。

这里的 ChannelInitializer 所对应的 DefaultChannelHandlerContext 的 inbound 与 outbound 属性分别是什么呢?先来看 ChannelInitializer 的类层次结构图，如下图所示。

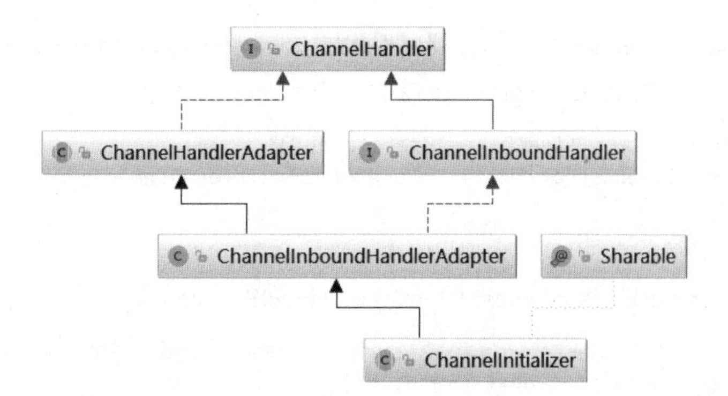

从上图可以清楚地看到，ChannelInitializer 仅仅实现了 ChannelInboundHandler 接口，因此这里实例化的 DefaultChannelHandlerContext 的 inbound = true，outbound = false。

为什么要这么大费周折地分析一番 inbound 和 outbound 两个属性呢？其实这两个属性关系到 Pipeline 事件的流向与分类，因此是十分关键的，但暂时先不分析这两个属性所起的作用。至此，我们先记住一个结论：ChannelInitializer 所对应的 DefaultChannelHandlerContext 的 inbound = true，outbound = false。

当创建好 Context 之后，就将这个 Context 插入 Pipeline 的双向链表中。

```
private void addLast0(AbstractChannelHandlerContext newCtx) {
      AbstractChannelHandlerContext prev = tail.prev;
      newCtx.prev = prev;
      newCtx.next = tail;
      prev.next = newCtx;
      tail.prev = newCtx;
}
```

添加完 ChannelInitializer 的 Pipeline 内部如下图所示。

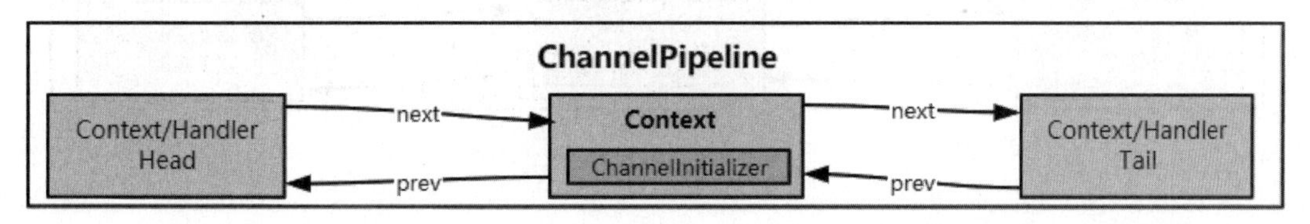

9.1.4　自定义 ChannelHandler 的添加过程

上一节我们已经分析了 ChannelInitializer 是如何插入 Pipeline 中的，接下来探讨 ChannelInitializer 在哪里被调用、ChannelInitializer 的作用及自定义的 ChannelHandler 是如何插入 Pipeline 中的。

我们自定义 ChannelHandler 的添加过程，发生在 AbstractUnsafe 的 register0()方法中，在这个方法中调用了 pipeline.fireChannelRegistered()方法，其代码实现如下。

```
public final ChannelPipeline fireChannelRegistered() {
        AbstractChannelHandlerContext.invokeChannelRegistered(head);
        return this;
}
```

再看 AbstractChannelHandlerContext 的 invokeChannelRegistered()方法。

```
static void invokeChannelRegistered(final AbstractChannelHandlerContext next) {
        EventExecutor executor = next.executor();
        if (executor.inEventLoop()) {
            next.invokeChannelRegistered();
        } else {
            executor.execute(new Runnable() {
                @Override
                public void run() {
                    next.invokeChannelRegistered();
                }
            });
        }
}
```

很显然，这个代码将从 Head 开始遍历 Pipeline 的双向链表，然后找到第一个属性 inbound 为 true 的 ChannelHandlerContext 实例。我们在分析 ChannelInitializer 时，花了大量的篇幅来分析 inbound 和 outbound 属性，现在这里就用上了。回想一下，ChannelInitializer 实现了 ChannelInboudHandler，因此它所对应的 ChannelHandlerContext 的 inbound 属性就是 true，因此这里返回的就是 ChannelInitializer 实例所对应的 ChannelHandlerContext 对象，如下图所示。

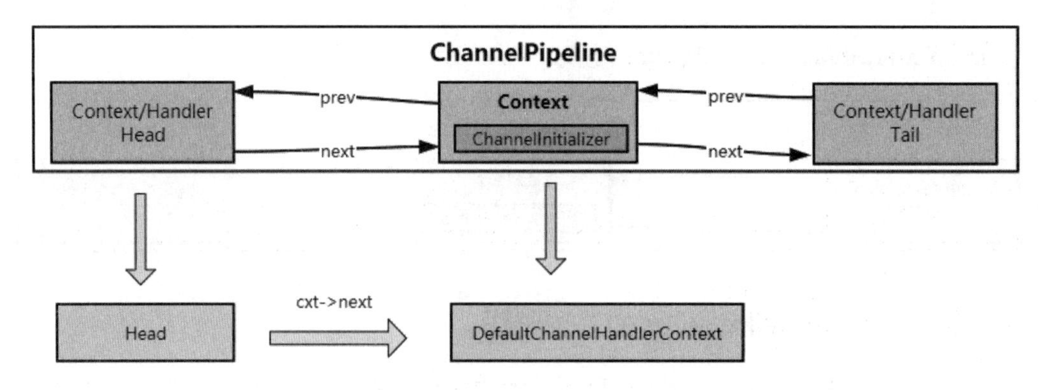

当获取 inbound 的 Context 后，就调用它的 invokeChannelRegistered()方法。

```java
private void invokeChannelRegistered() {
        if (invokeHandler()) {
            try {
                ((ChannelInboundHandler) handler()).channelRegistered(this);
            } catch (Throwable t) {
                notifyHandlerException(t);
            }
        } else {
            fireChannelRegistered();
        }
}
```

我们已经知道，每个 ChannelHandler 都和一个 ChannelHandlerContext 关联，可以通过 ChannelHandlerContext 获取对应的 ChannelHandler。很明显，这里 handler()方法返回的对象其实就是一开始实例化的 ChannelInitializer 对象，接着调用了 ChannelInitializer 的 channelRegistered()方法。ChannelInitializer 的 channelRegistered()方法我们在第 7 章已经接触了，但是并没有深入分析其调用过程。下面来看这个方法到底有什么玄机。继续看代码。

```java
    public final void channelRegistered(ChannelHandlerContext ctx) throws Exception {
        if (initChannel(ctx)) {
            ctx.pipeline().fireChannelRegistered();
        } else {
            ctx.fireChannelRegistered();
        }
    }
    private boolean initChannel(ChannelHandlerContext ctx) throws Exception {
        if (initMap.putIfAbsent(ctx, Boolean.TRUE) == null) { // Guard against
            try {
                initChannel((C) ctx.channel());
            } catch (Throwable cause) {
                exceptionCaught(ctx, cause);
            } finally {
                remove(ctx);
            }
            return true;
        }
        return false;
    }
```

initChannel((C) ctx.channel())方法我们也很熟悉，它就是在初始化 Bootstrap 时，调用 handler()方法传入的匿名内部类所实现的方法，代码如下。

```java
.handler(new ChannelInitializer<SocketChannel>() {
                 @Override
                 protected void initChannel(SocketChannel ch) throws Exception {
                     ChannelPipeline pipeline = ch.pipeline();
```

```
                pipeline.addLast(new ChatClientHandler(nickName));
            }
});
```

因此，在调用这个方法之后，我们自定义的 ChannelHandler 就插入 Pipeline 中了，此时 Pipeline 的状态如下图所示。

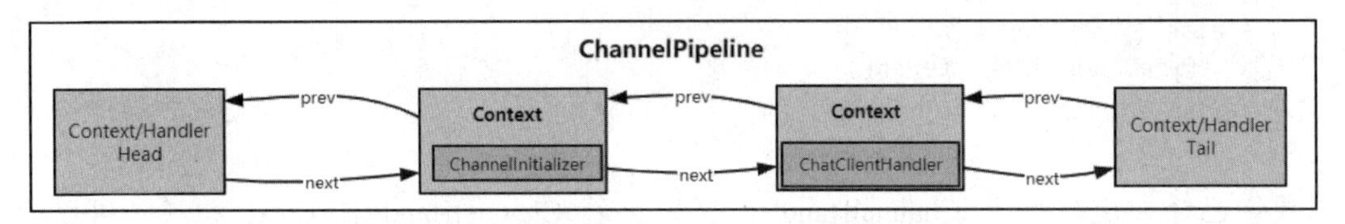

当添加完自定义的 ChannelHandler 后，在 finally 代码块会删除自定义的 ChannelInitializer，也就是 remove(ctx)，最终调用 ctx.pipeline().remove(this)，因此最后 Pipeline 的状态如下图所示。

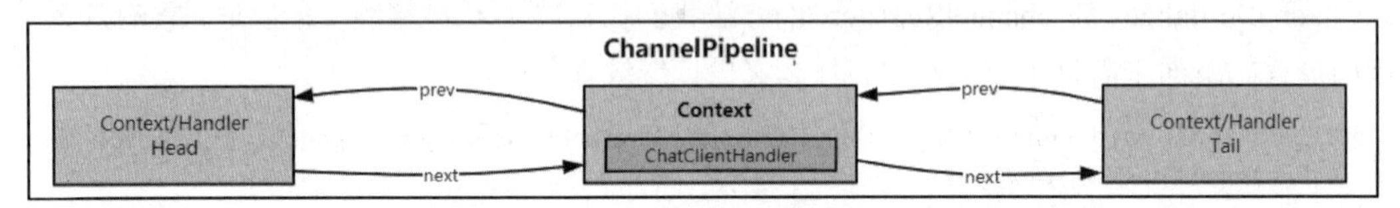

至此，自定义 ChannelHandler 的添加过程也分析完成了。

9.1.5 给 ChannelHandler 命名

不知道大家注意到没有，pipeline.addXXX()都有一个重载的方法，例如 addLast()有一个重载的版本，代码如下。

```
ChannelPipeline addLast(String name, ChannelHandler handler);
```

第一个参数指定添加的是 Handler 的名字（更准确地说是 ChannelHandlerContext 的名字，说成 Handler 的名字更便于理解）。那么 Handler 的名字有什么用呢？如果我们不设置 name，那么 Handler 默认的名字是怎样的呢？带着这些疑问，我们依旧去代码中寻找答案。还是以 addLast()方法为例，代码如下。

```
public final ChannelPipeline addLast(String name, ChannelHandler handler) {
     return addLast(null, name, handler);
}
```

这个方法会调用重载的 addLast()方法，代码如下。

```
    public final ChannelPipeline addLast(EventExecutorGroup group, String name, ChannelHandler
handler) {
     final AbstractChannelHandlerContext newCtx;
```

```
        synchronized (this) {
            checkMultiplicity(handler);

            newCtx = newContext(group, filterName(name, handler), handler);

            addLast0(newCtx);
            //略去 N 句代码
        }
        return this;
    }
```

第一个参数设置为 null，我们不用关心它。第二个参数就是 Handler 的名字。由代码可知，在添加一个 Handler 之前，需要调用 checkMultiplicity()方法来确定新添加的 Handler 名字是否与已添加的 Handler 名字重复。

9.1.6　ChannelHandler 的默认命名规则

如果我们调用如下的 addLast()方法：

```
ChannelPipeline addLast(ChannelHandler... handlers);
```

那么 Netty 就会调用 generateName()方法为新添加的 Handler 自动生成一个默认的名字。

```
    private String filterName(String name, ChannelHandler handler) {
        if (name == null) {
            return generateName(handler);
        }
        checkDuplicateName(name);
        return name;
    }
    private String generateName(ChannelHandler handler) {
        Map<Class<?>, String> cache = nameCaches.get();
        Class<?> handlerType = handler.getClass();
        String name = cache.get(handlerType);
        if (name == null) {
            name = generateName0(handlerType);
            cache.put(handlerType, name);
        }
        //此处省略 N 行代码
        return name;
    }
```

而 generateName()方法会接着调用 generateName0()方法来实际生成一个新的 Handler 名字。

```
private static String generateName0(Class<?> handlerType) {
     return StringUtil.simpleClassName(handlerType) + "#0";
}
```

默认命名的规则很简单，就是用反射获取 Handler 的 simpleName 加上“#0”，因此我们自定义 ChatClientHandler 的名字就是“ChatClientHandler#0”。

9.2 Pipeline 的事件传播机制

上一节中，我们已经知道 AbstractChannelHandlerContext 中有 Inbound 和 Outbound 两个 boolean 变量，分别用于标识 Context 所对应的 Handler 的类型。

（1）Inbound 为 true 时，表示其对应的 ChannelHandler 是 ChannelInboundHandler 的子类。

（2）Outbound 为 true 时，表示其对应的 ChannelHandler 是 ChannelOutboundHandler 的子类。

大家肯定还有很多疑惑，不知道这两个属性到底有什么作用？这还要从 ChannelPipeline 的事件传播类型说起。Netty 中的传播事件可以分为两种：Inbound 事件和 Outbound 事件。如下是 Netty 官网针对这两个事件的说明。

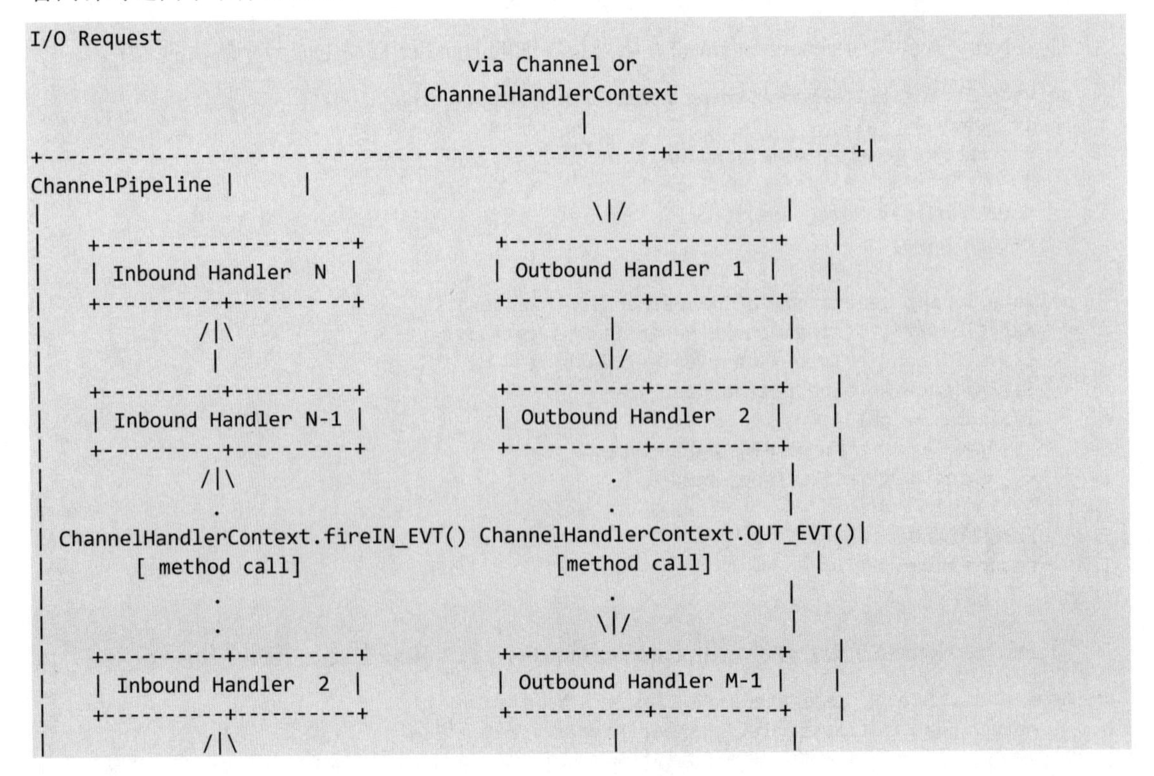

```
I/O Request
                                    via Channel or
                                 ChannelHandlerContext
                                          |
+---------------------------------------------------+---------------+|
ChannelPipeline |     |
|                                          \|/            |
|    +---------------------+         +-----------+----------+     |
|    | Inbound Handler  N  |         | Outbound Handler  1  |     |
|    +----------+----------+         +-----------+----------+     |
|              /|\                               |              |
|               |                                \|/             |
|    +----------+----------+         +-----------+----------+     |
|    | Inbound Handler N-1 |         | Outbound Handler  2  |     |
|    +----------+----------+         +-----------+----------+     |
|              /|\                               .              |
|               .                                .              |
| ChannelHandlerContext.fireIN_EVT() ChannelHandlerContext.OUT_EVT()|
|        [ method call]                     [method call]         |
|               .                                .              |
|               .                               \|/             |
|    +----------+----------+         +-----------+----------+     |
|    | Inbound Handler  2  |         | Outbound Handler M-1 |     |
|    +----------+----------+         +-----------+----------+     |
|              /|\                               |              |
```

```
|           |                                   \|/              |
|    +----------+----------+            +-----------+----------+    |
|    | Inbound Handler  1  |            | Outbound Handler  M  |    |
|    +----------+----------+            +-----------+----------+    |
|             /|\                                   |              |
+---------------+-----------------------------------+---------------+
                |                                  \|/
+---------------+-----------------------------------+---------------+|     |     |     |
|      [ Socket.read() ]                      [ Socket.write() ]     |
|                                                                   |
|  Netty Internal I/O Threads (Transport Implementation)            |
+-------------------------------------------------------------------+
```

由上可以看出，Inbound 事件和 Outbound 事件的流向是不一样的，Inbound 事件的流向是从下至上的，而 Outbound 恰好相反，是从上到下。并且 Inbound 方法是通过调用相应的 ChannelHandlerContext.fireIN_EVT() 方法来传递的，而 Outbound 方法是通过调用 ChannelHandlerContext.OUT_EVT() 方法来传递的。例如：ChannelHandlerContext 的 fireChannelRegistered() 调用会发送一个 ChannelRegistered 的 Inbound 给下一个 ChannelHandlerContext，而 ChannelHandlerContext 的 bind()方法调用时会发送一个 bind 的 Outbound 事件给下一个 ChannelHandlerContext。

Inbound 事件传播方法的代码如下。

```
public interface ChannelInboundHandler extends ChannelHandler {
    void channelRegistered(ChannelHandlerContext ctx) throws Exception;
    void channelUnregistered(ChannelHandlerContext ctx) throws Exception;
    void channelActive(ChannelHandlerContext ctx) throws Exception;
    void channelInactive(ChannelHandlerContext ctx) throws Exception;
    void channelRead(ChannelHandlerContext ctx, Object msg) throws Exception;
    void channelReadComplete(ChannelHandlerContext ctx) throws Exception;
    void userEventTriggered(ChannelHandlerContext ctx, Object evt) throws Exception;
    void channelWritabilityChanged(ChannelHandlerContext ctx) throws Exception;
    void exceptionCaught(ChannelHandlerContext ctx, Throwable cause) throws Exception;
}
```

Outbound 事件传播方法的代码如下。

```
public interface ChannelOutboundHandler extends ChannelHandler {
    void bind(ChannelHandlerContext ctx, SocketAddress localAddress, ChannelPromise promise)
throws Exception;
    void connect(
            ChannelHandlerContext ctx, SocketAddress remoteAddress,
            SocketAddress localAddress, ChannelPromise promise) throws Exception;
    void disconnect(ChannelHandlerContext ctx, ChannelPromise promise) throws Exception;
    void close(ChannelHandlerContext ctx, ChannelPromise promise) throws Exception;
```

```
    void deregister(ChannelHandlerContext ctx, ChannelPromise promise) throws Exception;
    void read(ChannelHandlerContext ctx) throws Exception;
    void flush(ChannelHandlerContext ctx) throws Exception;
}
```

我们发现：Inbound 类似于事件回调（响应请求的事件），而 Outbound 类似于主动触发（发起请求的事件）。注意，如果我们捕获了一个事件，并且想让这个事件继续传递下去，那么需要调用 Context 对应的传播方法 fireXXX()方法。

```
public class MyInboundHandler extends ChannelInboundHandlerAdapter {
    @Override
    public void channelActive(ChannelHandlerContext ctx) throws Exception {
        System.out.println("连接成功");
        ctx.fireChannelActive();
    }
}

public class MyOutboundHandler extends ChannelOutboundHandlerAdapter {
    @Override
    public void close(ChannelHandlerContext ctx, ChannelPromise promise) throws Exception {
        System.out.println("客户端关闭");
        ctx.close(promise);
    }
}
```

如上面的代码所示，MyInboundHandler 收到了一个 channelActive 事件，它在处理后，如果希望将事件继续传播下去，那么需要接着调用 ctx.fireChannelActive()方法。

接下来我们可以用一个代码案例来了解一下 Pipeline 的传播机制。我们分别编写 InboundHandlerA、InboundHandlerB、InboundHandlerC 和 OutboundHandlerA、OutboundHandlerB、OutboundHandlerC 类。

InboundHandlerA 的代码如下。

```
public class InboundHandlerA extends ChannelInboundHandlerAdapter {

    public void channelRead(ChannelHandlerContext ctx, Object msg) throws Exception {
        System.out.println("InboundHandlerA");
        ctx.fireChannelRead(msg);
    }
}
```

InboundHandlerB 的代码如下。

```
public class InboundHandlerB extends ChannelInboundHandlerAdapter {
```

```
    @Override
    //读取 Client 发送的信息，并打印出来
    public void channelRead(ChannelHandlerContext ctx, Object msg) throws Exception {
        System.out.println("InboundHandlerB");
        ctx.fireChannelRead(msg);
    }
}
```

InboundHandlerC 的代码如下：

```
public class InboundHandlerC extends ChannelInboundHandlerAdapter {

    public void channelRead(ChannelHandlerContext ctx, Object msg) throws Exception {
        System.out.println("InboundHandlerC");
        ctx.fireChannelRead(msg);
    }
}
```

以上代码中 InboundHandlerA 、 InboundHandlerB 、 InboundHandlerC 都调用了 ctx.fireChannelRead()方法向下传播。

OutboundHandlerA 的代码如下。

```
public class OutboundHandlerA extends ChannelOutboundHandlerAdapter {
    @Override
    //向 Client 发送消息
    public void write(ChannelHandlerContext ctx, Object msg, ChannelPromise promise) throws
Exception {
        System.out.println("OutboundHandlerA.write");

        ctx.write(msg,promise);
    }
}
```

OutboundHandlerB 的代码如下。

```
public class OutboundHandlerB extends ChannelOutboundHandlerAdapter {

    @Override
    public void write(ChannelHandlerContext ctx, Object msg, ChannelPromise promise) throws
Exception {
        System.out.println("OutboundHandlerB.write");
        //执行下一个 OutboundHandler
        ctx.write(msg,promise);
    }

    @Override
    public void handlerAdded(final ChannelHandlerContext ctx) throws Exception {
```

```
        ctx.executor().schedule(new Runnable() {
            public void run() {
                ctx.channel().write("say hello");
            }
        },3, TimeUnit.SECONDS);
    }
}
```

OutboundHandlerC 的代码如下。

```
public class OutboundHandlerC extends ChannelOutboundHandlerAdapter {

    @Override
    public void write(ChannelHandlerContext ctx, Object msg, ChannelPromise promise) throws
Exception {
        System.out.println("OutboundHandlerC.write");
        //执行下一个 OutboundHandler
        ctx.write(msg,promise);
    }
}
```

以上代码中 OutboundHandlerA、OutboundHandlerB、OutboundHandlerC 都调用了 ctx.write() 方法输出。

下面我们编写测试代码，来了解其传播顺序。先是编写服务端代码。

PipelineServer 类主要完成 Pipeline 的注册工作，代码如下。

```
public class PipelineServer {
    public void start(int port) throws Exception {
        EventLoopGroup bossGroup = new NioEventLoopGroup();
        EventLoopGroup workerGroup = new NioEventLoopGroup();
        try {
            ServerBootstrap b = new ServerBootstrap();
            b.group(bossGroup, workerGroup).channel(NioServerSocketChannel.class)
                    .childHandler(new ChannelInitializer<SocketChannel>() {
                                @Override
                                public void initChannel(SocketChannel ch) throws Exception {

                                    //InboundHandler 的执行顺序为注册顺序，应该是 A→B→C
                                    ch.pipeline().addLast(new InboundHandlerA());
                                    ch.pipeline().addLast(new InboundHandlerB());
                                    ch.pipeline().addLast(new InboundHandlerC());

                                    // OutboundHandler 的执行顺序为注册顺序的逆序，应该是 C→B→A
                                    ch.pipeline().addLast(new OutboundHandlerA());
                                    ch.pipeline().addLast(new OutboundHandlerB());
```

```
                            ch.pipeline().addLast(new OutboundHandlerC());

                        }
                    }).option(ChannelOption.SO_BACKLOG, 128)
            .childOption(ChannelOption.SO_KEEPALIVE, true);

         ChannelFuture f = b.bind(port).sync();

         f.channel().closeFuture().sync();
      } finally {
         workerGroup.shutdownGracefully();
         bossGroup.shutdownGracefully();
      }
   }

   public static void main(String[] args) throws Exception {
   PipelineServer server = new PipelineServer();
      server.start(8000);
   }
}
```

PipelineClient 类，与服务端建立连接并向服务端发送数据，代码如下。

```
public class PipelineClient {
   public void connect(String host, int port) throws Exception {
      EventLoopGroup workerGroup = new NioEventLoopGroup();

      try {
         Bootstrap b = new Bootstrap();
         b.group(workerGroup);
         b.channel(NioSocketChannel.class);
         b.option(ChannelOption.SO_KEEPALIVE, true);
         b.handler(new ChannelInitializer<SocketChannel>() {
            @Override
            public void initChannel(SocketChannel ch) throws Exception {
               ch.pipeline().addLast(new ClientIntHandler());
            }
         });

         //Start the Client.
         ChannelFuture f = b.connect(host, port).sync();
         f.channel().closeFuture().sync();
      } finally {
         workerGroup.shutdownGracefully();
      }
   }
```

```
    public static void main(String[] args) throws Exception {
        PipelineClient client = new PipelineClient();
        client.connect("127.0.0.1", 8000);
    }
}
```

ClientInHandler 类，完成向服务端发送数据的动作，代码如下。

```
public class ClientIntHandler extends ChannelInboundHandlerAdapter {
    @Override
    //读取服务端的信息
    public void channelRead(ChannelHandlerContext ctx, Object msg) throws Exception {
        System.out.println("ClientIntHandler.channelRead");
        ByteBuf result = (ByteBuf) msg;
        byte[] result1 = new byte[result.readableBytes()];
        result.readBytes(result1);
        result.release();
        ctx.close();
        System.out.println("Server said:" + new String(result1));
    }
    @Override
    //当连接建立的时候向服务端发送消息，channelActive 事件在连接建立的时候会被触发
    public void channelActive(ChannelHandlerContext ctx) throws Exception {
     System.out.println("ClientIntHandler.channelActive");
        String msg = "Are you ok?";
        ByteBuf encoded = ctx.alloc().buffer(4 * msg.length());
        encoded.writeBytes(msg.getBytes());
        ctx.write(encoded);
        ctx.flush();
    }
}
```

接下来，我们运行测试代码，分别启动 PipelineServer 和 PipelineClient，得到的运行结果如下图所示。

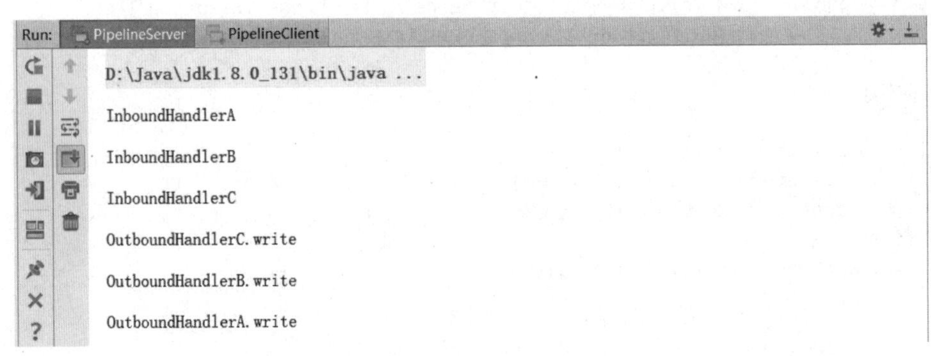

我们再尝试调整 PipelineServer 中 Handler 的注册顺序，代码如下。

```
public class PipelineServer {
    public void start(int port) throws Exception {
        EventLoopGroup bossGroup = new NioEventLoopGroup();
        EventLoopGroup workerGroup = new NioEventLoopGroup();
        try {
            ServerBootstrap b = new ServerBootstrap();
            b.group(bossGroup, workerGroup).channel(NioServerSocketChannel.class)
                    .childHandler(new ChannelInitializer<SocketChannel>() {
                                @Override
                                public void initChannel(SocketChannel ch) throws Exception {

                                    ch.pipeline().addLast(new OutboundHandlerC());
                                    ch.pipeline().addLast(new InboundHandlerB());
                                    ch.pipeline().addLast(new OutboundHandlerA());
                                    ch.pipeline().addLast(new InboundHandlerA());
                                    ch.pipeline().addLast(new InboundHandlerC());
                                    ch.pipeline().addLast(new OutboundHandlerB());
                                                  ......

                                }
                            }).option(ChannelOption.SO_BACKLOG, 128)
                    .childOption(ChannelOption.SO_KEEPALIVE, true);

            ChannelFuture f = b.bind(port).sync();

            f.channel().closeFuture().sync();
        } finally {
            workerGroup.shutdownGracefully();
            bossGroup.shutdownGracefully();
        }
    }
    ......
}
```

调整代码后，执行结果如下图所示。

从执行结果上看，我们已经知道了 Handler 的传播顺序：从 Inbound 开始顺序执行，然后从 Outbound 逆序执行。

9.2.1 Outbound 事件传播方式

Outbound 事件都是请求事件（Request Event），即请求某件事情的发生，然后通过 Outbound 事件进行通知。

Outbound 事件的传播方向是从 Tail 到 customContext 再到 Head。

下面我们以 Connect 事件为例，分析一下 Outbound 事件的传播机制。

首先，当用户调用了 Bootstrap 的 connect()方法时，就会触发一个 Connect 请求事件，此调用会触发调用链，如下图所示。

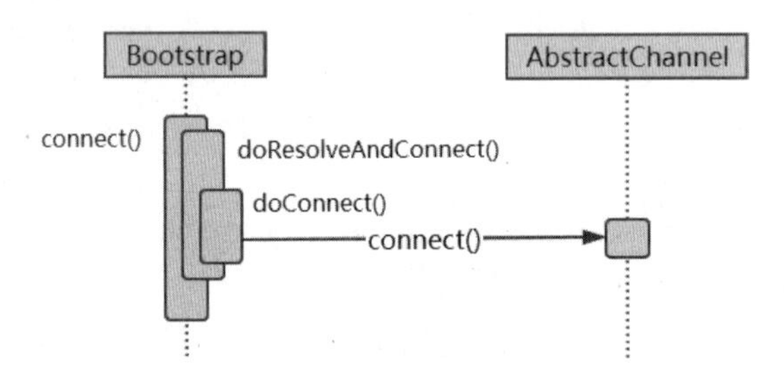

继续跟踪，我们就发现 AbstractChannel 的 connect()方法其实又调用了 DefaultChannelPipeline 的 connect()方法，代码如下。

```
public ChannelFuture connect(SocketAddress remoteAddress, ChannelPromise promise) {
    return pipeline.connect(remoteAddress, promise);
}
```

而 pipeline.connect()方法的实现代码如下。

```
public final ChannelFuture connect(SocketAddress remoteAddress, ChannelPromise promise)
{
    return tail.connect(remoteAddress, promise);
}
```

可以看到，当 Outbound 事件（这里是 Connect 事件）传递到 Pipeline 后，其实是以 Tail 为起点开始传播的。

而 tail.connect()调用的是 AbstractChannelHandlerContext 的 connect()方法。

```
public ChannelFuture connect(
        final SocketAddress remoteAddress,
```

```
        final SocketAddress localAddress, final ChannelPromise promise) {
    //此处省略N句
    final AbstractChannelHandlerContext next = findContextOutbound();
    EventExecutor executor = next.executor();
        next.invokeConnect(remoteAddress, localAddress, promise);
    //此处省略N句
    return promise;
}
```

顾名思义，findContextOutbound()方法的作用是以当前 Context 为起点，向 Pipeline 中 Context 双向链表的前端寻找第一个 Outbound 属性为 true 的 Context（即关联 ChannelOutboundHandler 的 Context），然后返回。findContextOutbound()方法的实现代码如下。

```
private AbstractChannelHandlerContext findContextOutbound() {
    AbstractChannelHandlerContext ctx = this;
    do {
        ctx = ctx.prev;
    } while (!ctx.outbound);
    return ctx;
}
```

当我们找到了一个 Outbound 的 Context 后，就调用它的 invokeConnect()方法，这个方法会调用 Context 关联的 ChannelHandler 的 connect()方法，代码如下。

```
    private void invokeConnect(SocketAddress remoteAddress, SocketAddress localAddress,
ChannelPromise promise) {
        if (invokeHandler()) {
            try {
                ((ChannelOutboundHandler) handler()).connect(this, remoteAddress,
localAddress, promise);
            } catch (Throwable t) {
                notifyOutboundHandlerException(t, promise);
            }
        } else {
            connect(remoteAddress, localAddress, promise);
        }
    }
```

如果用户没有重写 ChannelHandler 的 connect() 方法，那么会调用 ChannelOutboundHandlerAdapter 的 connect()方法，代码如下。

```
    public void connect(ChannelHandlerContext ctx, SocketAddress remoteAddress,
            SocketAddress localAddress, ChannelPromise promise) throws Exception {
        ctx.connect(remoteAddress, localAddress, promise);
    }
```

我们看到，ChannelOutboundHandlerAdapter 的 connect()方法仅仅调用了 ctx.connect()，而这个调用又回到了 Context.connect()方法调用 Connect.findContextOutbound()方法，然后调用 next.invokeConnect()方法，其次调用 handler.connect()方法，最后又调用 Context.connect()方法，如此循环下去，直到 Connect 事件传递到 DefaultChannelPipeline 的双向链表的头节点，即 Head 中。为什么会传递到 Head 中呢？回想一下，Head 实现了 ChannelOutboundHandler，因此它的 Outbound 属性是 true。

因为 Head 本身既是一个 ChannelHandlerContext，又实现了 ChannelOutboundHandler 接口，所以当 connect()消息传递到 Head 后，会将消息传递到对应的 ChannelHandler 中处理，而 Head 的 handler()方法返回的就是 Head 本身，代码如下。

```
public ChannelHandler handler() {
        return this;
}
```

因此最终 Connect 事件是在 Head 中被处理的。Head 的 Connect 事件处理逻辑的代码如下。

```
public void connect(
            ChannelHandlerContext ctx,
            SocketAddress remoteAddress, SocketAddress localAddress,
            ChannelPromise promise) throws Exception {
        unsafe.connect(remoteAddress, localAddress, promise);
}
```

到这里，整个 Connect 请求事件就结束了。下图描述了整个 Connect 请求事件的处理过程。

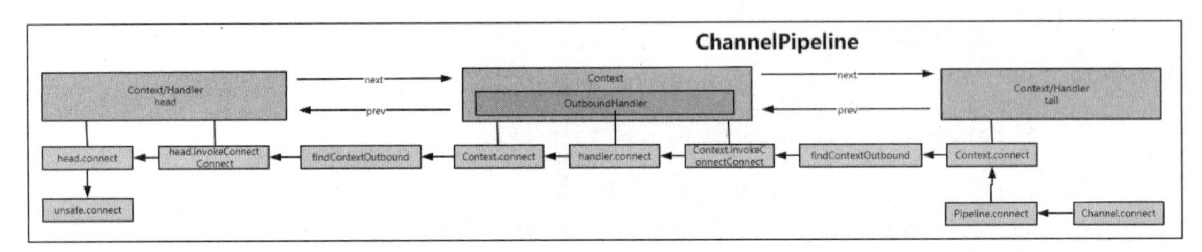

我们仅仅以 Connect 请求事件为例，分析了 Outbound 事件的传播过程，但是其实所有的 Outbound 的事件传播都遵循着一样的传播规律，小伙伴们可以试着分析一下其他 Outbound 事件，体会一下它们的传播过程。

9.2.2 Inbound 事件传播方式

Inbound 事件和 Outbound 事件的处理过程是类似的，只是传播方向不同。

Inbound 事件是一个通知事件，即某件事已经发生了，然后通过 Inbound 事件进行通知。Inbound

通常发生在 Channel 的状态改变或 I/O 事件就绪时。

Inbound 的特点是其传播方向从 Head 到 customContext 再到 Tail。

上面我们分析了 connect()方法其实是一个 Outbound 事件，那么接着分析 Connect()事件后会发生什么 Inbound 事件，并最终找到 Outbound 和 Inbound 事件之间的联系。当 Connect()事件传播到 Unsafe 后，其实是在 AbstractNioUnsafe 的 connect()方法中进行处理的，代码如下。

```
public final void connect(
            final SocketAddress remoteAddress,
            final SocketAddress localAddress, final ChannelPromise promise) {

            if (doConnect(remoteAddress, localAddress)) {
                fulfillConnectPromise(promise, wasActive);
            } else {
                ...
            }
}
```

在 AbstractNioUnsafe 的 connect()方法中，先调用 doConnect()方法进行实际的 Socket 连接，当连接后会调用 fulfillConnectPromise()方法，代码如下。

```
private void fulfillConnectPromise(ChannelPromise promise, boolean wasActive) {
         if (!wasActive && active) {
             pipeline().fireChannelActive();
         }
}
```

我们看到，在 fulfillConnectPromise()方法中，会通过调用 pipeline().fireChannelActive()方法将通道激活的消息（即 Socket 连接成功）发送出去。而这里，当调用 pipeline.fireXXX 后，就是 Inbound 事件的起点。因此当调用 pipeline().fireChannelActive()方法时，就产生了一个 ChannelActive Inbound 事件，接下来看一下 Inbound 事件是怎么传播的，代码如下。

```
public final ChannelPipeline fireChannelActive() {
      AbstractChannelHandlerContext.invokeChannelActive(head);
      return this;
}
```

果然，在 fireChannelActive()方法中调用了 head.invokeChannelActive()方法，因此可以证明 Inbound 事件在 Pipeline 中传输的起点是 Head。head.invokeChannelActive()方法的代码如下。

```
static void invokeChannelActive(final AbstractChannelHandlerContext next) {
      EventExecutor executor = next.executor();
      if (executor.inEventLoop()) {
         next.invokeChannelActive();
```

```
        } else {
            executor.execute(new Runnable() {
                @Override
                public void run() {
                    next.invokeChannelActive();
                }
            });
        }
}
```

上面的代码应该很熟悉了。回想一下在 Outbound 事件（例如 Connect 事件）的传输过程中，我们也有类似的如下操作。

（1）首先调用 findContextInbound()，从 Pipeline 的双向链表中找到第一个 Inbound 属性为 true 的 Context，然后将其返回。

（2）调用 Context 的 invokeChannelActive()方法，invokeChannelActive()方法的代码如下。

```
    private void invokeChannelActive() {
        if (invokeHandler()) {
            try {
                ((ChannelInboundHandler) handler()).channelActive(this);
            } catch (Throwable t) {
                notifyHandlerException(t);
            }
        } else {
            fireChannelActive();
        }
    }
```

这个方法和 Outbound 的对应方法（如 invokeConnect()方法）如出一辙。与 Outbound 一样，如果用户没有重写 channelActive()方法，就会调用 ChannelInboundHandlerAdapter 的 channelActive()方法，代码如下。

```
public void channelActive(ChannelHandlerContext ctx) throws Exception {
        ctx.fireChannelActive();
}
```

同样地，在 ChannelInboundHandlerAdapter 的 channelActive()方法中，仅仅调用了 ctx.fireChannelActive()方法，因此就调用 Context.fireChannelActive()方法，其次调用 Connect.findContextInbound()方法，然后调用 nextContext.invokeChannelActive()方法，再然后调用 nextHandler.channelActive()方法，最后调用 nextContext.fireChannelActive()方法，如此循环。同理，Tail 本身既实现了 ChannelInboundHandler 接口，又实现了 ChannelHandlerContext 接口，因此当 channelActive()消息传递到 Tail 后，会将消息传递到对应的 ChannelHandler 中处理，而 Tail 的

handler()方法返回的就是 Tail 本身，代码如下。

```
public ChannelHandler handler() {
        return this;
}
```

因此 ChannelActive Inbound 事件最终是在 Tail 中处理的，我们看一下它的处理方法。

```
public void channelActive(ChannelHandlerContext ctx) throws Exception {
}
```

TailContext 的 channelActive()方法是空的。大家自行查看 TailContext 的 Inbound 处理方法时会发现，它们的实现都是空的。可见，如果是 Inbound，当用户没有实现自定义的处理器时，那么默认是不处理的。下图描述了 Inbound 事件的传输过程。

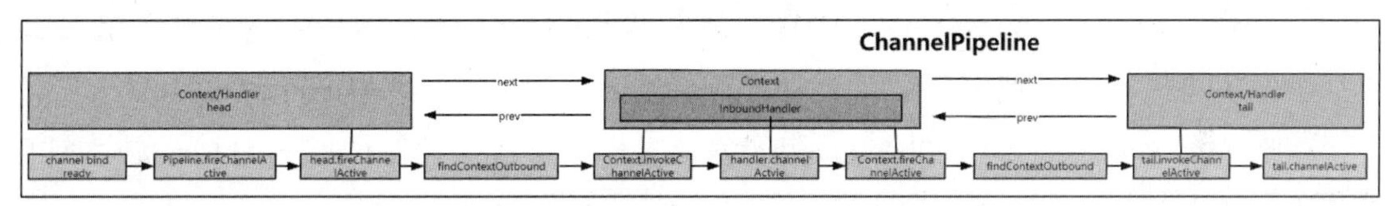

9.2.3　小结

Outbound 事件为传播过程总结如下。

（1）Outbound 事件是请求事件（由 Connect 发起一个请求，并最终由 Unsafe 处理这个请求）。

（2）Outbound 事件的发起者是 Channel。

（3）Outbound 事件的处理者是 Unsafe。

（4）Outbound 事件在 Pipeline 中的传输方向是从 Tail 到 Head。

（5）在 ChannelHandler 中处理事件时，如果这个 Handler 不是最后一个 Handler，则需要调用 ctx 的方法（如 ctx.connect()方法）将此事件继续传播下去。如果不这样做，那么此事件的传播会提前终止。

（6）Outbound 事件的传播方向是，从 Context.OUT_EVT()方法到 Connect.findContextOutbound()方法，再到 nextContext.invokeOUT_EVT()方法，再到 nextHandler.OUT_EVT()方法，最后到 nextContext.OUT_EVT()方法。

Inbound 事件传播过程总结如下。

（1）Inbound 事件为通知事件，当某件事情已经就绪后，会通知上层。

（2）Inbound 事件的发起者是 Unsafe。

（3）Inbound 事件的处理者是 Channel，如果用户没有实现自定义的处理方法，那么 Inbound 事件默认的处理者是 TailContext，并且其处理方法是空实现。

（4）Inbound 事件在 Pipeline 中的传输方向是从 Head 到 Tail。

（5）在 ChannelHandler 中处理事件时，如果这个 Handler 不是最后一个 Handler，则需要调用 ctx.fireIN_EVT()方法（如 ctx.fireChannelActive()方法）将此事件继续传播下去。如果不这样做，那么此事件的传播会提前终止。

（6）Intbound 事件的传播方向是，从 Context.fireIN_EVT()方法到 Connect.findContextInbound()方法，再到 nextContext.invokeIN_EVT()方法，再到 nextHandler.IN_EVT()方法，最后到 nextContext.fireIN_EVT()方法。

由此可知，Outbound 事件和 Inbound 事件在设计上十分相似，并且 Context 与 Handler 之间的调用关系也容易混淆，因此我们在阅读这里的代码时，需要特别注意。

9.3 Handler 的各种“姿势”

9.3.1 ChannelHandlerContext

每个 ChannelHandler 被添加到 ChannelPipeline 后，都会创建一个 ChannelHandlerContext，并与 ChannelHandler 关联绑定。ChannelHandlerContext 允许 ChannelHandler 与其他的 ChannelHandler 进行交互。ChannelHandlerContext 不会改变添加到其中的 ChannelHandler，因此它是安全的。ChannelHandlerContext、ChannelHandler 和 ChannelPipeline 的关系如下图所示。

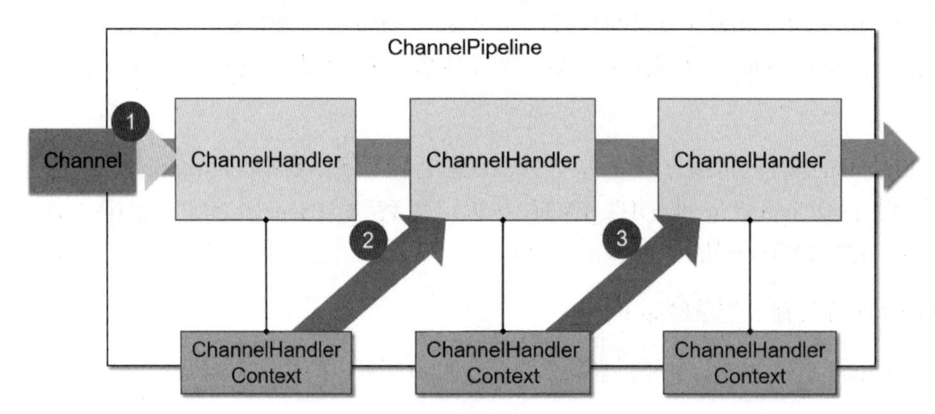

9.3.2　Channel 的生命周期

Netty 有一个简单但强大的状态模型，能完美映射到 ChannelInboundHandler 的各个方法。如下表所示是 Channel 生命周期中四个不同的状态。

状　　态	描　　述
channelUnregistered()	Channel已创建，还未注册到一个EventLoop
channelRegistered()	Channel已经注册到一个EventLoop
channelActive()	Channel是活跃状态（连接到某个远端），可以收发数据
channelInactive()	Channel未连接到远端

一个 Channel 正常的生命周期如下图所示。随着状态发生变化产生相应的事件。这些事件被转发到 ChannelPipeline 中的 ChannelHandler 来触发相应的操作。

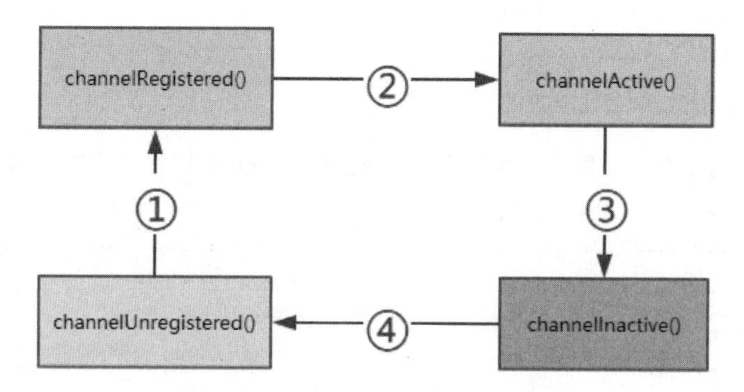

9.3.3　ChannelHandler 常用的 API

先看一下 Netty 中整个 Handler 体系的类关系图，如下图所示。

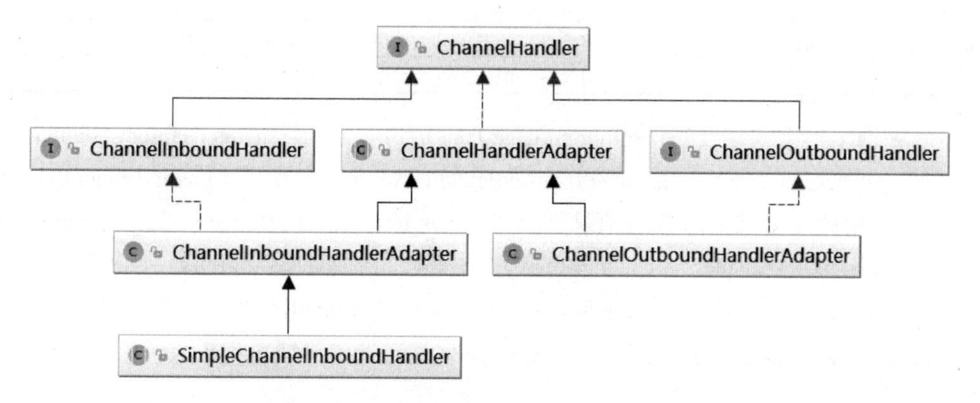

Netty 定义了良好的类型层次结构来表示不同的处理程序类型，所有类型的父类是 ChannelHandler。ChannelHandler 提供了在其生命周期内添加或从 ChannelPipeline 中删除的方法，如下表所示。

方　法	描　述
handlerAdded()	ChannelHandler添加到实际上下文中准备处理事件
handlerRemoved()	将ChannelHandler从实际上下文中删除，不再处理事件
exceptionCaught()	处理抛出的异常

Netty 还提供了一个实现了 ChannelHandler 的抽象类 ChannelHandlerAdapter。ChannelHandlerAdapter 实现了父类的所有方法，主要功能就是将请求从一个 ChannelHandler 往下传递到下一个 ChannelHandler，直到全部 ChannelHandler 传递完毕。也可以直接继承于 ChannelHandlerAdapter，然后重写里面的方法。

9.3.4 ChannelInboundHandler

ChannelInboundHandler 提供了一些在接收数据或 Channel 状态改变时被调用的方法。下面是 ChannelInboundHandler 的一些方法。

方　法	描　述
channelRegistered()	ChannelHandlerContext的Channel被注册到EventLoop
channelUnregistered()	ChannelHandlerContext的Channel从EventLoop中注销
channelActive()	ChannelHandlerContext的Channel已激活
channelInactive()	ChannelHandlerContxt的Channel结束生命周期
channelRead()	从当前Channel的对端读取消息
channelReadComplete()	消息读取完成后执行
userEventTriggered()	一个用户事件被触发
channelWritabilityChanged()	改变通道的可写状态，可以使用Channel.isWritable()检查
exceptionCaught()	重写父类ChannelHandler的方法，处理异常

Netty 提供了一个实现 ChannelInboundHandler 接口并继承 ChannelHandlerAdapter 的类：ChannelInboundHandlerAdapter。ChannelInboundHandlerAdapter 实现了 ChannelInboundHandler 的所有方法，作用就是处理消息并将消息转发到 ChannelPipeline 中的下一个 ChannelHandler。ChannelInboundHandlerAdapter 的 channelRead()方法处理完消息后不会自动释放消息，若想自动释放收到的消息，可以使用 SimpleChannelInboundHandler。ChannelRead()方法的实现代码如下。

```
public class UnreleaseHandler extends ChannelInboundHandlerAdapter {
    @Override
    public void channelRead(ChannelHandlerContext ctx, Object msg) throws Exception {
```

```
        //手动释放消息
        ReferenceCountUtil.release(msg);
    }
}
```

SimpleChannelInboundHandler 会自动释放消息，代码如下。

```
public class ReleaseHandler extends SimpleChannelInboundHandler<Object> {
    @Override
    protected void channelRead0(ChannelHandlerContext ctx, Object msg) throws Exception {
        //不需要手动释放
    }
}
```

ChannelInitializer 主要用来提供给用户初始化 ChannelHandler 的入口，将自定义的各种 ChannelHandler 添加到 ChannelPipe。

10

第 10 章 异步处理双子星 Future 与 Promise

10.1 异步结果 Future

java.util.concurrent.Future 是 Java 原生 API 中提供的接口，用来记录异步执行的状态，Future 的 get()方法会判断任务是否执行完成，如果完成立即返回执行结果，否则阻塞线程，直到任务完成再返回。

Netty 扩展了 Java 的 Future，在 Future 的基础上拓展了监听器（Listener）接口，通过监听器可以让异步执行更加有效率，不需要通过调用 get()方法来等待异步执行结束，而是通过监听器回调来精确地控制异步执行结束时间。

```
public interface Future<V> extends java.util.concurrent.Future<V> {
    boolean isSuccess();
    boolean isCancellable();
    Throwable cause();
    Future<V> addListener(GenericFutureListener<? extends Future<? super V>> listener);
    Future<V> addListeners(GenericFutureListener<? extends Future<? super V>>... listeners);
```

```
    Future<V> removeListener(GenericFutureListener<? extends Future<? super V>> listener);
    Future<V> removeListeners(GenericFutureListener<? extends Future<? super V>>...
listeners);
    Future<V> sync() throws InterruptedException;

    Future<V> syncUninterruptibly();
    Future<V> await() throws InterruptedException;
    Future<V> awaitUninterruptibly();
    boolean await(long timeout, TimeUnit unit) throws InterruptedException;
    boolean await(long timeoutMillis) throws InterruptedException;
    boolean awaitUninterruptibly(long timeout, TimeUnit unit);
    boolean awaitUninterruptibly(long timeoutMillis);
    V getNow();
    boolean cancel(boolean mayInterruptIfRunning);
}
```

ChannelFuture 接口又扩展了 Netty 的 Future 接口，表示一种没有返回值的异步调用，同时和一个 Channel 进行绑定。

```
public interface ChannelFuture extends Future<Void> {
    Channel channel();
    ChannelFuture addListener(GenericFutureListener<? extends Future<? super Void>> listener);
    ChannelFuture addListeners(GenericFutureListener<? extends Future<? super Void>>...
listeners);
    ChannelFuture removeListener(GenericFutureListener<? extends Future<? super Void>>
listener);
    ChannelFuture removeListeners(GenericFutureListener<? extends Future<? super Void>>...
listeners);
    ChannelFuture sync() throws InterruptedException;
    ChannelFuture syncUninterruptibly();
    ChannelFuture await() throws InterruptedException;
    ChannelFuture awaitUninterruptibly();
    boolean isVoid();
}
```

10.2 异步执行 Promise

Promise 接口也是 Future 的扩展接口，它表示一种可写的 Future，可以自定义设置异步执行的结果。

```
public interface Promise<V> extends Future<V> {
    Promise<V> setSuccess(V result);
    boolean trySuccess(V result);
    Promise<V> setFailure(Throwable cause);
    boolean tryFailure(Throwable cause);
```

```
    boolean setUncancellable();
    Promise<V> addListener(GenericFutureListener<? extends Future<? super V>> listener);
    Promise<V> addListeners(GenericFutureListener<? extends Future<? super V>>... listeners);
    Promise<V> removeListener(GenericFutureListener<? extends Future<? super V>> listener);
    Promise<V> removeListeners(GenericFutureListener<? extends Future<? super V>>...
listeners);

    Promise<V> await() throws InterruptedException;
    Promise<V> awaitUninterruptibly();
    Promise<V> sync() throws InterruptedException;
    Promise<V> syncUninterruptibly();
}
```

ChannelPromise 接口扩展了 Promise 和 ChannelFuture，绑定了 Channel，既可以写异步执行结果，又具备了监听者的功能，是 Netty 实际编程中使用的表示异步执行的接口。

```
public interface ChannelPromise extends ChannelFuture, Promise<Void> {
    Channel channel();
    ChannelPromise setSuccess(Void result);
    ChannelPromise setSuccess();
    boolean trySuccess();
    ChannelPromise setFailure(Throwable cause);
    ChannelPromise addListener(GenericFutureListener<? extends Future<? super Void>>
listener);
    ChannelPromise addListeners(GenericFutureListener<? extends Future<? super Void>>...
listeners);
    ChannelPromise removeListener(GenericFutureListener<? extends Future<? super Void>>
listener);
    ChannelPromise removeListeners(GenericFutureListener<? extends Future<? super Void>>...
listeners);
    ChannelPromise sync() throws InterruptedException;
    ChannelPromise syncUninterruptibly();
    ChannelPromise await() throws InterruptedException;
    ChannelPromise awaitUninterruptibly();
    ChannelPromise unvoid();
}
```

DefaultChannelPromise 是 ChannelPromise 的实现类，它是实际运行时的 Promise 实例。Netty 使用 addListener()方法来回调异步执行的结果。DefaultPromise 的 addListener()方法的代码如下。

```
public Promise<V> addListener(GenericFutureListener<? extends Future<? super V>> listener) {
        checkNotNull(listener, "listener");

        synchronized (this) {
            addListener0(listener);
        }

        if (isDone()) {
```

```
            notifyListeners();
        }

        return this;
}

private void addListener0(GenericFutureListener<? extends Future<? super V>> listener) {
        if (listeners == null) {
            listeners = listener;
        } else if (listeners instanceof DefaultFutureListeners) {
            ((DefaultFutureListeners) listeners).add(listener);
        } else {
            listeners = new DefaultFutureListeners((GenericFutureListener<? extends Future<V>>)
listeners, listener);
        }
}

private void notifyListeners() {
        EventExecutor executor = executor();
        if (executor.inEventLoop()) {
            final InternalThreadLocalMap threadLocals = InternalThreadLocalMap.get();
            final int stackDepth = threadLocals.futureListenerStackDepth();
            if (stackDepth < MAX_LISTENER_STACK_DEPTH) {
                threadLocals.setFutureListenerStackDepth(stackDepth + 1);
                try {
                    notifyListenersNow();
                } finally {
                    threadLocals.setFutureListenerStackDepth(stackDepth);
                }
                return;
            }
        }

        safeExecute(executor, new Runnable() {
            @Override
            public void run() {
                notifyListenersNow();
            }
        });
}
```

从上述代码中看到，DefaultChannelPromise 会判断异步任务执行的状态，如果执行完成就立即通知监听者，否则加入监听者队列。通知监听者就是找一个线程来执行调用监听的回调函数。

再来看监听者的接口，其实就是一个方法，即等待异步任务执行完成后，获得 Future 结果，执行回调的逻辑，代码如下。

```
public interface GenericFutureListener<F extends Future<?>> extends EventListener {
    void operationComplete(F future) throws Exception;
}
```

第 11 章 Netty 内存分配 ByteBuf

11.1 初识 ByteBuf

ByteBuf 是 Netty 整个结构里面最为底层的模块，主要负责把数据从底层 I/O 读到 ByteBuf，然后传递给应用程序，应用程序处理完成之后再把数据封装成 ByteBuf 写回 I/O。所以，ByteBuf 是直接与底层打交道的一层抽象。相对于 Netty 其他模块来说，这部分内容是非常复杂的。笔者会把这部分内容拆解，从不同角度来分析 ByteBuf 的分配和回收。本章主要从内存与内存管理器的抽象、不同规格大小和不同类别的内存的分配策略及内存的回收过程等内容来展开分析。

11.1.1 ByteBuf 的基本结构

我们来看 Netty 官方对 ByteBuf 的描述，具体如下。

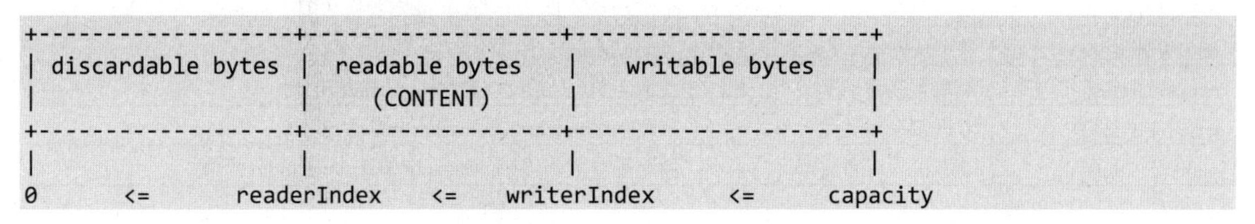

```
+-------------------+------------------+------------------+
| discardable bytes |  readable bytes  |  writable bytes  |
|                   |     (CONTENT)    |                  |
+-------------------+------------------+------------------+
|                   |                  |                  |
0      <=      readerIndex   <=   writerIndex    <=    capacity
```

从上面 ByteBuf 的结构来看，我们发现 ByteBuf 有三个非常重要的指针，分别是 readerIndex（记录读指针的开始位置）、writerIndex（记录写指针的开始位置）和 capacity（缓冲区的总长度），三者的关系是 readerIndex<=writerIndex<=capacity。从 0 到 readerIndex 为 discardable bytes，表示是无效的；从 readerIndex 到 writerIndex 为 readable bytes，表示可读数据区；从 writerIndex 到 capacity 为 writable bytes，表示这段区间空闲，可以往里面写数据。除了这三个指针，ByteBuf 里面其实还有一个指针 maxCapacity，它相当于 ByteBuf 扩容的最大阈值，相应的代码如下。

```
/**
 * Returns the maximum allowed capacity of this buffer.  If a user attempts to increase the
 * capacity of this buffer beyond the maximum capacity using {@link #capacity(int)} or
 * {@link #ensureWritable(int)}, those methods will raise an
 * {@link IllegalArgumentException}.
 */
public abstract int maxCapacity();
```

这个 maxCapacity 指针可以看作是指向 capactiy 之后的这段区间，当 Netty 发现 writable bytes 写数据超出空间大小时，ByteBuf 会提前自动扩容，扩容之后，就有了足够的空间来写数据，同时 capactiy 也会同步更新，maxCapacity 就是扩容后 capactiy 的最大值。

11.1.2 ByteBuf 的重要 API

我们来看 ByteBuf 的基本 API，主要包括 read()、write()、set()、mark()及 reset()等方法。我们对 ByteBuf 最重要的 API 做一个详细说明，如下表所示。

方　法	解　释
readByte()	从当前readerIndex指针开始往后读1字节的数据并移动readerIndex，将数据存储单位转化为Byte
readUnsignedByte()	读取一个无符号的Byte数据
readShort()	从当前readerIndex指针开始往后读2字节的数据并移动readerIndex，将数据类型转化为short
readInt()	从当前readerIndex指针开始往后读4字节的数据并移动readerIndex，将数据类型转化为int
readLong()	从当前readerIndex指针开始往后读8字节的数据并移动readerIndex，将数据类型转化为long
writeByte()	从当前writerIndex指针开始往后写1字节的数据并移动writerIndex
setByte()	将Byte数据写入指定位置，不移动writerIndex
markReaderIndex()	在读数据之前，将readerIndex的状态保存起来，方便在读完数据之后将readerIndex复原
resetReaderIndex()	将readerIndex复原到调用markReaderIndex()之后的状态
markWriterIndex()	在写数据之前，将writerIndex的状态保存起来，方便在读完数据之后将writerIndex复原
resetWriterIndex()	将writerIndex复原到调用markWriterIndex()之后的状态
readableBytes()	获取可读数据区大小，相当于获取当前writerIndex减去readerIndex的值
writableBytes()	获取可写数据区大小，相当于获取当前capactiy减去writerIndex的值
maxWritableBytes()	获取最大可写数据区的大小，相当于获取当前maxCapactiy减去writerIndex的值

在 Netty 中，ByteBuf 的大部分功能是在 AbstractByteBuf 中实现的。

```
public abstract class AbstractByteBuf extends ByteBuf {

    ...

    int readerIndex;   //读指针
    int writerIndex;   //写指针
    private int markedReaderIndex;  //mark 之后的读指针
    private int markedWriterIndex;  //mark 之后的写指针
private int maxCapacity;   //最大容量

...

}
```

最重要的几个属性 readerIndex、writerIndex、markedReaderIndex、markedWriterIndex、maxCapacity 被定义在 AbstractByteBuf 抽象类中，下面来看基本读写的骨架代码实现。例如，几个基本的判断读写区间的 API，具体实现代码如下。

```
public abstract class AbstractByteBuf extends ByteBuf {

    ...

    @Override
    public boolean isReadable() {
        return writerIndex > readerIndex;
    }

    @Override
    public boolean isReadable(int numBytes) {
        return writerIndex - readerIndex >= numBytes;
    }

    @Override
    public boolean isWritable() {
        return capacity() > writerIndex;
    }

    @Override
    public boolean isWritable(int numBytes) {
        return capacity() - writerIndex >= numBytes;
    }

    @Override
    public int readableBytes() {
        return writerIndex - readerIndex;
```

```
    }

    @Override
    public int writableBytes() {
        return capacity() - writerIndex;
    }

    @Override
    public int maxWritableBytes() {
        return maxCapacity() - writerIndex;
}

@Override
public ByteBuf markReaderIndex() {
        markedReaderIndex = readerIndex;
        return this;
}

@Override
    public ByteBuf resetReaderIndex() {
        readerIndex(markedReaderIndex);
        return this;
        }

    @Override
    public ByteBuf markWriterIndex() {
        markedWriterIndex = writerIndex;
        return this;
    }

    @Override
    public ByteBuf resetWriterIndex() {
        writerIndex = markedWriterIndex;
        return this;
    }

    ...

}
```

上面代码已经介绍了这些 API 的功能。再来看几个读写操作的 API，具体代码如下。

```
public abstract class AbstractByteBuf extends ByteBuf {

    ...

    @Override
public byte readByte() {
        checkReadableBytes0(1);
```

```
        int i = readerIndex;
        byte b = _getByte(i);
        readerIndex = i + 1;
        return b;
}

@Override
    public ByteBuf writeByte(int value) {
        ensureAccessible();
        ensureWritable0(1);
        _setByte(writerIndex++, value);
        return this;
}

    @Override
    public byte getByte(int index) {
        checkIndex(index);
        return _getByte(index);
    }


protected abstract void _setByte(int index, int value);

protected abstract byte _getByte(int index);

    ...
}
```

可以看到，上面代码中 readByte()方法和 getByte()方法都调用了一个抽象的 getByte()方法，这个方法在 AbstractByteBuf 的子类中实现。在 writeByte()方法中调用了一个抽象的 setByte()方法，这个方法同样也在子类中实现。

11.1.3　ByteBuf 的基本分类

AbstractByteBuf 有众多子类，大致可以从三个维度来进行分类，分别如下。

- Pooled：池化内存，就是从预先分配好的内存空间中提取一段连续内存封装成一个 ByteBuf，分给应用程序使用。
- Unsafe：是 JDK 底层的一个负责 I/O 操作的对象，可以直接获得对象的内存地址，基于内存地址进行读写操作。
- Direct：堆外内存，直接调用 JDK 的底层 API 进行物理内存分配，不在 JVM 的堆内存中，需要手动释放。

综上所述，其实 ByteBuf 共会有六种组合：Pooled（池化内存）和 Unpooled（非池化内存）；Unsafe 和非 Unsafe；Heap（堆内内存）和 Direct（堆外内存）。下图是 ByteBuf 最重要的继承关系类结构图，通过命名就能一目了然。

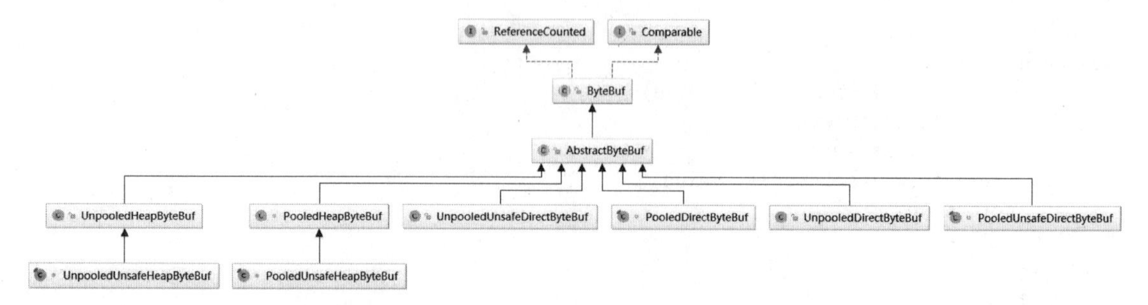

ByteBuf 最基本的读写 API 操作在 AbstractByteBuf 中已经实现了，其众多子类采用不同的策略来分配内存空间，下表是对重要的几个子类的总结。

类	解　释
PooledHeapByteBuf	池化的堆内缓冲区
PooledUnsafeHeapByteBuf	池化的Unsafe堆内缓冲区
PooledDirectByteBuf	池化的直接（堆外）缓冲区
PooledUnsafeDirectByteBuf	池化的Unsafe直接（堆外）缓冲区
UnpooledHeapByteBuf	非池化的堆内缓冲区
UnpooledUnsafeHeapByteBuf	非池化的Unsafe堆内缓冲区
UnpooledDirectByteBuf	非池化的直接（堆外）缓冲区
UnpooledUnsafeDirectByteBuf	非池化的Unsafe直接（堆外）缓冲区

11.2 ByteBufAllocator 内存管理器

Netty 中内存分配有一个顶层的抽象就是 ByteBufAllocator，负责分配所有 ByteBuf 类型的内存。功能其实不是很多，主要有几个重要的 API，如下表所示。

方　法	解　释
buffer()	分配一块内存，自动判断是否分配堆内内存或者堆外内存
ioBuffer()	尽可能地分配一块堆外直接内存，如果系统不支持则分配堆内内存
heapBuffer()	分配一块堆内内存
directBuffer()	分配一块堆外内存
compositeBuffer()	组合分配，把多个ByteBuf组合到一起变成一个整体

可能小伙伴会有疑问，以上 API 中为什么没有前面提到的 8 种类型的内存分配 API？下面我们来看 ByteBufAllocator 的基本实现类 AbstractByteBufAllocator，重点分析主要 API 的基本实现，比如 buffer()方法的代码如下。

```
public abstract class AbstractByteBufAllocator implements ByteBufAllocator {
    ...

    public ByteBuf buffer() {
        if (directByDefault) {
            return directBuffer();
        }
        return heapBuffer();
    }

    ...
}
```

我们发现 buffer()方法中对是否默认支持 directBuffer 做了判断，如果支持则分配 directBuffer，否则分配 heapBuffer。

下面分别来看 directBuffer()方法和 heapBuffer()方法的实现，先来看 directBuffer()方法的代码。

```
public abstract class AbstractByteBufAllocator implements ByteBufAllocator {
    ...

    @Override
    public ByteBuf directBuffer() {
        return directBuffer(DEFAULT_INITIAL_CAPACITY, Integer.MAX_VALUE);
    }

    @Override
    public ByteBuf directBuffer(int initialCapacity) {
        return directBuffer(initialCapacity, Integer.MAX_VALUE);
    }

    @Override
    public ByteBuf directBuffer(int initialCapacity, int maxCapacity) {
        if (initialCapacity == 0 && maxCapacity == 0) {
            return emptyBuf;
        }
        validate(initialCapacity, maxCapacity);
        return newDirectBuffer(initialCapacity, maxCapacity);
}

    ...
}
```

directBuffer()方法有多个重载方法，最终会调用 newDirectBuffer()方法，继续跟进 newDirectBuffer()方法的代码。

```
public abstract class AbstractByteBufAllocator implements ByteBufAllocator {
    ...

protected abstract ByteBuf newDirectBuffer(int initialCapacity, int maxCapacity);
    ...
}
```

我们发现 newDirectBuffer()方法其实是一个抽象方法，最终，交给 AbstractByteBufAllocator 的子类来实现。同理，我们再来看 heapBuffer()方法的代码。

```
public abstract class AbstractByteBufAllocator implements ByteBufAllocator {
    ...

        @Override
    public ByteBuf heapBuffer() {
        return heapBuffer(DEFAULT_INITIAL_CAPACITY, Integer.MAX_VALUE);
    }

    @Override
    public ByteBuf heapBuffer(int initialCapacity) {
        return heapBuffer(initialCapacity, Integer.MAX_VALUE);
    }

    @Override
    public ByteBuf heapBuffer(int initialCapacity, int maxCapacity) {
        if (initialCapacity == 0 && maxCapacity == 0) {
            return emptyBuf;
        }
        validate(initialCapacity, maxCapacity);
        return newHeapBuffer(initialCapacity, maxCapacity);
    }

protected abstract ByteBuf newHeapBuffer(int initialCapacity, int maxCapacity);
    ...
}
```

我们发现 heapBuffer()方法最终是调用 newHeapBuffer()方法，而 newHeapBuffer()方法也是抽象方法，具体交给 AbstractByteBufAllocator 的子类实现。AbstractByteBufAllocator 的子类主要有两个：PooledByteBufAllocator 和 UnpooledByteBufAllocator。AbstractByteBufAllocator 子类实现的类结构图如下图所示。

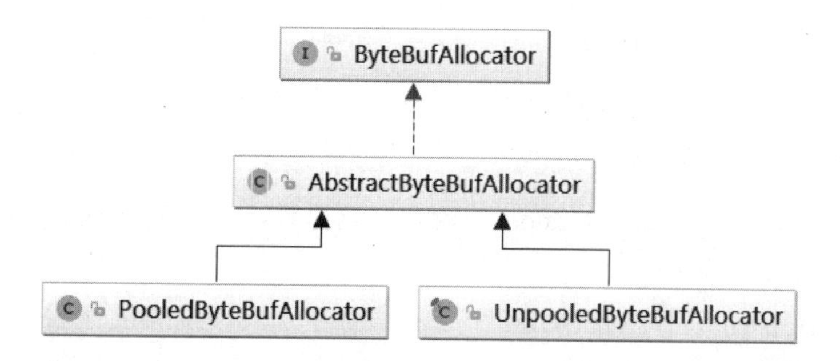

分析到这里，其实我们还只知道 directBuffer、heapBuffer 和 Pooled、Unpooled 的分配规则，那么 Unsafe 和非 Unsafe 是如何判别的呢？其实是 Netty 自动判别的。如果操作系统底层支持 Unsafe 那就采用 Unsafe 读写，否则采用非 Unsafe 读写。我们可以从 UnpooledByteBufAllocator 的源码中验证，代码如下。

```
public final class UnpooledByteBufAllocator extends AbstractByteBufAllocator {

    ...

@Override
    protected ByteBuf newHeapBuffer(int initialCapacity, int maxCapacity) {
        return PlatformDependent.hasUnsafe() ? new UnpooledUnsafeHeapByteBuf(this,
initialCapacity, maxCapacity)
                : new UnpooledHeapByteBuf(this, initialCapacity, maxCapacity);
    }

    @Override
    protected ByteBuf newDirectBuffer(int initialCapacity, int maxCapacity) {
        ByteBuf buf = PlatformDependent.hasUnsafe() ?
                UnsafeByteBufUtil.newUnsafeDirectByteBuf(this, initialCapacity,
maxCapacity) :
                new UnpooledDirectByteBuf(this, initialCapacity, maxCapacity);

        return disableLeakDetector ? buf : toLeakAwareBuffer(buf);
    }

    ...
}
```

我们发现在 newHeapBuffer()方法和 newDirectBuffer()方法中，分配内存判断 PlatformDependent 是否支持 Unsafe，如果支持则创建 Unsafe 类型的 Buffer，否则创建非 Unsafe 类型的 Buffer，由 Netty 自动判断。

11.3 非池化内存分配

11.3.1 堆内内存的分配

现在来看 UnpooledByteBufAllocator 的内存分配原理。首先是 heapBuffer 的分配逻辑，newHeapBuffer()方法的代码如下。

```
public final class UnpooledByteBufAllocator extends AbstractByteBufAllocator {

    ...

@Override
   protected ByteBuf newHeapBuffer(int initialCapacity, int maxCapacity) {
       return PlatformDependent.hasUnsafe() ? new UnpooledUnsafeHeapByteBuf(this,
initialCapacity, maxCapacity)
                : new UnpooledHeapByteBuf(this, initialCapacity, maxCapacity);
    }

    ...
}
```

通过调用 PlatformDependent.hasUnsafe()方法来判断操作系统是否支持 Unsafe，如果支持 Unsafe 则创建 UnpooledUnsafeHeapByteBuf 类，否则创建 UnpooledHeapByteBuf 类。我们先进入 UnpooledUnsafeHeapByteBuf 的构造器看看会进行哪些操作？代码如下。

```
final class UnpooledUnsafeHeapByteBuf extends UnpooledHeapByteBuf {

    UnpooledUnsafeHeapByteBuf(ByteBufAllocator alloc, int initialCapacity, int maxCapacity)
{
        super(alloc, initialCapacity, maxCapacity);
    }
}
```

我们发现 UnpooledUnsafeHeapByteBuf 继承了 UnpooledHeapByte，并且在 UnpooledUnsafeHeapByteBuf 的构造器中直接调用了 super()方法，也就是其父类 UnpooledHeapByte 的构造方法。我们看 UnpooledHeapByte 的构造器代码。

```
public class UnpooledHeapByteBuf extends AbstractReferenceCountedByteBuf {

    private final ByteBufAllocator alloc;
    byte[] array;
    private ByteBuffer tmpNioBuf;
    protected UnpooledHeapByteBuf(ByteBufAllocator alloc, int initialCapacity, int maxCapacity)
{
        this(alloc, new byte[initialCapacity], 0, 0, maxCapacity);
```

```
    }

    protected UnpooledHeapByteBuf(ByteBufAllocator alloc, byte[] initialArray, int maxCapacity)
{
        this(alloc, initialArray, 0, initialArray.length, maxCapacity);
    }

    private UnpooledHeapByteBuf(
            ByteBufAllocator alloc, byte[] initialArray, int readerIndex, int writerIndex, int
maxCapacity) {

        super(maxCapacity);

        ...

        this.alloc = alloc;
        setArray(initialArray);
        setIndex(readerIndex, writerIndex);
}

    ...
}
```

其中调用了一个关键方法就是 setArray()方法。这个方法的功能非常简单，就是把默认分配的数组 new byte[initialCapacity]赋值给全局变量 initialArray 数组。

```
private void setArray(byte[] initialArray) {
    array = initialArray;
    tmpNioBuf = null;
}
```

紧接着就是调用 setIndex()方法。

```
public ByteBuf setIndex(int readerIndex, int writerIndex) {
      if (readerIndex < 0 || readerIndex > writerIndex || writerIndex > capacity()) {
          throw new IndexOutOfBoundsException(String.format(
                  "readerIndex: %d, writerIndex: %d (expected: 0 <= readerIndex <= writerIndex
<= capacity(%d))",
                  readerIndex, writerIndex, capacity()));
      }
      setIndex0(readerIndex, writerIndex);
      return this;
}

final void setIndex0(int readerIndex, int writerIndex) {
      this.readerIndex = readerIndex;
      this.writerIndex = writerIndex;
  }
```

最终在 setIndex0()方法中初始化 readerIndex 属性和 writerIndex 属性。

既然 UnpooledUnsafeHeapByteBuf 和 UnpooledHeapByteBuf 调用的都是 UnpooledHeapByteBuf 的构造方法，那么它们之间到底有什么区别呢？其实根本区别在于 I/O 的读写，我们分别来看它们的 getByte()方法，了解二者的区别。先来看 UnpooledHeapByteBuf 的 getByte()方法的实现代码。

```
public byte getByte(int index) {
    ensureAccessible();
    return _getByte(index);
}

@Override
protected byte _getByte(int index) {
    return HeapByteBufUtil.getByte(array, index);
}
```

可以看到最终调用的是 HeapByteBufUtil 的 getByte()方法。

```
final class HeapByteBufUtil {

    static byte getByte(byte[] memory, int index) {
        return memory[index];
    }
        ...
}
```

getByte()这个方法中的处理逻辑也非常简单，就是根据 index 索引直接从数组中取值。接着来看 UnpooledUnsafeHeapByteBuf 的 getByte()方法中的代码实现。

```
public byte getByte(int index) {
        checkIndex(index);
        return _getByte(index);
    }

    @Override
    protected byte _getByte(int index) {
        return UnsafeByteBufUtil.getByte(array, index);
    }
```

可以看到，最终调用的是 UnsafeByteBufUtil 的 getByte()方法。

```
final class UnsafeByteBufUtil{

    static byte getByte(byte[] array, int index) {
        return PlatformDependent.getByte(array, index);
    }
```

```
    ...
}
```

通过这样对比代码，我们已经基本了解 UnpooledUnsafeHeapByteBuf 和 UnpooledHeapByteBuf 的区别了。

11.3.2　堆外内存的分配

再回到 UnpooledByteBufAllocator 的 newDirectBuffer()方法，代码如下。

```
public final class UnpooledByteBufAllocator extends AbstractByteBufAllocator {

    ...

    @Override
    protected ByteBuf newDirectBuffer(int initialCapacity, int maxCapacity) {
        ByteBuf buf = PlatformDependent.hasUnsafe() ?
                UnsafeByteBufUtil.newUnsafeDirectByteBuf(this, initialCapacity,
maxCapacity) :
                new UnpooledDirectByteBuf(this, initialCapacity, maxCapacity);

        return disableLeakDetector ? buf : toLeakAwareBuffer(buf);
    }

    ...
}
```

从上面代码可以看出，如果支持 Unsafe 则调用 UnsafeByteBufUtil.newUnsafeDirectByteBuf()方法，否则创建 UnpooledDirectByteBuf 类。来看 UnpooledDirectByteBuf 构造器。

```
public class UnpooledDirectByteBuf extends AbstractReferenceCountedByteBuf {

    private final ByteBufAllocator alloc;

    private ByteBuffer buffer;
    private ByteBuffer tmpNioBuf;
    private int capacity;
    private boolean doNotFree;

    protected UnpooledDirectByteBuf(ByteBufAllocator alloc, int initialCapacity, int
maxCapacity) {
        super(maxCapacity);
            ...
        this.alloc = alloc;
        setByteBuffer(ByteBuffer.allocateDirect(initialCapacity));
}
```

```
    ...
}
```

首先调用 ByteBuffer.allocateDirect.allocateDirect()通过 JDK 底层分配一个直接缓冲区，然后传给 setByteBuffer()方法，继续跟进，代码如下。

```
    private void setByteBuffer(ByteBuffer buffer) {
        ByteBuffer oldBuffer = this.buffer;
        if (oldBuffer != null) {
            if (doNotFree) {
                doNotFree = false;
            } else {
                freeDirect(oldBuffer);
            }
        }

        this.buffer = buffer;
        tmpNioBuf = null;
        capacity = buffer.remaining();
    }
```

由上述代码可以看到，setByteBuffer()方法主要做了一次赋值。

继续看 UnsafeByteBufUtil.newUnsafeDirectByteBuf()方法的逻辑。

```
final class UnsafeByteBufUtil {
    ...
    static UnpooledUnsafeDirectByteBuf newUnsafeDirectByteBuf(
            ByteBufAllocator alloc, int initialCapacity, int maxCapacity) {
        if (PlatformDependent.useDirectBufferNoCleaner()) {
            return new UnpooledUnsafeNoCleanerDirectByteBuf(alloc, initialCapacity,
maxCapacity);
        }
        return new UnpooledUnsafeDirectByteBuf(alloc, initialCapacity, maxCapacity);
}
...
}
```

从上面代码中，我们看到这个方法中返回了一个 UnpooledUnsafeDirectByteBuf 对象，关于 UnpooledUnsafeNoCleanerDirectByteBuf，我们在后续章节再进行详细分析。下面我们继续来看 UnpooledUnsafeDirectByteBuf 构造器中的代码。

```
public class UnpooledUnsafeDirectByteBuf extends AbstractReferenceCountedByteBuf {
        protected UnpooledUnsafeDirectByteBuf(ByteBufAllocator alloc, int initialCapacity,
int maxCapacity) {
        super(maxCapacity);
        ...
```

```
        this.alloc = alloc;
        setByteBuffer(allocateDirect(initialCapacity), false);
    }

        ...
}
```

UnpooledUnsafeDirectByteBuf 构造器的逻辑和 UnpooledDirectByteBuf 构造器的逻辑是相似的，其 setByteBuffer()方法的实现代码如下。

```
final void setByteBuffer(ByteBuffer buffer, boolean tryFree) {
        if (tryFree) {
            ByteBuffer oldBuffer = this.buffer;
            if (oldBuffer != null) {
                if (doNotFree) {
                    doNotFree = false;
                } else {
                    freeDirect(oldBuffer);
                }
            }
        }
        this.buffer = buffer;
        memoryAddress = PlatformDependent.directBufferAddress(buffer);
        tmpNioBuf = null;
        capacity = buffer.remaining();
    }
```

同样还是先保存在 JDK 底层创建的 Buffer，接下来有个很重要的操作就是调用 PlatformDependent.directBufferAddress()方法获取 Buffer 真实的内存地址，并保存到 memoryAddress 变量中。PlatformDependent.directBufferAddress()方法的实现代码如下。

```
public static long directBufferAddress(ByteBuffer buffer) {
        return PlatformDependent0.directBufferAddress(buffer);
    }
```

PlatformDependent0 的 directBufferAddress()方法的实现代码如下。

```
static long directBufferAddress(ByteBuffer buffer) {
        return getLong(buffer, ADDRESS_FIELD_OFFSET);
    }
```

getLong()方法的实现代码如下。

```
        private static long getLong(Object object, long fieldOffset) {
        return UNSAFE.getLong(object, fieldOffset);
    }
```

可以看到，上述代码调用了 Unsafe 的 getLong()方法，这是一个 native 方法。它直接通过 Buffer 的内存地址加上一个偏移量去取数据。到这里，我们已经基本清楚 UnpooledUnsafeDirectByteBuf 和 UnpooledDirectByteBuf 的区别，非 Unsafe 通过数组的下标取数据，Unsafe 直接操作内存地址，相对于非 Unsafe 来说效率当然要更高。

11.4 池化内存分配

11.4.1 PooledByteBufAllocator 简述

现在开始，我们来分析池化内存的分配原理。首先找到 AbstractByteBufAllocator 的子类 PooledByteBufAllocator 实现分配内存的两个方法：newDirectBuffer()方法和 newHeapBuffer()方法。

```
public class PooledByteBufAllocator extends AbstractByteBufAllocator {
    ...

    @Override
    protected ByteBuf newHeapBuffer(int initialCapacity, int maxCapacity) {
        PoolThreadCache cache = threadCache.get();
        PoolArena<byte[]> heapArena = cache.heapArena;

        ByteBuf buf;
        if (heapArena != null) {
            buf = heapArena.allocate(cache, initialCapacity, maxCapacity);
        } else {
            buf = new UnpooledHeapByteBuf(this, initialCapacity, maxCapacity);
        }

        return toLeakAwareBuffer(buf);
    }

    @Override
    protected ByteBuf newDirectBuffer(int initialCapacity, int maxCapacity) {
        PoolThreadCache cache = threadCache.get();
        PoolArena<ByteBuffer> directArena = cache.directArena;

        ByteBuf buf;
        if (directArena != null) {
            buf = directArena.allocate(cache, initialCapacity, maxCapacity);
        } else {
            if (PlatformDependent.hasUnsafe()) {
                buf = UnsafeByteBufUtil.newUnsafeDirectByteBuf(this, initialCapacity,
maxCapacity);
            } else {
```

```
            buf = new UnpooledDirectByteBuf(this, initialCapacity, maxCapacity);
        }
    }

    return toLeakAwareBuffer(buf);
}

...
}
```

我们发现这两个方法大体结构是一样的，以 newDirectBuffer()方法为例，简单地分析一下。

首先，通过 threadCache.get()方法获得一个类型为 PoolThreadCache 的 cache 对象；然后，通过 cache 获得 directArena 对象；最后，调用 directArena.allocate()方法分配 ByteBuf。这里读者可能会有点看不懂，我们接下来详细分析一下。threadCache 对象其实是 PoolThreadLocalCache 类型的变量，PoolThreadLocalCache 的相关代码如下。

```
final class PoolThreadLocalCache extends FastThreadLocal<PoolThreadCache> {

    @Override
    protected synchronized PoolThreadCache initialValue() {
            //从 heapArenas 中获得一个使用率最少的 Arena
        final PoolArena<byte[]> heapArena = leastUsedArena(heapArenas);
            //从 directArenas 中获得一个使用率最少的 Arena
        final PoolArena<ByteBuffer> directArena = leastUsedArena(directArenas);

        return new PoolThreadCache(
                heapArena, directArena, tinyCacheSize, smallCacheSize, normalCacheSize,
                DEFAULT_MAX_CACHED_BUFFER_CAPACITY, DEFAULT_CACHE_TRIM_INTERVAL);
    }

        ...
}
```

从名字来看，我们发现 PoolThreadLocalCache 的 initialValue()方法就是用来初始化 PoolThreadLocalCache 的。首先调用 leastUsedArena()方法分别获得类型为 PoolArena 的 heapArena 和 directArena 对象。然后把 heapArena 和 directArena 对象作为参数传递到 PoolThreadCache 的构造器中。那么 heapArena 和 directArena 对象是在哪里初始化的呢？经过查找，发现在 PooledByteBufAllocator 的构造方法中调用 newArenaArray()方法给 heapArenas 和 directArenas 进行了赋值，代码如下。

```
public PooledByteBufAllocator(boolean preferDirect, int nHeapArena, int nDirectArena, int
pageSize, int maxOrder,
                              int tinyCacheSize, int smallCacheSize, int normalCacheSize) {
```

```
...
        if (nHeapArena > 0) {
            heapArenas = newArenaArray(nHeapArena);

                    ...

        } else {
            heapArenas = null;
            heapArenaMetrics = Collections.emptyList();
        }

        if (nDirectArena > 0) {
            directArenas = newArenaArray(nDirectArena);

            ...

        }

            ...
}
```

newArenaArray()方法的实现代码如下。

```
    private static <T> PoolArena<T>[] newArenaArray(int size) {
        return new PoolArena[size];
    }
```

其实就是创建了一个固定大小的 PoolArena 数组，数组大小由传入的参数 nHeapArena 和 nDirectArena 决定。再回到 PooledByteBufAllocator 的构造器源码，看 nHeapArena 和 nDirectArena 是怎么初始化的，PooledByteBufAllocator 的重载构造器的代码如下。

```
    public PooledByteBufAllocator(boolean preferDirect) {
        this(preferDirect, DEFAULT_NUM_HEAP_ARENA, DEFAULT_NUM_DIRECT_ARENA,
DEFAULT_PAGE_SIZE, DEFAULT_MAX_ORDER);
    }
```

我们发现，nHeapArena 和 nDirectArena 是通过 DEFAULT_NUM_HEAP_ARENA 和 DEFAULT_NUM_DIRECT_ARENA 这两个常量默认赋值的。相关常量的定义代码如下。

```
final int defaultMinNumArena = runtime.availableProcessors() * 2;

    DEFAULT_NUM_HEAP_ARENA = Math.max(0,
            SystemPropertyUtil.getInt(
                    "io.netty.allocator.numHeapArenas",
                    (int) Math.min(
```

```
                            defaultMinNumArena,
                            runtime.maxMemory() / defaultChunkSize / 2 / 3)));
    DEFAULT_NUM_DIRECT_ARENA = Math.max(0,
            SystemPropertyUtil.getInt(
                    "io.netty.allocator.numDirectArenas",
                    (int) Math.min(
                            defaultMinNumArena,
                            PlatformDependent.maxDirectMemory() / defaultChunkSize / 2 /
3)));
```

到这里为止，我们才知道 nHeapArena 和 nDirectArena 的默认值就是 CPU 核数×2，也就是把 defaultMinNumArena 的值赋给 nHeapArena 和 nDirectArena。大家对于 CPU 核数×2 应该有印象，我们在第 8 章介绍 EventLoopGroup 分配线程时，默认线程数也是 CPU 核数×2。那么，Netty 为什么要这样设计呢？其实，主要目的就是保证 Netty 中的每一个任务线程都可以有一个独享的 Arena，保证在每个线程分配内存的时候不用加锁。

基于上面的分析，我们知道 Arena 有 heapArena 和 directArena，这里统称为 Arena。假设有四个线程，那么对应会分配四个 Arena。在创建 ByteBuf 的时候，首先通过 PoolThreadCache 获取 Arena 对象并赋值给其成员变量，然后每个线程通过 PoolThreadCache 调用 get()方法的时候会获得它底层的 Arena，也就是说通过 EventLoop1 获得 Arena1，通过 EventLoop2 获得 Arena2，依此类推，如下图所示。

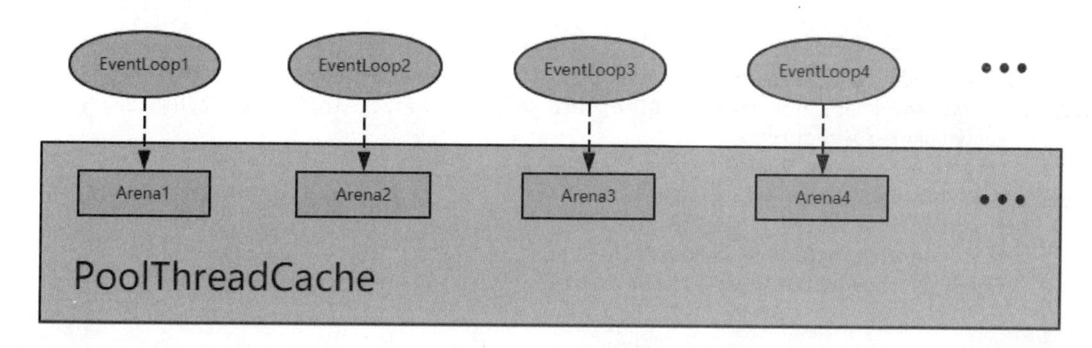

那么 PoolThreadCache 除了可以在 Arena 上进行内存分配，还可以在它底层维护的 ByteBuf 缓存列表进行分配。举个例子：我们通过 PooledByteBufAllocator 创建了一个 1 024 字节的 ByteBuf，当用完释放后，可能在其他地方会继续分配 1 024 字节的 ByteBuf。这时，其实不需要在 Arena 上进行内存分配，而是直接通过 PoolThreadCache 中维护的 ByteBuf 的缓存列表直接拿过来返回。在 PooledByteBufAllocator 中维护着三种规格大小的缓存列表，分别是三个值 tinyCacheSize、smallCacheSize、normalCacheSize，相关代码如下。

```
public class PooledByteBufAllocator extends AbstractByteBufAllocator {

        ...

   DEFAULT_TINY_CACHE_SIZE =
                SystemPropertyUtil.getInt("io.netty.allocator.tinyCacheSize", 512);
       DEFAULT_SMALL_CACHE_SIZE =
                         SystemPropertyUtil.getInt("io.netty.allocator.smallCacheSize",
256);
       DEFAULT_NORMAL_CACHE_SIZE =
                          SystemPropertyUtil.getInt("io.netty.allocator.normalCacheSize",
64);


private final int tinyCacheSize;
   private final int smallCacheSize;
   private final int normalCacheSize;
       ...


  public PooledByteBufAllocator(boolean preferDirect, int nHeapArena, int nDirectArena, int
pageSize, int maxOrder) {
       this(preferDirect, nHeapArena, nDirectArena, pageSize, maxOrder,DEFAULT_TINY_CACHE
_SIZE,
                                    DEFAULT_SMALL_CACHE_SIZE, DEFAULT_NORMAL_CACHE_SIZE);
   }

   public PooledByteBufAllocator(boolean preferDirect, int nHeapArena, int nDirectArena, int
pageSize, int maxOrder,int tinyCacheSize, int smallCacheSize, int normalCacheSize) {
       super(preferDirect);
       threadCache = new PoolThreadLocalCache();
       this.tinyCacheSize = tinyCacheSize;
       this.smallCacheSize = smallCacheSize;
       this.normalCacheSize = normalCacheSize;
       final int chunkSize = validateAndCalculateChunkSize(pageSize, maxOrder);

       ...
   }

}
```

我们看到，在 PooledByteBufAllocator 的构造器中，分别赋值 tinyCacheSize=512，smallCacheSize=256，normalCacheSize=64。通过这种方式，Netty 预创建了固定规格的内存池，大大提高了内存分配的性能。

11.4.2　DirectArena 内存分配流程

Arena 分配内存的基本流程有三个步骤。

（1）优先从对象池里获得 PooledByteBuf 进行复用。

（2）然后在缓存中进行内存分配。

（3）最后考虑从内存堆里进行内存分配。

以 directBuffer 为例，首先来看从对象池里获得 PooledByteBuf 进行复用的情况，我们依旧跟进到 PooledByteBufAllocator 的 newDirectBuffer()方法，代码如下。

```
    @Override
    protected ByteBuf newDirectBuffer(int initialCapacity, int maxCapacity) {
        PoolThreadCache cache = threadCache.get();
        PoolArena<ByteBuffer> directArena = cache.directArena;

        ByteBuf buf;
        if (directArena != null) {
            buf = directArena.allocate(cache, initialCapacity, maxCapacity);
        } else {
            ...
        }

        return toLeakAwareBuffer(buf);
}
```

对于上面的 PoolArena 我们已经清楚，我们直接跟进 PoolArena 的 allocate()方法，代码如下。

```
    PooledByteBuf<T> allocate(PoolThreadCache cache, int reqCapacity, int maxCapacity) {
        PooledByteBuf<T> buf = newByteBuf(maxCapacity);
        allocate(cache, buf, reqCapacity);
        return buf;
    }
```

在这个地方其实思路就非常清晰了，首先调用 newByteBuf()方法获得一个 PooledByteBuf 对象，然后通过 allocate()方法在线程私有的 PoolThreadCache 中分配一块内存，再对 buf 里面的内存地址之类的值进行初始化。跟进 newByteBuf()方法，选择 DirectArena 对象。

```
protected PooledByteBuf<ByteBuffer> newByteBuf(int maxCapacity) {
    if (HAS_UNSAFE) {
            return PooledUnsafeDirectByteBuf.newInstance(maxCapacity);
    } else {
            return PooledDirectByteBuf.newInstance(maxCapacity);
    }
}
```

首先判断是否支持 Unsafe，默认情况下一般是支持 Unsafe 的，继续看 PooledUnsafeDirectByteBuf 的 newInstance()方法，代码如下。

```
final class PooledUnsafeDirectByteBuf extends PooledByteBuf<ByteBuffer> {
    private static final Recycler<PooledUnsafeDirectByteBuf> RECYCLER =
    new Recycler<PooledUnsafeDirectByteBuf>() {
        @Override
        protected PooledUnsafeDirectByteBuf newObject(Handle<PooledUnsafeDirectByteBuf>
handle) {
            return new PooledUnsafeDirectByteBuf(handle, 0);
        }
};

static PooledUnsafeDirectByteBuf newInstance(int maxCapacity) {
        PooledUnsafeDirectByteBuf buf = RECYCLER.get();
        buf.reuse(maxCapacity);
        return buf;
}
    ...
}
```

顾名思义，首先通过 RECYCLER（内存回收站）对象的 get()方法获得一个 buf。从上面的代码片段来看，RECYCLER 对象实现了一个 newObject()方法，当回收站里面没有可用的 buf 时就会创建一个新的 buf。因为获得的 buf 可能是回收站里取出来的，所以复用前需要重置。继续往下看就会调用 buf 的 reuse()方法，代码如下。

```
    final void reuse(int maxCapacity) {
        maxCapacity(maxCapacity);
        setRefCnt(1);
        setIndex0(0, 0);
        discardMarks();
    }
```

我们发现 reuse()方法就是让所有的参数重新归为初始状态。到这里我们应该已经清楚从内存池获取 buf 对象的全过程。接下来，再回到 PoolArena 的 allocate()方法，看看真实的内存是如何分配出来的？buf 的内存分配主要有两种情况，分别是从缓存中进行内存分配和从内存堆里进行内存分配。我们来看 allocate()方法的具体逻辑代码。

```
    private void allocate(PoolThreadCache cache, PooledByteBuf<T> buf, final int reqCapacity)
{
        ...
        if (normCapacity <= chunkSize) {
            if (cache.allocateNormal(this, buf, reqCapacity, normCapacity)) {
                //was able to allocate out of the cache so move on
                return;
```

```
        }
        allocateNormal(buf, reqCapacity, normCapacity);
    } else {
        //Huge allocations are never served via the cache so just call allocateHuge
        allocateHuge(buf, reqCapacity);
    }
}
```

这段代码逻辑看上去非常复杂，其实我们省略掉的逻辑基本上都是判断不同规格大小，从其对应的缓存中获取内存。如果所有规格都不满足，那就直接调用 allocateHuge()方法进行真实的内存分配。

11.4.3　内存池的内存规格

在前面的源码分析过程中，关于内存规格大小我们应该还有印象。其实在 Netty 内存池中主要设置了四种规格大小的内存：tiny 指 0~512Byte 的规格大小，small 指 512Byte~8KB 的规格大小，normal 指 8KB~16MB 的规格大小，huge 指 16MB 以上的规格大小。为什么 Netty 会选择这些值作为分界点呢？其实在 Netty 底层还有一个内存单位的封装，为了更高效地管理内存，避免内存浪费，把每一个区间的内存规格又做了细分。默认情况下，Netty 将内存规格划分为四个部分。Netty 中所有的内存申请是以 Chunk 为单位向系统申请的，每个 Chunk 大小为 16MB，后续的所有内存分配都是在这个 Chunk 里的操作。一个 Chunk 会以 Page 为单位进行切分，8KB 对应的是一个 Page，而一个 Chunk 被划分为 2 048 个 Page。小于 8KB 的是 SubPage。例如，我们申请的一段内存空间只有 1KB，却给我们分配了一个 Page，显然另外 7KB 就会被浪费，所以就继续把 Page 进行划分，以节省空间。内存规格大小如下图所示。

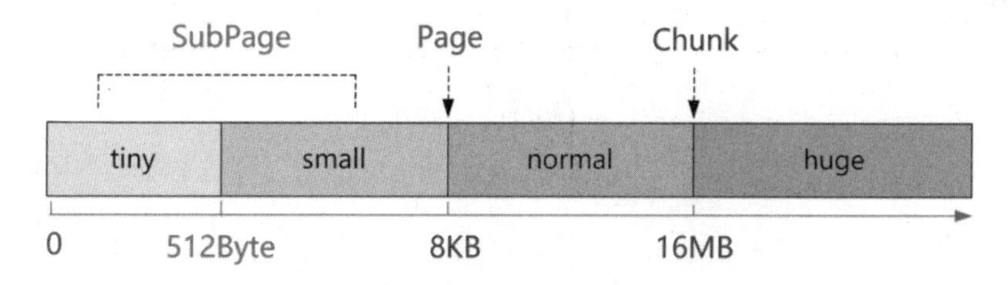

至此，小伙伴们应该已经非常清楚 Netty 的内存池缓存管理机制了。

11.4.4　命中缓存的分配

前面我们简单分析了 DirectArena 内存分配的大概流程，知道其先命中缓存，如果命中不到，则去分配一款连续内存。现在剖析命中缓存的相关逻辑。前面讲到 PoolThreadCache 中维护了三个缓存数组（实际上是六个，这里仅以 Direct 为例，Heap 类型的逻辑是一样的）：tinySubPageDirectCaches、smallSubPageDirectCaches 和 normalDirectCaches，分别代表 tiny 类型、small 类型和 normal 类型的缓存数组。这三个数组保存在 PoolThreadCache 的成员变量中，代码如下。

```
final class PoolThreadCache {
...
private final MemoryRegionCache<ByteBuffer>[] tinySubPageDirectCaches;
private final MemoryRegionCache<ByteBuffer>[] smallSubPageDirectCaches;
private final MemoryRegionCache<ByteBuffer>[] normalDirectCaches;
    ...
}
```

其实在构造方法中进行了初始化，代码如下。

```
final class PoolThreadCache {
...
    PoolThreadCache(PoolArena<byte[]> heapArena, PoolArena<ByteBuffer> directArena,
                    int tinyCacheSize, int smallCacheSize, int normalCacheSize,
                    int maxCachedBufferCapacity, int freeSweepAllocationThreshold) {
        ...

        if (directArena != null) {
            tinySubPageDirectCaches = createSubPageCaches(
                    tinyCacheSize, PoolArena.numTinySubpagePools, SizeClass.Tiny);
            smallSubPageDirectCaches = createSubPageCaches(
                    smallCacheSize, directArena.numSmallSubpagePools, SizeClass.Small);

            numShiftsNormalDirect = log2(directArena.pageSize);
            normalDirectCaches = createNormalCaches(
                    normalCacheSize, maxCachedBufferCapacity, directArena);

            directArena.numThreadCaches.getAndIncrement();
     }
          ...
    }
    ...
}
```

以 tiny 类型为例，具体分析一下 SubPage 的缓存结构，实现代码如下。

```
private static <T> MemoryRegionCache<T>[] createSubPageCaches(
        int cacheSize, int numCaches, SizeClass sizeClass) {
    if (cacheSize > 0) {
        @SuppressWarnings("unchecked")
        MemoryRegionCache<T>[] cache = new MemoryRegionCache[numCaches];
```

```
        for (int i = 0; i < cache.length; i++) {
            cache[i] = new SubPageMemoryRegionCache<T>(cacheSize, sizeClass);
        }
        return cache;
    } else {
        return null;
    }
}
```

从以上代码中看出，createSubPageCaches()方法中的操作其实就是创建了一个缓存数组，这个缓存数组的长度是 numCaches。

在 PoolThreadCache 给数组 tinySubPageDirectCaches 赋值前，需要设定的数组长度就是对应每一种规格的固定值。以 tinySubPageDirectCaches[1]为例（下标选择 1 是因为下标为 0 代表的规格是 0Byte，其实就代表一个空的缓存，这里不进行举例）。在 tinySubPageDirectCaches[1]的缓存对象中所缓存的 ByteBuf 的缓冲区大小是 16Byte，在 tinySubPageDirectCaches[2]中缓存的 ByteBuf 大小为 32Byte，依此类推，tinySubPageDirectCaches[31]中缓存的 ByteBuf 大小为 496Byte。具体类型规则的配置如下。

我们知道，不同类型的缓存数组规格不一样，tiny 类型的数组长度是 32，small 类型的数组长度是 4，normal 类型的数组长度是 3。缓存数组中的每一个元素都是 MemoryRegionCache 类型，代表一个缓存对象。每个 MemoryRegionCache 对象中维护了一个队列，队列的容量大小由 PooledByteBufAllocator 类中定义的 tinyCacheSize、smallCacheSize、normalCacheSize 的值来决定。

MemoryRegionCache 对象的队列中的元素为 ByteBuf 类型，ByteBuf 的大小也是固定的。这样，Netty 就将每种 ByteBuf 的容量大小划分成了不同的规格。同一个队列中，每个 ByteBuf 的容量大小是相同的规格。比如，在 tiny 类型中，Netty 将其长度分成 32 种规格，每种规格都是 16 的整数倍，也就是包含 0Byte、16Byte、32Byte、48Byte、64Byte、80Byte、96Byte……496Byte，总共 32 种规格。

- tiny：共 32 种规格，均是 16 的整数倍，0Byte、16Byte、32Byte、48Byte、64Byte、80Byte、96Byte……496Byte。
- small：4 种规格，512Byte、1KB、2KB、4KB。
- normal：3 种规格，8KB、16KB、32KB。

由此，我们得出结论，PoolThreadCache 中缓存数组的数据结构如下图所示。

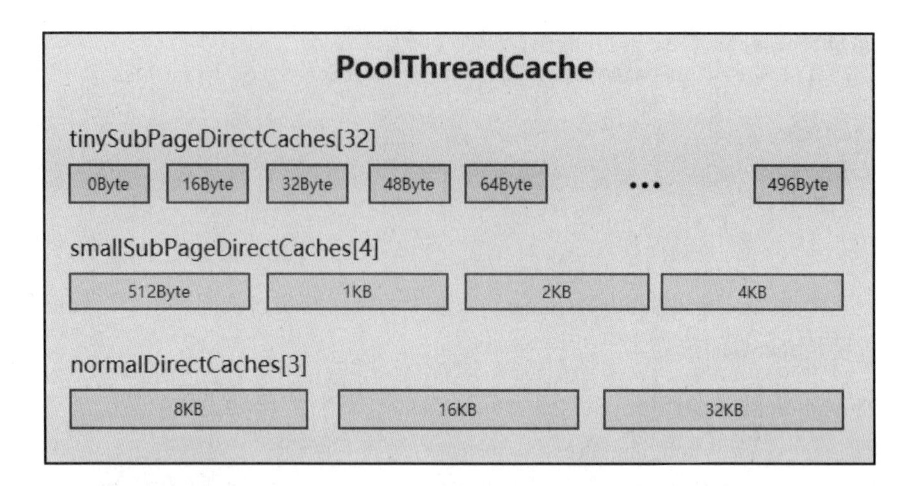

在基本了解缓存数组的数据结构之后，继续剖析在缓存中分配内存的逻辑，回到 PoolArena 的 allocate()方法，代码如下。

```
private void allocate(PoolThreadCache cache, PooledByteBuf<T> buf, final int reqCapacity) {
    //规格化
    final int normCapacity = normalizeCapacity(reqCapacity);
    if (isTinyOrSmall(normCapacity)) {
        int tableIdx;
        PoolSubpage<T>[] table;
        //判断是不是 tiny 类型
        boolean tiny = isTiny(normCapacity);
        if (tiny) { // < 512Byte
            //缓存分配
            if (cache.allocateTiny(this, buf, reqCapacity, normCapacity)) {
                return;
            }
            //通过 tinyIdx 获取 tableIdx
            tableIdx = tinyIdx(normCapacity);
            //SubPage 的数组
            table = tinySubpagePools;
        } else {
            if (cache.allocateSmall(this, buf, reqCapacity, normCapacity)) {
                return;
            }
            tableIdx = smallIdx(normCapacity);
            table = smallSubpagePools;
        }

        //获取对应的节点
        final PoolSubpage<T> head = table[tableIdx];
```

```
        synchronized (head) {
            final PoolSubpage<T> s = head.next;
            //默认情况下，Head 的 next 也是自身
            if (s != head) {
                assert s.doNotDestroy && s.elemSize == normCapacity;
                long handle = s.allocate();
                assert handle >= 0;
                s.chunk.initBufWithSubpage(buf, handle, reqCapacity);

                if (tiny) {
                    allocationsTiny.increment();
                } else {
                    allocationsSmall.increment();
                }
                return;
            }
        }
        allocateNormal(buf, reqCapacity, normCapacity);
        return;
    }
    if (normCapacity <= chunkSize) {
        //首先在缓存上进行内存分配
        if (cache.allocateNormal(this, buf, reqCapacity, normCapacity)) {
            //分配成功，返回
            return;
        }
        //分配不成功，做实际的内存分配
        allocateNormal(buf, reqCapacity, normCapacity);
    } else {
        //大于这个值，就不在缓存上分配
        allocateHuge(buf, reqCapacity);
    }
}
```

首先通过 normalizeCapacity()方法进行内存规格化，代码如下。

```
int normalizeCapacity(int reqCapacity) {
    if (reqCapacity < 0) {
        throw new IllegalArgumentException("capacity: " + reqCapacity + " (expected: 0+)");
    }
    if (reqCapacity >= chunkSize) {
        return reqCapacity;
    }
    //如果大于 tiny 类型的大小
    if (!isTiny(reqCapacity)) { // >= 512 KB
        //找一个 2 的 n 次方的数值，确保数值大于等于 reqCapacity
        int normalizedCapacity = reqCapacity;
```

```
        normalizedCapacity --;
        normalizedCapacity |= normalizedCapacity >>>  1;
        normalizedCapacity |= normalizedCapacity >>>  2;
        normalizedCapacity |= normalizedCapacity >>>  4;
        normalizedCapacity |= normalizedCapacity >>>  8;
        normalizedCapacity |= normalizedCapacity >>> 16;
        normalizedCapacity ++;

        if (normalizedCapacity < 0) {
            normalizedCapacity >>>= 1;
        }

        return normalizedCapacity;
    }
    //如果是 16 的倍数
    if ((reqCapacity & 15) == 0) {
        return reqCapacity;
    }
    //不是 16 的倍数，变成最小大于当前值的值+16
    return (reqCapacity & ~15) + 16;
}
```

上面代码中 if (!isTiny(reqCapacity))的作用是，如果分配的缓存空间大于 tiny 类型的大小（也就是大于等于 512 Byte），则会找一个 2 的 n 次方的数值，以便确保这个数值大于等于 reqCapacity。如果是 tiny 类型，则继续往下执行 if ((reqCapacity & 15) == 0)，这里判断如果是 16 的倍数，则直接返回。如果不是 16 的倍数，则返回(reqCapacity & ~15) + 16，也就是变成最小大于当前值的 16 的倍数值。从上面规格化逻辑可以看出，这里将缓存大小规格化成固定大小，确保每个缓存对象缓存的 ByteBuf 容量统一。allocate()方法中的 if(isTinyOrSmall(normCapacity))则是根据规格化后的大小判断类型是 tiny 还是 small。

```
boolean isTinyOrSmall(int normCapacity) {
    return (normCapacity & subpageOverflowMask) == 0;
}
```

这个方法通过判断 normCapacity 是否小于一个 Page 的大小（8KB）来判断类型（tiny 或者 small）。

继续看 allocate()方法，如果当前大小类型是 tiny 或者 small，则通过 isTiny(normCapacity)判断是否是 tiny 类型，代码如下。

```
static boolean isTiny(int normCapacity) {
    return (normCapacity & 0xFFFFFE00) == 0;
}
```

这个方法是判断如果小于 512Byte，则认为是 tiny 类型。

再继续看 allocate()方法，如果是 tiny 类型，则通过 cache.allocateTiny(this, buf, reqCapacity, normCapacity)在缓存上进行分配。我们就以 tiny 类型为例，分析在缓存上分配 ByteBuf 的流。allocateTiny 是缓存分配的入口，PoolThreadCache 的 allocateTiny()方法的实现代码如下。

```
boolean allocateTiny(PoolArena<?> area, PooledByteBuf<?> buf, int reqCapacity, int
normCapacity) {
    return allocate(cacheForTiny(area, normCapacity), buf, reqCapacity);
}
```

这里有个方法 cacheForTiny(area, normCapacity)，其作用是根据 normCapacity 找到 tiny 类型缓存数组中的一个缓存对象。cacheForTiny()方法的代码如下。

```
    private MemoryRegionCache<?> cacheForTiny(PoolArena<?> area, int normCapacity) {
        int idx = PoolArena.tinyIdx(normCapacity);
        if (area.isDirect()) {
            return cache(tinySubPageDirectCaches, idx);
        }
        return cache(tinySubPageHeapCaches, idx);
    }
```

PoolArena.tinyIdx(normCapacity)是找到 tiny 类型缓存数组的下标。继续看 tinyIdx()方法的代码。

```
static int tinyIdx(int normCapacity) {
    return normCapacity >>> 4;
}
```

这里相当于直接将 normCapacity 除以 16，通过前面的内容我们已经知道，tiny 类型缓存数组中每个元素规格化的数据都是 16 的倍数，所以通过这种方式可以找到其下标，如果是 16Byte 会获得下标为 1 的元素，如果是 32Byte 则会获得下标为 2 的元素。

在 cacheForTiny()方法中，通过 if (area.isDirect())判断是否分配堆外内存，因为我们是按照堆外内存进行举例的，所以这里为 true。cache(tinySubPageDirectCaches, idx)方法的实现代码如下。

```
private static <T> MemoryRegionCache<T> cache(MemoryRegionCache<T>[] cache, int idx) {
    if (cache == null || idx > cache.length - 1) {
        return null;
    }
    return cache[idx];
}
```

我们看到，直接通过下标的方式获取了缓存数组中的对象，回到 PoolThreadCache 的

allocateTiny()方法，代码如下。

```
boolean allocateTiny(PoolArena<?> area, PooledByteBuf<?> buf, int reqCapacity, int
normCapacity) {
    return allocate(cacheForTiny(area, normCapacity), buf, reqCapacity);
}
```

获取缓存对象之后，来看 allocate(cacheForTiny(area, normCapacity), buf, reqCapacity)方法的实现代码。

```
private boolean allocate(MemoryRegionCache<?> cache, PooledByteBuf buf, int reqCapacity) {
    if (cache == null) {
        return false;
    }
    boolean allocated = cache.allocate(buf, reqCapacity);
    if (++ allocations >= freeSweepAllocationThreshold) {
        allocations = 0;
        trim();
    }
    return allocated;
}
```

分析上面代码，看到 cache.allocate(buf, reqCapacity)继续进行分配。我们来看一下内部类 MemoryRegionCache 的 allocate(PooledByteBuf<T> buf, int reqCapacity)方法的具体代码。

```
public final boolean allocate(PooledByteBuf<T> buf, int reqCapacity) {
    Entry<T> entry = queue.poll();
    if (entry == null) {
        return false;
    }
    initBuf(entry.chunk, entry.handle, buf, reqCapacity);
    entry.recycle();
    ++ allocations;
    return true;
}
```

在这个方法中，首先通过 queue.poll()方法弹出一个 Entry，我们知道 MemoryRegionCache 内部维护着一个队列，而队列中的每一个值是一个 Entry。来看 Entry 类的实现代码。

```
static final class Entry<T> {
    final Handle<Entry<?>> recyclerHandle;
    PoolChunk<T> chunk;
    long handle = -1;

    //代码省略
}
```

我们重点来看 Chunk 和 Handle 的这两个属性，Chunk 代表一块连续的内存。Netty 是以 Chunk 为单位进行内存分配的，后面我们会对 Chunk 进行详细剖析。Handle 相当于一个指针，可以唯一定位 Chunk 里的一块连续内存，之后也会详细分析。这样，通过 Chunk 和 Handle 就可以定位 ByteBuf 中指定的一块连续内存，有关 ByteBuf 相关的读写操作，都会在这块内存中进行。MemoryRegionCache 的 allocate(PooledByteBuf<T> buf, int reqCapacity)方法的具体代码如下。

```
public final boolean allocate(PooledByteBuf<T> buf, int reqCapacity) {
    Entry<T> entry = queue.poll();
    if (entry == null) {
        return false;
    }
    initBuf(entry.chunk, entry.handle, buf, reqCapacity);
    entry.recycle();
    ++ allocations;
    return true;
}
```

弹出 Entry 之后，通过 initBuf(entry.chunk, entry.handle，buf, reqCapacity)方法初始化 ByteBuf，这里参数传入 Entry 的 Chunk 和 Handle。因为我们分析的 tiny 类型的缓存对象是 SubPageMemoryRegionCache 类型，所以继续看 SubPageMemoryRegionCache 类的 initBuf(entry.chunk, entry.handle, buf, reqCapacity)方法，代码如下。

```
protected void initBuf(
        PoolChunk<T> chunk, long handle, PooledByteBuf<T> buf, int reqCapacity) {
    chunk.initBufWithSubpage(buf, handle, reqCapacity);
}
```

这里的 Chunk 调用了 initBufWithSubpage(buf, handle, reqCapacity)方法，其实就是 PoolChunk 类中的方法。继续看 initBufWithSubpage()方法，代码如下。

```
void initBufWithSubpage(PooledByteBuf<T> buf, long handle, int reqCapacity) {
    initBufWithSubpage(buf, handle, bitmapIdx(handle), reqCapacity);
}
```

上面代码中调用了 bitmapIdx()方法。有关 bitmapIdx(handle)相关的逻辑，会在后续的章节进行剖析。这里继续看 initBufWithSubpage()的逻辑代码。

```
private void initBufWithSubpage(PooledByteBuf<T> buf, long handle, int bitmapIdx, int
reqCapacity) {
    assert bitmapIdx != 0;
    int memoryMapIdx = memoryMapIdx(handle);
    PoolSubpage<T> subpage = subpages[subpageIdx(memoryMapIdx)];
    assert subpage.doNotDestroy;
    assert reqCapacity <= subpage.elemSize;
```

```
    buf.init(
        this, handle,
        runOffset(memoryMapIdx) + (bitmapIdx & 0x3FFFFFFF) * subpage.elemSize, reqCapacity,
subpage.elemSize,
        arena.parent.threadCache());
}
```

我们先看 init()方法，因为我们是以 PooledUnsafeDirectByteBuf 为例的，所以这里用的是 PooledUnsafeDirectByteBuf 的 init()方法。init()方法的代码如下。

```
void init(PoolChunk<ByteBuffer> chunk, long handle, int offset, int length, int maxLength,
        PoolThreadCache cache) {
    super.init(chunk, handle, offset, length, maxLength, cache);
    initMemoryAddress();
}
```

首先调用了父类的 init()方法，代码如下。

```
void init(PoolChunk<T> chunk, long handle, int offset, int length, int maxLength,
PoolThreadCache cache) {
    //初始化
    assert handle >= 0;
    assert chunk != null;
    //指定当前 ByteBuf 分配的内存块
    this.chunk = chunk;
    //指定当前 ByteBuf 连续内存指向的位置
    this.handle = handle;
    memory = chunk.memory;
    this.offset = offset;
    this.length = length;
    this.maxLength = maxLength;
    tmpNioBuf = null;
    this.cache = cache;
}
```

上面的代码将 PooledUnsafeDirectByteBuf 的各个属性进行了初始化。

- this.chunk = chunk 初始化了 chunk，指定当前 ByteBuf 分配的内存块。
- this.handle = handle 初始化了 handle，指定当前 ByteBuf 连续内存指向的位置。

有关 offset 和 length，我们会在之后再分析，在这里只需要知道，通过缓存分配 ByteBuf，只要通过一个 chunk 和 handle 就可以确定一块内存。以上就是通过缓存分配 ByteBuf 对象的全过程。

现在，回到 MemoryRegionCache 的 allocate(PooledByteBuf<T> buf, int reqCapacity)方法。

```
public final boolean allocate(PooledByteBuf<T> buf, int reqCapacity) {
    Entry<T> entry = queue.poll();
```

```
    if (entry == null) {
        return false;
    }
    initBuf(entry.chunk, entry.handle, buf, reqCapacity);
    entry.recycle();
    ++ allocations;
    return true;
}
```

分析完了 initBuf()方法，再继续往下看。entry.recycle()这步将 Entry 对象回收，因为 Entry 对象弹出之后没有再被引用，所以可能 GC 会将 Entry 对象回收。Netty 为了将对象循环利用，将其放在对象回收站进行回收。跟进 recycle()方法，代码如下。

```
void recycle() {
    chunk = null;
    handle = -1;
    recyclerHandle.recycle(this);
}
```

chunk = null 和 handle = -1 表示当前 Entry 不指向任何一块内存。recyclerHandle.recycle(this) 将当前 Entry 回收。

以上就是命中缓存的流程，因为是假设在缓存中有值的情况下进行分配的，所以如果是第一次分配，缓存中是没有值的，那么在缓存中没有值的情况下，Netty 是如何进行分配的呢?我们后面再详细分析。

最后，简单总结一下 MemoryRegionCache 对象的基本结构，如下图所示。

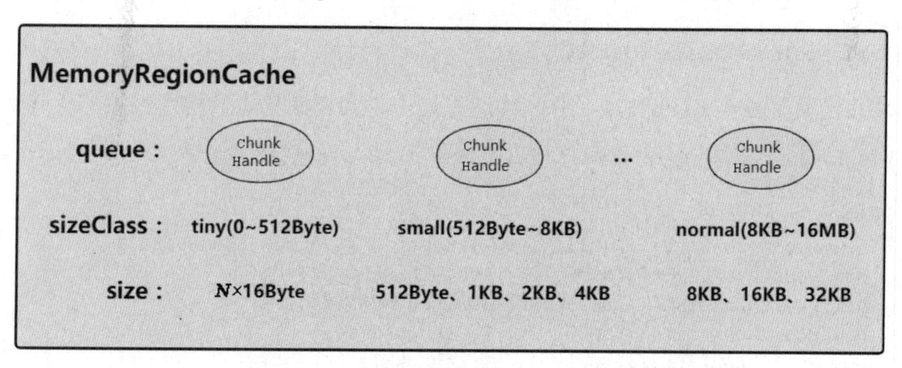

11.4.5 Page 级别的内存分配

11.4.3 节中介绍过，Netty 内存分配的单位是 Chunk，一个 Chunk 的大小是 16MB，每个 Chunk 都以双向链表的形式保存在一个 ChunkList 中。同样，多个 ChunkList 也是以双向链表的形式进行

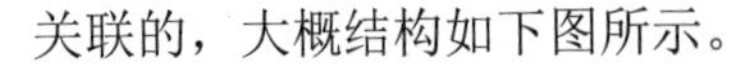

关联的，大概结构如下图所示。

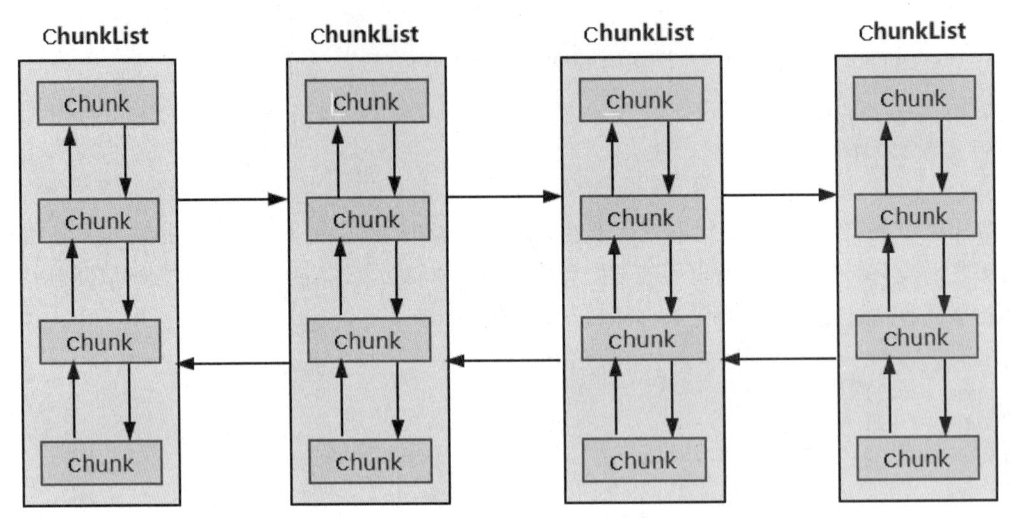

在 ChunkList 中，根据 Chunk 的内存使用率划分 ChunkList。这样，在内存分配时，会根据百分比找到相应的 ChunkList，在 ChunkList 中选择一个 Chunk 进行内存分配。我们来看 PoolArena 中有关 ChunkList 的成员变量。

```
private final PoolChunkList<T> q050;
private final PoolChunkList<T> q025;
private final PoolChunkList<T> q000;
private final PoolChunkList<T> qInit;
private final PoolChunkList<T> q075;
private final PoolChunkList<T> q100;
```

这里总共定义了 6 个 ChunkList，并在构造方法中将其初始化，其构造方法的代码如下。

```
protected PoolArena(PooledByteBufAllocator parent, int pageSize, int maxOrder, int pageShifts,
int chunkSize) {
    //代码省略
    q100 = new PoolChunkList<T>(null, 100, Integer.MAX_VALUE, chunkSize);
    q075 = new PoolChunkList<T>(q100, 75, 100, chunkSize);
    q050 = new PoolChunkList<T>(q075, 50, 100, chunkSize);
    q025 = new PoolChunkList<T>(q050, 25, 75, chunkSize);
    q000 = new PoolChunkList<T>(q025, 1, 50, chunkSize);
    qInit = new PoolChunkList<T>(q000, Integer.MIN_VALUE, 25, chunkSize);

    //用双向链表的形式进行连接
    q100.prevList(q075);
    q075.prevList(q050);
    q050.prevList(q025);
```

```
    q025.prevList(q000);
    q000.prevList(null);
    qInit.prevList(qInit);
    //代码省略
}
```

我们通过 new PoolChunkList()创建每个 ChunkList，以 q050 = new PoolChunkList<T>(q075, 50, 100, chunkSize) 为例进行简单的介绍。q075 表示当前 q050 的下一个节点。参数 50 和 100 表示当前 ChunkList 中存储的 Chunk 的内存使用率都在 50%~100%，ChunkSize 为其设置大小。创建完 ChunkList 后，再设置其上一个节点，以 q050.prevList(q025)为例，表示当前 ChunkList 的上一个节点是 q025。ChunkList 的节点关系如下图所示。

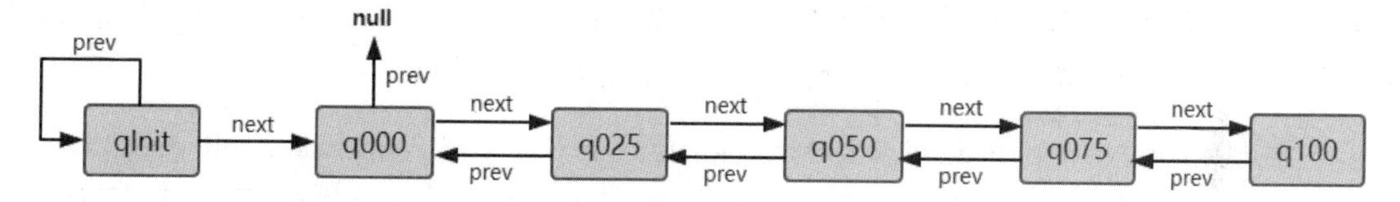

Netty 中，Chunk 又包含了多个 Page，每个 Page 的大小为 8KB，如果要分配 16KB 的内存，则在 Chunk 中找到连续的两个 Page 就可以，对应关系如下图所示。

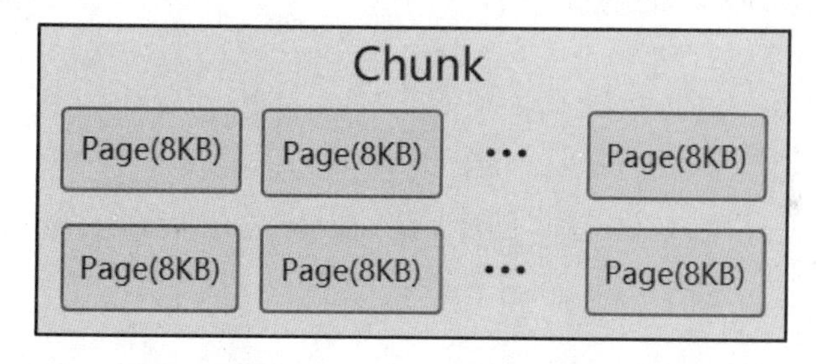

在很多场景下，为缓冲区分配 8KB 的内存也是一种浪费，比如只需要分配 2KB 的缓冲区，如果使用 8KB 会造成 6KB 的浪费。这种情况下，Netty 又会将 Page 切分成多个 SubPage，每个 SubPage 大小要根据分配的缓冲区大小而定，比如要分配 2KB 的内存，就会将一个 Page 切分成 4 个 SubPage，每个 SubPage 的大小为 2KB，如下图所示。

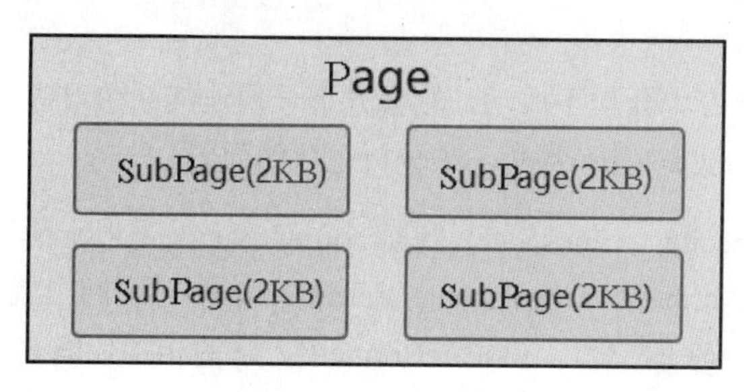

PoolSubpage 的基本结构的代码如下。

```
final PoolChunk<T> chunk;
private final int memoryMapIdx;
private final int runOffset;
private final int pageSize;
private final long[] bitmap;
PoolSubpage<T> prev;
PoolSubpage<T> next;
boolean doNotDestroy;
int elemSize;
```

chunk 代表其子页属于哪个 Chunk；bitmap 用于记录子页的内存分配情况；prev 和 next 代表子页是按照双向链表进行关联的，分别指向上一个节点和下一个节点；elemSize 属性代表的是子页是按照多大内存进行划分的，如果按照 1KB 划分，则可以划分出 8 个子页。

下面开始剖析 Netty 在 Page 级别上分配内存的流程，还是回到 PoolArena 的 allocate()方法。

```
private void allocate(PoolThreadCache cache, PooledByteBuf<T> buf, final int reqCapacity) {
    //规格化
    final int normCapacity = normalizeCapacity(reqCapacity);
    if (isTinyOrSmall(normCapacity)) {
        //此处省略处理逻辑
    ...
    }
    if (normCapacity <= chunkSize) {
        //首先在缓存上进行内存分配
        if (cache.allocateNormal(this, buf, reqCapacity, normCapacity)) {
            //分配成功，返回
            return;
        }
        //分配不成功，做实际的内存分配
        allocateNormal(buf, reqCapacity, normCapacity);
    } else {
        //大于这个值，就不在缓存上分配
        allocateHuge(buf, reqCapacity);
    }
}
```

之前讲过，如果在缓存中分配不成功，则会开辟一块连续的内存进行缓冲区分配。这里我们先跳过 isTinyOrSmall(normCapacity)判断，来分析之后的代码。

首先执行 if(normCapacity<=chunkSize) 判断 normCapacity 是否小于等于 chunkSize，如果条件满足，会执行到 allocateNormal(buf,reqCapacity,normCapacity)，实际上就是在 Page 级别上进行分配，分配一个或者多个 Page 的空间。allocateNormal()方法的相关代码如下。

```
private synchronized void allocateNormal(PooledByteBuf<T> buf, int reqCapacity, int
normCapacity) {
    //在原有的 Chunk 上进行内存分配(1)
    if (q050.allocate(buf, reqCapacity, normCapacity) || q025.allocate(buf, reqCapacity,
normCapacity) ||
        q000.allocate(buf, reqCapacity, normCapacity) || qInit.allocate(buf, reqCapacity,
normCapacity) ||
        q075.allocate(buf, reqCapacity, normCapacity)) {
        ++allocationsNormal;
        return;
    }

    //创建 Chunk 进行内存分配(2)
    PoolChunk<T> c = newChunk(pageSize, maxOrder, pageShifts, chunkSize);
    long handle = c.allocate(normCapacity);
    ++allocationsNormal;
    assert handle > 0;
    //初始化 ByteBuf(3)
    c.initBuf(buf, handle, reqCapacity);
    qInit.add(c);
}
```

上面代码主要拆解了如下步骤。

（1）优先在原有的 Chunk 中进行内存分配。

（2）如果是首次分配，那就创建 Chunk 进行内存分配。

（3）最后初始化 ByteBuf。

首先看第一步，在原有的 Chunk 中进行内存分配，代码如下。

```
if (q050.allocate(buf, reqCapacity, normCapacity) || q025.allocate(buf, reqCapacity,
normCapacity) ||
    q000.allocate(buf, reqCapacity, normCapacity) || qInit.allocate(buf, reqCapacity,
normCapacity) ||
    q075.allocate(buf, reqCapacity, normCapacity)) {
    ++allocationsNormal;
    return;
}
```

我们之前讲过，ChunkList 是存储不同内存使用量的 Chunk 集合，每个 ChunkList 以双向链表的形式进行关联，这里的 q050.allocate(buf,reqCapacity,normCapacity)代表首先在 q050 这个 ChunkList 上进行内存分配。以 q050 为例进行分析，q050.allocate(buf,reqCapacity,normCapacity)方法的代码如下。

```
boolean allocate(PooledByteBuf<T> buf, int reqCapacity, int normCapacity) {
    if (head == null || normCapacity > maxCapacity) {
        return false;
    }
    //从 Head 节点往下遍历
    for (PoolChunk<T> cur = head;;) {
        long handle = cur.allocate(normCapacity);
        if (handle < 0) {
            cur = cur.next;
            if (cur == null) {
                return false;
            }
        } else {
            cur.initBuf(buf, handle, reqCapacity);
            if (cur.usage() >= maxUsage) {
                remove(cur);
                nextList.add(cur);
            }
            return true;
        }
    }
}
```

我们从 Head 节点往下遍历：long handle=cur.allocate(normCapacity)表示对每个 Chunk 都尝试去分配。

- if (handle < 0)说明没有分配到，则通过 cur=cur.next 找到下一个节点继续进行分配，我们知道 Chunk 也是通过双向链表进行关联的，所以对这块逻辑应该不会陌生。如果 Handle 大于 0 说明已经分配到了内存，则通过 cur.initBuf(buf,handle,reqCapacity)对 ByteBuf 进行初始化。
- if(cur.usage()>=maxUsage)代表当前 Chunk 的内存使用率大于其最大使用率，则通过 remove(cur)从当前的 ChunkList 中移除，再通过 nextList.add(cur)添加到下一个 ChunkList 中。

我们再回到 PoolArena 的 allocateNormal()方法中，看第二步 PoolChunk<T>c= newChunk(pageSize,maxOrder,pageShifts,chunkSize)，其中参数 pageSize 是 8 192，也就是 8KB；maxOrder 为 11；pageShifts 为 13，2 的 13 次方正好是 8 192，也就是 8KB；chunkSize 为 16 777 216，也就是 16MB。

因为我们分析的是堆外内存，所以 newChunk(pageSize,maxOrder,pageShifts,chunkSize)调用的是 DirectArena 的 newChunk()方法。

```
protected PoolChunk<ByteBuffer> newChunk(int pageSize, int maxOrder, int pageShifts, int
chunkSize) {
    return new PoolChunk<ByteBuffer>(
            this, allocateDirect(chunkSize),
            pageSize, maxOrder, pageShifts, chunkSize);
}
```

这里直接通过构造函数创建了一个 Chunk。allocateDirect(chunkSize)通过 JDK 的 API 申请了一块直接内存，PoolChunk 的构造函数的代码如下。

```
PoolChunk(PoolArena<T> arena, T memory, int pageSize, int maxOrder, int pageShifts, int
chunkSize) {
    unpooled = false;
    this.arena = arena;
    //memeory 为一个 ByteBuf
    this.memory = memory;
    //8kB
    this.pageSize = pageSize;
    //13
    this.pageShifts = pageShifts;
    //11
    this.maxOrder = maxOrder;
    this.chunkSize = chunkSize;
    unusable = (byte) (maxOrder + 1);
    log2ChunkSize = log2(chunkSize);
    subpageOverflowMask = ~(pageSize - 1);
    freeBytes = chunkSize;

    assert maxOrder < 30 : "maxOrder should be < 30, but is: " + maxOrder;
    maxSubpageAllocs = 1 << maxOrder;

    //节点数量为 4 096
    memoryMap = new byte[maxSubpageAllocs << 1];
    //也是 4 096 个节点
    depthMap = new byte[memoryMap.length];
    int memoryMapIndex = 1;
    //d 相当于一个深度，赋值的内容代表当前节点的深度
    for (int d = 0; d <= maxOrder; ++ d) {
        int depth = 1 << d;
        for (int p = 0; p < depth; ++ p) {
            memoryMap[memoryMapIndex] = (byte) d;
            depthMap[memoryMapIndex] = (byte) d;
            memoryMapIndex ++;
        }
    }
```

```
    subpages = newSubpageArray(maxSubpageAllocs);
}
```

上述代码中将参数传入的值进行赋值 this.memory = memory，即将参数中创建的堆外内存进行保存，这里的内存是指 Chunk 所指向的那块连续的内存，在这个 Chunk 中所分配的 ByteBuf 都会在这块内存中进行读写。

我们重点关注 memoryMap=new byte[maxSubpageAllocs<<1] 和 depthMap=new byte[memoryMap.length]这两步：memoryMap = new byte[maxSubpageAllocs << 1]初始化了一个字节数组 memoryMap，大小为 maxSubpageAllocs << 1，也就是 4 096；depthMap = new byte[memoryMap.length] 同样初始化了一个字节数组，大小为 memoryMap 的大小，也是 4 096。继续往下分析之前，先来看一下 Chunk 的层级关系，如下图所示。

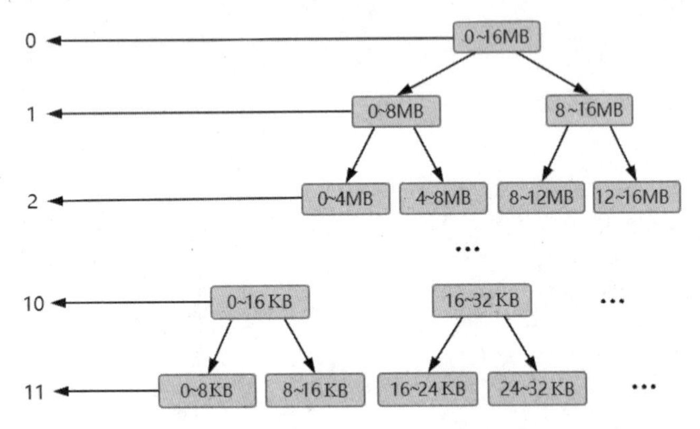

上图是一个二叉树的结构，左侧的数字代表层级，右侧代表一块连续的内存，每个父节点下又拆分成多个子节点，顶层表示的内存范围为 0~16MB，其下又分为两层，范围为 0~8MB、8~16MB，依此类推，最后到 11 层，以 8KB 的大小划分，也就是一个 Page 的大小。

如果我们分配一个 8MB 的缓冲区，则会将第二层的第一个节点，也就是 0~8MB 这个连续的内存进行分配。分配完成之后，会将这个节点设置为不可用。结合上面的图，我们再看构造方法中的 for 循环代码。

```
for (int d = 0; d <= maxOrder; ++ d) {
    int depth = 1 << d;
    for (int p = 0; p < depth; ++ p) {
        memoryMap[memoryMapIndex] = (byte) d;
        depthMap[memoryMapIndex] = (byte) d;
        memoryMapIndex ++;
    }
}
```

实际上，这个 for 循环就是将上面的结构包装成一个字节数组 memoryMap，外层循环用于控制层数，内层循环用于控制里面每层的节点。经过循环之后，memoryMap 和 depthMap 内容为以下表现形式。

```
[0, 0, 1, 1, 2, 2, 2, 2, 3, 3, 3, 3, 3, 3, 3, 3, 4, 4, 4, 4, 4, 4, 4, 4, 4, 4, 4, 4, 4, 4, 4,
4..........]
```

需要注意的是，因为程序中数组的下标是从 1 开始设置的，所以第 0 个节点元素为默认值 0。这里数字代表层级，同时也代表了当前层级的节点，相同的数字个数就是这一层级的节点数。其中 0 为 2 个（因为分配时下标是从 1 开始的，所以第 0 个位置是默认值 0，实际上第 0 层元素只有一个，就是头节点），1 为 2 个，2 为 4 个，3 为 8 个，4 为 16 个，*n* 为 2 的 *n* 次方个，直到 11，也就是 11 有 2 的 11 次方个。

我们再回到 PoolArena 的 allocateNormal()方法。

```
private synchronized void allocateNormal(PooledByteBuf<T> buf, int reqCapacity, int
normCapacity) {

    //此处省略部分逻辑
    …

    //创建 Chunk 进行内存分配(2)
    PoolChunk<T> c = newChunk(pageSize, maxOrder, pageShifts, chunkSize);
    long handle = c.allocate(normCapacity);
    ++allocationsNormal;
    assert handle > 0;
    //初始化 ByteBuf(3)
    c.initBuf(buf, handle, reqCapacity);
    qInit.add(c);
}
```

继续剖析 long handle = c.allocate(normCapacity) 这步，看 allocate(normCapacity)的代码。

```
long allocate(int normCapacity) {
    if ((normCapacity & subpageOverflowMask) != 0) {
        return allocateRun(normCapacity);
    } else {
        return allocateSubpage(normCapacity);
    }
}
```

如果以 Page 为单位分配，则执行到 allocateRun(normCapacity)方法中，代码如下。

```
private long allocateRun(int normCapacity) {
    int d = maxOrder - (log2(normCapacity) - pageShifts);
    int id = allocateNode(d);
```

```
    if (id < 0) {
        return id;
    }
    freeBytes -= runLength(id);
    return id;
}
```

- int d = maxOrder - (log2(normCapacity) - pageShifts) 表示根据 normCapacity 计算出第几层。
- int id = allocateNode(d) 表示根据层级关系去分配一个节点，其中 id 代表 memoryMap 中的下标。

allocateNode()方法的具体实现代码如下。

```
private int allocateNode(int d) {
    //下标初始值为 1
    int id = 1;
    //代表当前层级第一个节点的初始下标
    int initial = - (1 << d);
    //获取第一个节点的值
    byte val = value(id);
    //如果值大于层级，说明 Chunk 不可用
    if (val > d) {
        return -1;
    }
    //当前下标对应的节点值如果小于层级，或者当前下标小于层级的初始下标
    while (val < d || (id & initial) == 0) {
        //当前下标乘以 2，代表当前节点的子节点的起始位置
        id <<= 1;
        //获得 id 位置的值
        val = value(id);
        //如果当前节点值大于层数（节点不可用）
        if (val > d) {
            //id 为偶数则+1，id 为奇数则-1（用的是其兄弟节点）
            id ^= 1;
            //获取 id 的值
            val = value(id);
        }
    }
    byte value = value(id);
    assert value == d && (id & initial) == 1 << d : String.format("val = %d, id & initial = %d,
d = %d",
            value, id & initial, d);
    //将找到的节点设置为不可用
    setValue(id, unusable);
    //逐层往上标记被使用
    updateParentsAlloc(id);
```

```
    return id;
}
```

上述代码实际上是从第一个节点往下找的，找到层级为 d、未被使用的节点，可以通过注释体会其逻辑。找到相关节点后通过 setValue 将当前节点设置为不可用，其中 id 是当前节点的下标，unusable 代表一个不可用的值，这里是 12，因为我们的层级只有 12 层，所以设置为 12 之后就相当于标记不可用。设置为不可用之后，通过 updateParentsAlloc(id)逐层设置为缓存被使用的状态。updateParentsAlloc()方法的代码如下。

```
private void updateParentsAlloc(int id) {
    while (id > 1) {
        //获取当前节点的父节点的 id
        int parentId = id >>> 1;
        //获取当前节点的值
        byte val1 = value(id);
        //找到当前节点的兄弟节点
        byte val2 = value(id ^ 1);
        //如果当前节点值小于兄弟节点，则保存当前节点值到 val，否则，保存兄弟节点值到 val
        //如果当前节点不可用，则当前节点值是 12，大于兄弟节点的值，将兄弟节点的值进行保存
        byte val = val1 < val2 ? val1 : val2;
        //将 val 的值设置为父节点下标所对应的值
        setValue(parentId, val);
        //id 设置为父节点 id，继续循环
        id = parentId;
    }
}
```

这里其实是将循环兄弟节点的值替换成父节点的值，可以通过注释仔细地进行逻辑分析。如果实在理解有困难，可以通过画图帮助理解，简单起见，这里只设置三层，如下图所示。

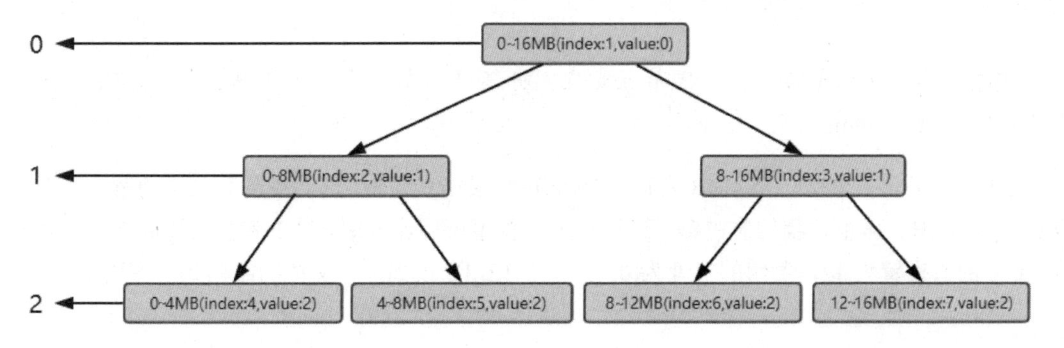

我们模拟其分配场景，假设只有三层，其中 index 代表数组 memoryMap 的下标，value 代表其值，memoryMap 中的值就为[0, 0, 1, 1, 2, 2, 2, 2]。我们要分配一个 4MB 的 ByteBuf，在调用 allocateNode(int d)中传入的 d 是 2，也就是第二层。根据上面分析的逻辑，这里会找到第二层的第

一个节点，也就是 0~4MB 这个节点，找到之后将其设置为不可用，这样 memoryMap 中的值就为[0, 0, 1, 1, 12, 2, 2, 2]，二叉树的结构就会如下图所示。

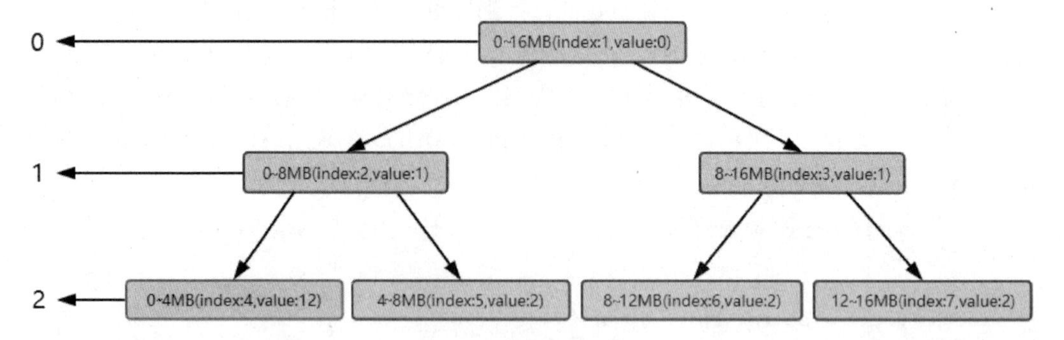

注意深色节点部分，将 index 为 4 的节点设置为不可用，之后则将向上设置不可用，循环将兄弟节点数值较小的节点替换到父节点，也就是将 index 为 2 的节点的值替换成了 index 为 5 的节点的值，这样数组的值就会变为[0, 1, 2, 1, 12, 2, 2, 2]，二叉树的结构如下图所示。

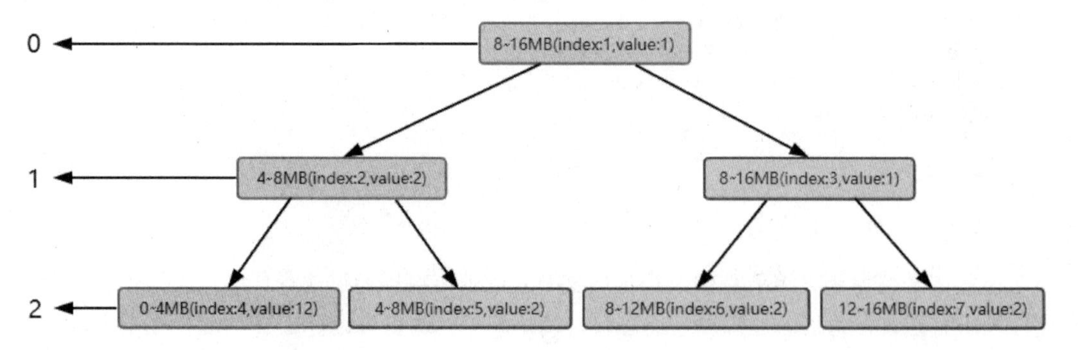

> **注意：**这里深色节点仅仅代表节点变化，并不是当前节点为不可用状态，不可用状态的真正判断依据是 value 值为 12。

这样，如果再次分配一个 4MB 内存的 ByteBuf，根据其逻辑，则会找到第二层的第二个节点，也就是 4~8MB。再根据我们的逻辑，通过向上设置不可用，index 为 2 就会设置成不可用状态，将 value 的值设置为 12，数组的值变为[0, 1, 12, 1, 12, 12, 2, 2]，二叉树的结构如下图所示。

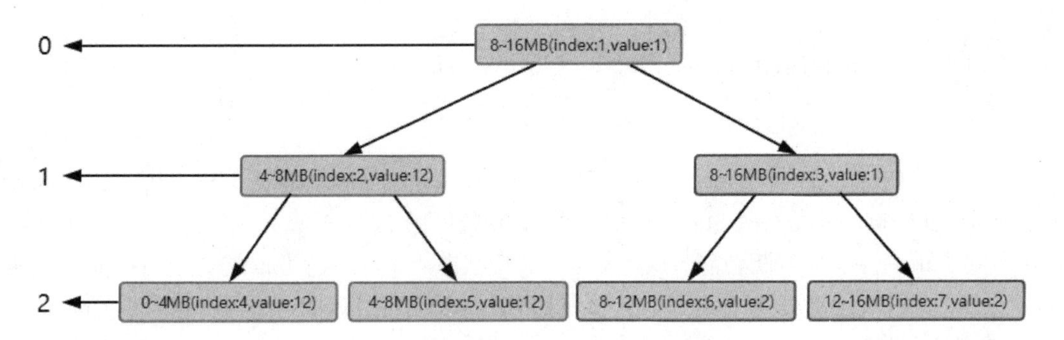

可以看到，分配两个 4MB 的 ByteBuf 之后，当前节点和其父节点都会设置成不可用状态，当 index=2 的节点设置为不可用之后，将不会再找这个节点下的子节点。依此类推，直到所有的内存分配完毕，index 为 1 的节点也会变成不可用状态，这样所有的 Page 就都分配完毕，Chunk 中再无可用节点。现在再回到 PoolArena 的 allocateNormal()方法。

```
private synchronized void allocateNormal(PooledByteBuf<T> buf, int reqCapacity, int
normCapacity) {

    //此处省略部分逻辑

    //创建 Chunk 进行内存分配(2)
    PoolChunk<T> c = newChunk(pageSize, maxOrder, pageShifts, chunkSize);
    long handle = c.allocate(normCapacity);
    ++allocationsNormal;
    assert handle > 0;
    //初始化 ByteBuf(3)
    c.initBuf(buf, handle, reqCapacity);
    qInit.add(c);
}
```

通过以上逻辑，我们知道 long handle = c.allocate(normCapacity)这一步其实返回的就是 memoryMap 的一个下标。通过这个下标，我们能唯一地定位一块内存。继续往下跟，通过 c.initBuf(buf, handle, reqCapacity)初始化 ByteBuf 后，通过 qInit.add(c)将新创建的 Chunk 添加到 ChunkList 中，我们看 initBuf()方法的代码。

```
void initBuf(PooledByteBuf<T> buf, long handle, int reqCapacity) {
    int memoryMapIdx = memoryMapIdx(handle);
    int bitmapIdx = bitmapIdx(handle);
    if (bitmapIdx == 0) {
        byte val = value(memoryMapIdx);
        assert val == unusable : String.valueOf(val);
        buf.init(this, handle, runOffset(memoryMapIdx), reqCapacity, runLength(memoryMapIdx),
              arena.parent.threadCache());
```

```
    } else {
        initBufWithSubpage(buf, handle, bitmapIdx, reqCapacity);
    }
}
```

从上面代码中看出，通过 memoryMapIdx(handle)找到 memoryMap 的下标，其实就是 Handle 的值。bitmapIdx(handle)是有关 SubPage 中使用到的逻辑，如果是 Page 级别的分配，只返回 0，进入 if 块中。if 块中首先判断当前节点是不是不可用状态，然后通过 init()方法进行初始化。其中 runOffset(memoryMapIdx)表示偏移量，偏移量相当于分配给缓冲区的这块内存相对于 Chunk 中申请的内存的首地址偏移了多少。参数 runLength(memoryMapIdx)表示根据下标获取可分配的最大长度。跟进 init()方法，PooledByteBuf 的 init()方法的代码如下。

```
void init(PoolChunk<T> chunk, long handle, int offset, int length, int maxLength,
PoolThreadCache cache) {
    //初始化
    assert handle >= 0;
    assert chunk != null;

    //在哪一块内存上进行分配
    this.chunk = chunk;
    //这一块内存上的哪一块连续内存
    this.handle = handle;
    memory = chunk.memory;
    this.offset = offset;
    this.length = length;
    this.maxLength = maxLength;
    tmpNioBuf = null;
    this.cache = cache;
}
```

这段代码也是我们熟悉的部分，将属性进行了初始化。以上就是 DirectUnsafePooledByteBuf 在 Page 级别分配的完整流程，逻辑也非常复杂，想真正地熟练掌握，还需要读者们多下功夫进行调试和剖析。

11.4.6 SubPage 级别的内存分配

通过之前的学习我们知道，如果分配一个缓冲区大小远小于 Page，直接在一个 Page 上进行分配会造成内存浪费，所以需要将 Page 继续切分成多个子块进行分配，子块分配的个数根据要分配的缓冲区大小而定，比如只需要分配 1KB 的内存，就将一个 Page 分成 8 等分。简单起见，仅以 16 字节为例，讲解其分配逻辑。在分析其逻辑前，首先看 PoolArena 的一个属性。

```
private final PoolSubpage<T>[] tinySubpagePools;
```

这个属性是一个 PoolSubpage 的数组，有点类似于一个 SubPage 的缓存。我们创建一个 SubPage 之后，会将创建好的 SubPage 与 PoolArena 中 tinySubpagePools 数组的每一个元素进行关联，下次再分配的时候可以直接通过 tinySubpagePools 数组元素去找关联的 SubPage。而 tinySubpagePools 是在 PoolArena 的构造方法中初始化的，代码如下。

```
protected PoolArena(PooledByteBufAllocator parent, int pageSize, int maxOrder, int pageShifts,
int chunkSize) {
...
smallSubpagePools = newSubpagePoolArray(numSmallSubpagePools);
...
}
```

上述代码中 numTinySubpagePools 为 32，newSubpagePoolArray(numTinySubpagePools)方法的代码如下。

```
private PoolSubpage<T>[] newSubpagePoolArray(int size) {
    return new PoolSubpage[size];
}
```

这里直接创建了一个 PoolSubpage 数组，长度为 32，在构造方法中创建完毕之后，会通过循环为其赋值，代码如下。

```
for (int i = 0; i < tinySubpagePools.length; i ++) {
    tinySubpagePools[i] = newSubpagePoolHead(pageSize);
}
```

继续看 newSubpagePoolHead()方法。

```
private PoolSubpage<T> newSubpagePoolHead(int pageSize) {
    PoolSubpage<T> head = new PoolSubpage<T>(pageSize);
    head.prev = head;
    head.next = head;
    return head;
}
```

在 newSubpagePoolHead()方法中创建了一个 PoolSubpage 对象 Head。

```
head.prev = head;
head.next = head;
```

由上面代码知道，SubPage 其实也是一个双向链表，这里将 Head 的上一个节点和下一个节点都设置为自身，有关 PoolSubpage 的关联关系，我们稍后分析。通过循环创建 PoolSubpage，总共创建 32 个 SubPage，每个 SubPage 实际代表一块内存大小，如下图所示。

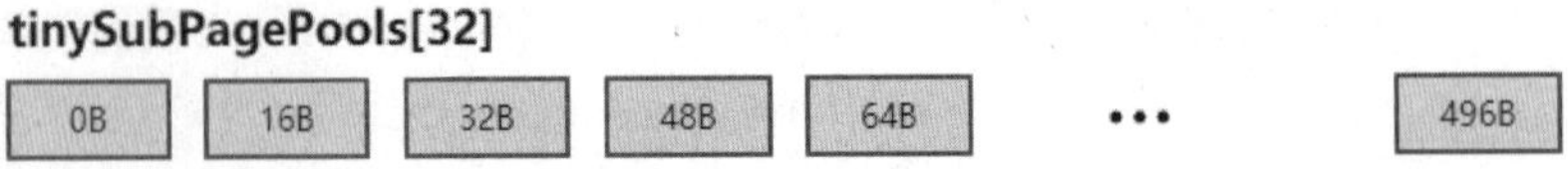

tinySubPagePools[32]

0B 16B 32B 48B 64B ••• 496B

tinySubpagePools 的结构有点类似于之前小节的缓存数组 tinySubPageDirectCaches 的结构。了解了 tinySubpagePools 的属性，我们看 PoolArena 的 allocate()方法，也就是缓冲区的入口方法。

```
private void allocate(PoolThreadCache cache, PooledByteBuf<T> buf, final int reqCapacity) {
    //规格化
    final int normCapacity = normalizeCapacity(reqCapacity);
    if (isTinyOrSmall(normCapacity)) {
        int tableIdx;
        PoolSubpage<T>[] table;
        //判断是不是 tiny 类型
        boolean tiny = isTiny(normCapacity);
        if (tiny) { // < 512
            //缓存分配
            if (cache.allocateTiny(this, buf, reqCapacity, normCapacity)) {
                return;
            }
            //通过 tinyIdx 获取 tableIdx
            tableIdx = tinyIdx(normCapacity);
            //SubPage 的数组
            table = tinySubpagePools;
        } else {
            if (cache.allocateSmall(this, buf, reqCapacity, normCapacity)) {
                return;
            }
            tableIdx = smallIdx(normCapacity);
            table = smallSubpagePools;
        }

        //获取对应的节点
        final PoolSubpage<T> head = table[tableIdx];

        synchronized (head) {
            final PoolSubpage<T> s = head.next;
            //默认情况下，Head 的 next 也是自身
            if (s != head) {
                assert s.doNotDestroy && s.elemSize == normCapacity;
                long handle = s.allocate();
                assert handle >= 0;
                s.chunk.initBufWithSubpage(buf, handle, reqCapacity);

                if (tiny) {
                    allocationsTiny.increment();
```

```
                } else {
                    allocationsSmall.increment();
                }
                return;
            }
        }
        allocateNormal(buf, reqCapacity, normCapacity);
        return;
    }
    if (normCapacity <= chunkSize) {
        //此处省略部分逻辑
        …
    } else {
        //此处省略部分逻辑
        …
    }
}
```

之前用这个方法剖析过 Page 级别内存分配的相关逻辑，现在来看 SubPage 级别分配的相关逻辑。假设分配 16 字节的缓冲区，isTinyOrSmall(normCapacity)就会返回 true，进入 if 块，同样 if (tiny)会返回 true，继续跟进 if (tiny)中的逻辑。首先会在缓存中分配缓冲区，如果分配不到，就开辟一块内存进行内存分配，先看这一步的代码。

```
tableIdx = tinyIdx(normCapacity);
```

通过 normCapacity 获得 tableIdx，继续看代码。

```
static int tinyIdx(int normCapacity) {
    return normCapacity >>> 4;
}
```

将 normCapacity 除以 16，其实就是 1。回到 PoolArena 的 allocate()方法，继续看代码。

```
table = tinySubpagePools
```

这里将 tinySubpagePools 赋值到局部变量 table 中，继续往下看。

```
final PoolSubpage<T> head = table[tableIdx]
```

这步是通过下标获得一个 PoolSubpage，因为我们以 16 字节为例，所以我们获得下标为 1 的 PoolSubpage，对应的内存大小就是 16Byte。再看 final PoolSubpage<T> s = head.next 这一步，根据之前了解的 tinySubpagePools 属性，默认情况下 head.next 也是自身，所以 if (s != head)会返回 false，继续往下看，allocateNormal(buf, reqCapacity,normCapacity)方法的逻辑我们已经剖析过。先在原来的 Chunk 中分配，如果分配不成功，则会创建 Chunk 进行分配。

我们看这一步 long handle = c.allocate(normCapacity)，allocate(normCapacity)方法的代码如下。

```
long allocate(int normCapacity) {
    if ((normCapacity & subpageOverflowMask) != 0) {
        return allocateRun(normCapacity);
    } else {
        return allocateSubpage(normCapacity);
    }
}
```

上一小节我们分析 Page 级别分配的时候，剖析的是 allocateRun(normCapacity)方法。因为这里是以 16 字节举例，所以这次剖析 allocateSubpage(normCapacity)方法，也就是在 SubPage 级别进行内存分配，代码如下。

```
private long allocateSubpage(int normCapacity) {
    PoolSubpage<T> head = arena.findSubpagePoolHead(normCapacity);
    synchronized (head) {
        int d = maxOrder;
        //表示在第 11 层分配节点
        int id = allocateNode(d);
        if (id < 0) {
            return id;
        }

        //获取初始化的 SubPage
        final PoolSubpage<T>[] subpages = this.subpages;
        final int pageSize = this.pageSize;

        freeBytes -= pageSize;
        //表示第几个 subpageIdx
        int subpageIdx = subpageIdx(id);
        PoolSubpage<T> subpage = subpages[subpageIdx];
        if (subpage == null) {
            //如果 SubPage 为空
            subpage = new PoolSubpage<T>(head, this, id, runOffset(id), pageSize,
normCapacity);
            //则将当前的下标赋值为 SubPage
            subpages[subpageIdx] = subpage;
        } else {
            subpage.init(head, normCapacity);
        }
        //取出一个 SubPage
        return subpage.allocate();
    }
}
```

上述代码中，通过 PoolSubpage<T> head = arena.findSubpagePoolHead(normCapacity)这种方式找到 Head 节点，实际上这里 Head 就是之前分析的 tinySubpagePools 属性的第一个节点，也就是

对应 16Byte 的那个节点。int d = maxOrder 是将 11 赋值给 d，也就是在内存树的第 11 层取节点，这部分上一小节已经剖析过。int id = allocateNode(d)获取的是上一小节分析过的字节数组 memoryMap 的下标，这里指向一个 Page，如果是第一次分配，指向的是 0~8kB 的那个 Page。final PoolSubpage<T>[] subpages = this.subpages 这一步是获得 PoolChunk 中成员变量 SubPages 的值，也是个 PoolSubpage 的数组，在 PoolChunk 进行初始化的时候，也会初始化该数组，长度为 2 048。也就是说每个 Chunk 都维护着一个 SubPage 的列表。如果每一个 Page 级别的内存都需要被切分成 SubPage，则会将这个 Page 放入该列表中，专门用于分配 SubPage。所以这个列表中的 SubPage 其实就是一个用于切分的 Page。SubPages 如下图所示。

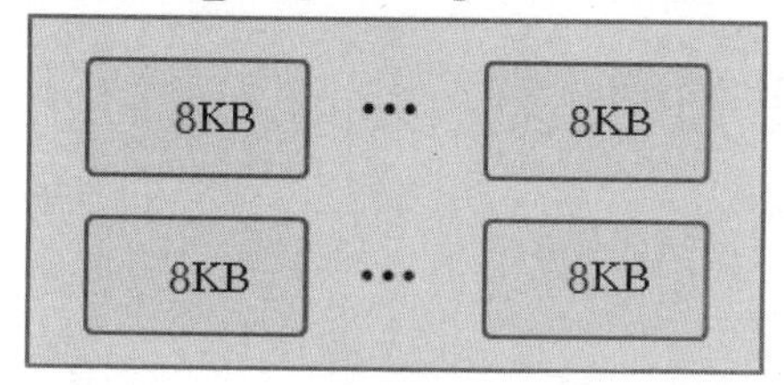

int subpageIdx = subpageIdx(id)这一步是通过 id 获得 PoolSubpage 数组的下标，如果 id 对应的 Page 是 0~8KB 的节点，这里获得的下标就是 0。在 if (subpage == null)中，因为默认 SubPages 只是创建一个数组，并没有往数组中赋值，所以第一次执行到这里会返回 true，跟到 if 块中，代码如下。

```
    subpage = new PoolSubpage<T>(head, this, id, runOffset(id), pageSize, normCapacity);
```

通过 new PoolSubpage 创建一个新的 SubPage 后，通过 subpages[subpageIdx] = subpage 将新创建的 SubPage 根据下标赋值到 SubPages 中的元素。在 new PoolSubpage 的构造方法中，传入 Head，就是之前提到的 tinySubpagePools 属性中的节点，如果分配 16 字节的缓冲区，这里对应的就是第一个节点，PoolSubpage 构造方法的代码如下。

```
PoolSubpage(PoolSubpage<T> head, PoolChunk<T> chunk, int memoryMapIdx, int runOffset, int
pageSize, int elemSize) {
    this.chunk = chunk;
    this.memoryMapIdx = memoryMapIdx;
    this.runOffset = runOffset;
    this.pageSize = pageSize;
    bitmap = new long[pageSize >>> 10];
    init(head, elemSize);
}
```

这里重点关注 bitmap 属性，这是一个 long 类型的数组，初始大小为 8，这只是初始化的大小，真正的大小要根据 Page 切分成多少块确定，这里将属性进行了赋值，我们看 init()方法的代码。

```
void init(PoolSubpage<T> head, int elemSize) {
    doNotDestroy = true;
    this.elemSize = elemSize;
    if (elemSize != 0) {
        maxNumElems = numAvail = pageSize / elemSize;
        nextAvail = 0;
        bitmapLength = maxNumElems >>> 6;
        if ((maxNumElems & 63) != 0) {
            bitmapLength ++;
        }

        for (int i = 0; i < bitmapLength; i ++) {
            //bitmap 标识哪个 SubPage 被分配
            //0 表示未分配，1 表示已分配
            bitmap [i] = 0;
        }
    }
    //加到 Arena 里面
    addToPool(head);
}
```

this.elemSize = elemSize 表示保存当前分配的缓冲区大小，因为以 16 字节为例，所以这里是 16。maxNumElems = numAvail = pageSize / elemSize 初始化了两个属性 maxNumElems 和 numAvail，值都为 pageSize / elemSize，表示一个 Page 大小除以分配的缓冲区大小，也就是表示当前 Page 被划分了多少份。numAvail 表示剩余可用的块数，由于第一次分配都是可用的，所以 numAvail=maxNumElems。bitmapLength 表示 bitmap 的实际大小，已经分析过，bitmap 初始化大小为 8，但实际上并不一定需要 8 个元素，元素个数要根据 Page 切分的子块而定，这里的大小是所切分的子块数除以 64。

再往下看，代码 if ((maxNumElems & 63) != 0)用于判断 maxNumElems 也就是当前配置所切分的子块是不是 64 的倍数，如果不是，则 bitmapLength 加 1，最后通过循环，将对应位置的子块标记为 0。

这里详细分析一下 bitmap，它是个 long 类型的数组，long 数组中的每一个值，也都是 long 类型的数字，其中每一个比特位都标记着 Page 中每一个子块的内存是否已分配，如果比特位是 1，表示该子块已分配；如果比特位是 0，表示该子块未分配，标记顺序是其二进制数从低位到高位进行排列。我们应该知道为什么 bitmap 大小要设置为子块数量除以 64，因为 long 类型的数字是 64 位，每一个元素都能记录 64 个子块的数量，这样就可以通过 SubPage 个数除以 64 的方式决定 bitmap 中元素的数量。如果子块不能整除 64，则通过元素数量+1 的方式，除以 64 之后剩余的子块通过 long 中比特位由低到高进行排列记录，其逻辑结构如下图所示。

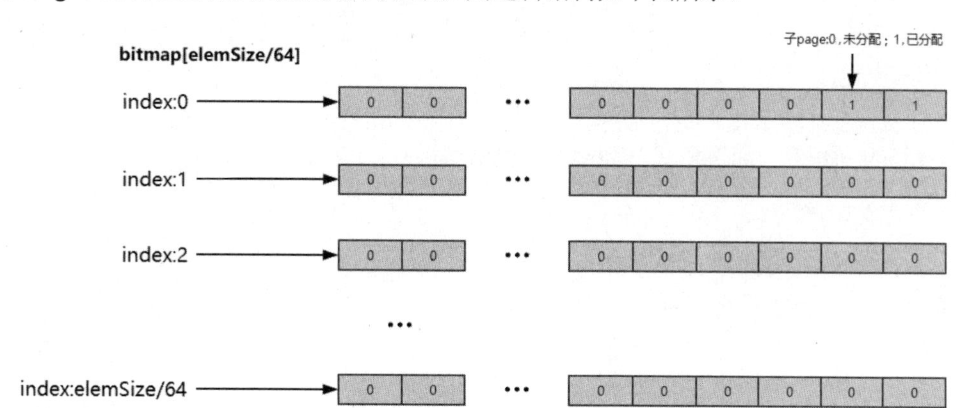

PoolSubpage 的 addToPool(head)方法的代码如下。

```
private void addToPool(PoolSubpage<T> head) {
    assert prev == null && next == null;
    prev = head;
    next = head.next;
    next.prev = this;
    head.next = this;
}
```

上面代码里的 Head 是 Arena 中数组 tinySubpagePools 中的元素，通过以上逻辑，就会将新创建的 SubPage 以双向链表的形式关联到 tinySubpagePools 中的元素，我们以 16 字节为例，关联关系如下图所示。

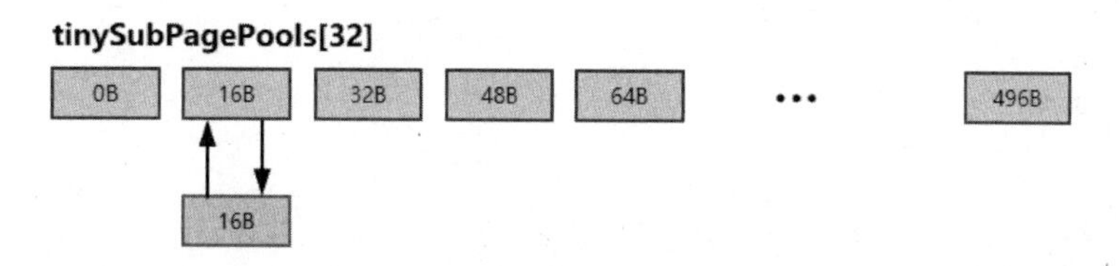

这样，下次如果还需要分配 16 字节的内存，就可以通过 tinySubpagePools 找到其元素关联的 SubPage 进行分配了。我们再回到 PoolChunk 的 allocateSubpage()方法。

```
private long allocateSubpage(int normCapacity) {
    PoolSubpage<T> head = arena.findSubpagePoolHead(normCapacity);
    synchronized (head) {
        //省略部分逻辑
            …

        //获取初始化的 SubPage
```

```
        final PoolSubpage<T>[] subpages = this.subpages;
        final int pageSize = this.pageSize;

        freeBytes -= pageSize;
        //表示第几个 subpageIdx
        int subpageIdx = subpageIdx(id);
        PoolSubpage<T> subpage = subpages[subpageIdx];
        …
            //省略部分逻辑
            …
        //取出一个子 SubPage
        return subpage.allocate();
    }
}
```

创建完一个 SubPage，就可以通过 subpage.allocate()方法进行内存分配了。我们看 allocate()方法，代码如下。

```
long allocate() {
    if (elemSize == 0) {
        return toHandle(0);
    }

    if (numAvail == 0 || !doNotDestroy) {
        return -1;
    }
    //取一个 bitmap 中可用的 id（绝对 id）
    final int bitmapIdx = getNextAvail();
    //除以 64（bitmap 的相对下标）
    int q = bitmapIdx >>> 6;
    //除以 64 取余，其实就是当前绝对 id 的偏移量
    int r = bitmapIdx & 63;
    assert (bitmap[q] >>> r & 1) == 0;

    //当前位标记为 1
    bitmap[q] |= 1L << r;
    //如果可用的 SubP 为 0
    //可用的 SubPage-1
    if (-- numAvail == 0) {
        //则移除相关 SubPage
        removeFromPool();
    }
    //bitmapIdx 转换成 handle
    return toHandle(bitmapIdx);
}
```

上述代码的逻辑看起来比较复杂，笔者带大家一点点来剖析，来看下面的代码。

```
final int bitmapIdx = getNextAvail();
```

其中 bitmapIdx 表示从 bitmap 中找到一个可用的比特位的下标，注意，这里是比特位的下标，并不是数组的下标，我们之前分析过，由于每一比特位都代表一个子块的内存分配情况，通过这个下标就可以知道哪个比特位是未分配状态，继续看 getNextAvail()分层的代码。

```
private int getNextAvail() {
    //nextAvail=0
    int nextAvail = this.nextAvail;
    if (nextAvail >= 0) {
        //一个子 SubPage 被释放之后，会记录当前 SubPage 的 bitmapIdx 的位置，下次分配可以直接通过
bitmapIdx 获取一个 SubPage
        this.nextAvail = -1;
        return nextAvail;
    }
    return findNextAvail();
}
```

上述代码片段中的 nextAvail 表示下一个可用的 bitmapIdx，在释放的时候会被标记，标记被释放的子块对应 bitmapIdx 的下标，如果<0 代表没有被释放的子块，则通过 findNextAvail()方法进行查找，继续看 findNextAvail()方法。

```
private int findNextAvail() {
    //当前 long 数组
    final long[] bitmap = this.bitmap;
    //获取其长度
    final int bitmapLength = this.bitmapLength;
    for (int i = 0; i < bitmapLength; i ++) {
        //第 i 个
        long bits = bitmap[i];
        //!=-1 说明 64 位没有全部占满
        if (~bits != 0) {
            //找下一个节点
            return findNextAvail0(i, bits);
        }
    }
    return -1;
}
```

这里会遍历 bitmap 中的每一个元素，如果当前元素中所有的比特位没有全部标记被使用，则通过 findNextAvail0(i, bits)方法一个一个地往后找标记未使用的比特位。继续看 findNextAvail0()，代码如下。

```
private int findNextAvail0(int i, long bits) {
    //多少份
    final int maxNumElems = this.maxNumElems;
```

```
    //乘以64，代表当前long的第一个下标
    final int baseVal = i << 6;
    //循环64次（指代当前的下标）
    for (int j = 0; j < 64; j ++) {
        //第一位为0（如果是2的倍数，则第一位就是0）
        if ((bits & 1) == 0) {
            //这里相当于加，将i×64之后加上j，获取绝对下标
            int val = baseVal | j;
            //小于块数（不能越界）
            if (val < maxNumElems) {
                return val;
            } else {
                break;
            }
        }
        //当前下标不为0
        //右移一位
        bits >>>= 1;
    }
    return -1;
}
```

上述代码从当前元素的第一个比特位开始找，直到找到一个标记为 0 的比特位，并返回当前比特位的下标，大致流程如下图所示。

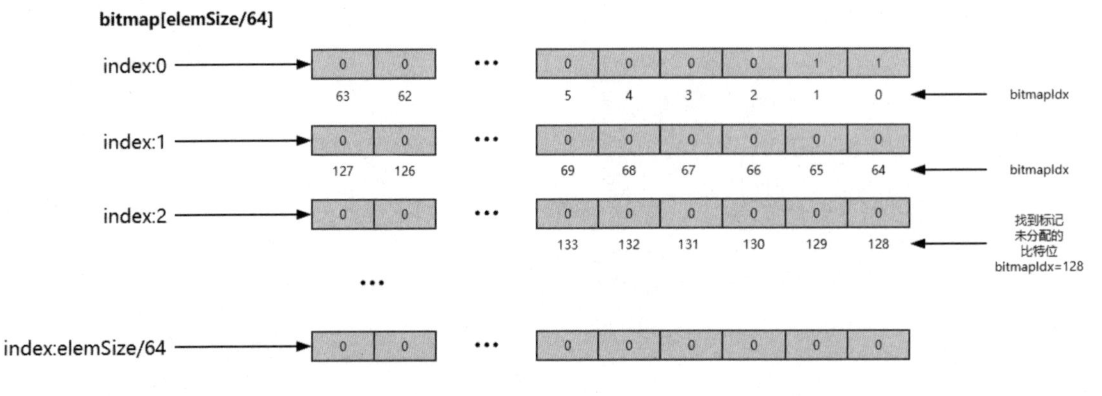

我们回到 allocate()方法中，代码如下。

```
long allocate() {
    if (elemSize == 0) {
        return toHandle(0);
    }

    if (numAvail == 0 || !doNotDestroy) {
        return -1;
```

```
    }
    //取一个 bitmap 中可用的 id（绝对 id）
    final int bitmapIdx = getNextAvail();
    //除以 64（bitmap 的相对下标）
    int q = bitmapIdx >>> 6;
    //除以 64 取余，其实就是当前绝对 id 的偏移量
    int r = bitmapIdx & 63;
    assert (bitmap[q] >>> r & 1) == 0;

    //当前位标记为 1
    bitmap[q] |= 1L << r;
    //如果可用的 SubPage 为 0
    //可用的 SubPage-1
    if (-- numAvail == 0) {
        //则移除相关 SubPage
        removeFromPool();
    }
    //bitmapIdx 转换成 handle
    return toHandle(bitmapIdx);
}
```

找到可用的 bitmapIdx 之后，首先通过 int q = bitmapIdx >>> 6 获取 bitmap 中 bitmapIdx 所属元素的数组下标。int r = bitmapIdx & 63 表示获取 bitmapIdx 的位置是从当前元素最低位开始的第几个比特位。bitmap[q] |= 1L << r 将 bitmap 的位置设置为不可用，也就是比特位设置为 1，表示已占用。然后将可用于配置的数量 numAvail 减 1。如果没有可用 SubPage 的数量，则会将 PoolArena 中的数组 tinySubpagePools 所关联的 SubPage 进行移除。最后通过 toHandle(bitmapIdx)获取当前子块的 Handle，上一小节我们知道 Handle 指向的是当前 Chunk 中唯一的一块内存，我们跟进 toHandle(bitmapIdx)，代码如下。

```
private long toHandle(int bitmapIdx) {
    return 0x4000000000000000L | (long) bitmapIdx << 32 | memoryMapIdx;
}
```

(long) bitmapIdx << 32 将 bitmapIdx 右移 32 位，而 32 位正好是一个 int 的长度，这样通过(long) bitmapIdx << 32 | memoryMapIdx 计算，就可以将 memoryMapIdx，也就是 Page 所属下标的二进制数保存在(long) bitmapIdx << 32 的低 32 位中。0x4000000000000000L 是一个最高位是 1 并且所有低位都是 0 的二进制数，通过按位或的方式可以将(long) bitmapIdx << 32 | memoryMapIdx 计算出来的结果保存在 0x4000000000000000L 的所有低位中，这样返回对的数字就可以指向 Chunk 中唯一的一块内存，回到 PoolArena 的 allocateNormal()方法中，代码如下。

```
private synchronized void allocateNormal(PooledByteBuf<T> buf, int reqCapacity, int
normCapacity) {
```

```
    //省略部分逻辑
    …

    //创建 Chunk 进行内存分配(2)
    PoolChunk<T> c = newChunk(pageSize, maxOrder, pageShifts, chunkSize);
    long handle = c.allocate(normCapacity);
    ++allocationsNormal;
    assert handle > 0;
    //初始化 ByteBuf(3)
    c.initBuf(buf, handle, reqCapacity);
    qInit.add(c);
}
```

分析完 long handle = c.allocate(normCapacity)，其中返回的 Handle 指向 Chunk 中某个 Page 的某个子块所对应的连续内存。通过 initBuf()方法初始化后，创建的 Chunk 被加到 ChunkList，跟到 initBuf()方法，代码如下。

```
void initBuf(PooledByteBuf<T> buf, long handle, int reqCapacity) {
    int memoryMapIdx = memoryMapIdx(handle);
    //bitmapIdx 在后面分配 SubPage 的时候使用到
    int bitmapIdx = bitmapIdx(handle);
    if (bitmapIdx == 0) {
        byte val = value(memoryMapIdx);
        assert val == unusable : String.valueOf(val);
        //runOffset(memoryMapIdx)：偏移量
        //runLength(memoryMapIdx)：当前节点的长度
        buf.init(this, handle, runOffset(memoryMapIdx), reqCapacity, runLength(memoryMapIdx),
                arena.parent.threadCache());
    } else {
        initBufWithSubpage(buf, handle, bitmapIdx, reqCapacity);
    }
}
```

这部分在前面已经剖析过，相信大家不会陌生，区别在于 if(bitmapIdx == 0)的判断，这里的 bitmapIdx 不会是 0，这样才会执行到 initBufWithSubpage(buf, handle, bitmapIdx, reqCapacity)方法中，跟进 initBufWithSubpage()方法，代码如下。

```
private void initBufWithSubpage(PooledByteBuf<T> buf, long handle, int bitmapIdx, int
reqCapacity) {
    assert bitmapIdx != 0;
    int memoryMapIdx = memoryMapIdx(handle);
    PoolSubpage<T> subpage = subpages[subpageIdx(memoryMapIdx)];
    assert subpage.doNotDestroy;
    assert reqCapacity <= subpage.elemSize;
    buf.init(
        this, handle,
```

```
        runOffset(memoryMapIdx) + (bitmapIdx & 0x3FFFFFFF) * subpage.elemSize, reqCapacity,
subpage.elemSize,
        arena.parent.threadCache());
}
```

首先获得 memoryMapIdx，这里会将之前计算的 Handle 传入，代码如下。

```
private static int memoryMapIdx(long handle) {
    return (int) handle;
}
```

这里将其强制转化为 int 类型，也就是去掉高 32 位，得到 memoryMapIdx，回到 initBufWithSubpage() 方法中。我们注意到 buf 调用 init() 方法中的一个参数：runOffset(memoryMapIdx) + (bitmapIdx & 0x3FFFFFFF) * subpage.elemSize，这里的偏移量就是原来 Page 的偏移量+子块的偏移量：bitmapIdx & 0x3FFFFFFF，代表当前分配的 SubPage 属于第几个 SubPage。(bitmapIdx & 0x3FFFFFFF) * subpage.elemSize 表示在当前 Page 的偏移量。这样，分配的 ByteBuf 在内存读写的时候，就会根据偏移量进行读写。跟到 init()方法中，代码如下。

```
void init(PoolChunk<T> chunk, long handle, int offset, int length, int maxLength,
PoolThreadCache cache) {

    //初始化
    assert handle >= 0;
    assert chunk != null;

    //在哪一块内存上进行分配
    this.chunk = chunk;

    //这一块内存上的哪一块连续内存
    this.handle = handle;
    memory = chunk.memory;

    //偏移量
    this.offset = offset;
    this.length = length;
    this.maxLength = maxLength;
    tmpNioBuf = null;
    this.cache = cache;
}
```

上面代码也是我们熟悉的逻辑，初始化属性之后，一个缓冲区分配完成。以上就是 SubPage 级别的缓冲区分配逻辑。

11.4.7 内存池 ByteBuf 的内存回收

在 11.1.3 节中我们提到，堆外内存是不受 JVM 垃圾回收机制控制的，所以我们分配一块堆外内存进行 ByteBuf 操作时，使用完毕要对对象进行回收，本节就以 PooledUnsafeDirectByteBuf 为例讲解有关内存分配的相关逻辑。PooledUnsafeDirectByteBuf 中内存释放的入口方法是其父类 AbstractReferenceCountedByteBuf 中的 release()方法，代码如下。

```
@Override
public boolean release() {
    return release0(1);
}
```

这里调用了 release0()方法，代码如下。

```
private boolean release0(int decrement) {
   for (;;) {
       int refCnt = this.refCnt;
       if (refCnt < decrement) {
           throw new IllegalReferenceCountException(refCnt, -decrement);
       }
       if (refCntUpdater.compareAndSet(this, refCnt, refCnt - decrement)) {
           if (refCnt == decrement) {
               deallocate();
               return true;
           }
           return false;
       }
   }
}
```

if(refCnt == decrement)中判断当前 ByteBuf 是否没有被引用，如果没有被引用，则通过 deallocate()方法进行释放。因为我们以 PooledUnsafeDirectByteBuf 为例，所以这里会调用其父类 PooledByteBuf 的 deallocate()方法，代码如下。

```
protected final void deallocate() {
   if (handle >= 0) {
       final long handle = this.handle;
           //表示当前的 ByteBuf 不再指向任何一块内存
       this.handle = -1;
           //这里将 memory 也设置为 null
       memory = null;
           //这一步将 ByteBuf 的内存进行释放
       chunk.arena.free(chunk, handle, maxLength, cache);
           //将对象放入对象回收站，循环利用
       recycle();
```

```
    }
}
```

首先分析 free()方法，代码如下。

```
void free(PoolChunk<T> chunk, long handle, int normCapacity, PoolThreadCache cache) {
    //是否为 Unpooled
    if (chunk.unpooled) {
        int size = chunk.chunkSize();
        destroyChunk(chunk);
        activeBytesHuge.add(-size);
        deallocationsHuge.increment();
    } else {
        //哪种级别的 Size
        SizeClass sizeClass = sizeClass(normCapacity);
        //加到缓存里
        if (cache != null && cache.add(this, chunk, handle, normCapacity, sizeClass)) {
            return;
        }
        //将缓存对象标记为未使用
        freeChunk(chunk, handle, sizeClass);
    }
}
```

先判断是不是 Unpooled，因为这里是 Pooled，所以会执行到 else 块中。

- sizeClass(normCapacity)计算是哪种级别的 Size，我们按照 tiny 级别进行分析。
- cache.add(this, chunk, handle, normCapacity, sizeClass)是将当前 ByteBuf 进行缓存。

之前讲过，在分配 ByteBuf 时首先在缓存上分配，而这一步就是将其缓存的过程，继续跟进，代码如下。

```
boolean add(PoolArena<?> area, PoolChunk chunk, long handle, int normCapacity, SizeClass
sizeClass) {
    //获取 MemoryRegionCache 节点
    MemoryRegionCache<?> cache = cache(area, normCapacity, sizeClass);
        …
    return cache.add(chunk, handle);
}
```

上述代码中根据类型获得相关类型缓存节点，这里会根据不同的内存规格去找不同的对象，我们简单回顾一下，每个缓存对象都包含一个 Queue，Queue 中每个节点都是 Entry，每一个 Entry 中都包含一个 Chunk 和 Handle，可以指向唯一连续的内存，我们看 Cache 的代码。

```
private MemoryRegionCache<?> cache(PoolArena<?> area, int normCapacity, SizeClass sizeClass)
{
    switch (sizeClass) {
```

```
    case Normal:
        return cacheForNormal(area, normCapacity);
    case Small:
        return cacheForSmall(area, normCapacity);
    case Tiny:
        return cacheForTiny(area, normCapacity);
    default:
        throw new Error();
    }
}
```

假设是 tiny 类型，就会执行到 cacheForTiny(area, normCapacity)方法中，代码如下。

```
private MemoryRegionCache<?> cacheForTiny(PoolArena<?> area, int normCapacity) {
    int idx = PoolArena.tinyIdx(normCapacity);
    if (area.isDirect()) {
        return cache(tinySubPageDirectCaches, idx);
    }
    return cache(tinySubPageHeapCaches, idx);
}
```

这个方法我们之前剖析过，就是根据大小找到对应的缓存，获得下标之后，通过 Cache 找相对应的缓存对象，代码如下。

```
private static <T>  MemoryRegionCache<T> cache(MemoryRegionCache<T>[] cache, int idx) {
    if (cache == null || idx > cache.length - 1) {
        return null;
    }
    return cache[idx];
}
```

可以看到，上面代码是直接通过下标获取的缓存对象，回到 add()方法，代码如下。

```
boolean add(PoolArena<?> area, PoolChunk chunk, long handle, int normCapacity, SizeClass
sizeClass) {
    //获取 MemoryRegionCache 节点
    MemoryRegionCache<?> cache = cache(area, normCapacity, sizeClass);
    if (cache == null) {
        return false;
    }
    //将 Chunk 和 Handle 封装成实体加到 Queue 里面
    return cache.add(chunk, handle);
}
```

这里的 Cache 对象调用了一个 add()方法，这个方法就是将 Chunk 和 Handle 封装成一个 Entry 加到 Queue 里面，add()方法的代码如下。

```
public final boolean add(PoolChunk<T> chunk, long handle) {
    Entry<T> entry = newEntry(chunk, handle);
    boolean queued = queue.offer(entry);
    if (!queued) {
        entry.recycle();
    }
    return queued;
}
```

我们之前介绍过，在缓存中分配的时候从 Queue 弹出一个 Entry，会放到一个对象池里面，而 Entry<T> entry = newEntry(chunk, handle)就是从对象池里取一个 Entry 对象，然后将 Chunk 和 Handle 赋值，通过 queue.offer(entry)加到 Queue，回到 free()方法，代码如下。

```
void free(PoolChunk<T> chunk, long handle, int normCapacity, PoolThreadCache cache) {
    //是否为 Unpooled
    if (chunk.unpooled) {
        int size = chunk.chunkSize();
        destroyChunk(chunk);
        activeBytesHuge.add(-size);
        deallocationsHuge.increment();
    } else {
        //哪种级别的 Size
        SizeClass sizeClass = sizeClass(normCapacity);
        //加到缓存里
        if (cache != null && cache.add(this, chunk, handle, normCapacity, sizeClass)) {
            return;
        }
        freeChunk(chunk, handle, sizeClass);
    }
}
```

这里加到缓存里后，如果成功，就会 return，如果不成功，就会调用 freeChunk(chunk, handle, sizeClass)方法。这个方法的意义是将原先给 ByteBuf 分配的内存区段标记为未使用，跟进 freeChunk()方法。

```
void freeChunk(PoolChunk<T> chunk, long handle, SizeClass sizeClass) {
    final boolean destroyChunk;
    synchronized (this) {
        switch (sizeClass) {
        case Normal:
            ++deallocationsNormal;
            break;
        case Small:
            ++deallocationsSmall;
            break;
        case Tiny:
```

```
            ++deallocationsTiny;
            break;
        default:
            throw new Error();
        }
        destroyChunk = !chunk.parent.free(chunk, handle);
    }
    if (destroyChunk) {
        destroyChunk(chunk);
    }
}
```

再看 free()方法，代码如下。

```
boolean free(PoolChunk<T> chunk, long handle) {
    chunk.free(handle);
    if (chunk.usage() < minUsage) {
        remove(chunk);
        return move0(chunk);
    }
    return true;
}
```

chunk.free(handle)的意思是通过 Chunk 释放一段连续的内存，再看 free()方法的代码。

```
void free(long handle) {
    int memoryMapIdx = memoryMapIdx(handle);
    int bitmapIdx = bitmapIdx(handle);

    if (bitmapIdx != 0) {
        PoolSubpage<T> subpage = subpages[subpageIdx(memoryMapIdx)];
        assert subpage != null && subpage.doNotDestroy;
        PoolSubpage<T> head = arena.findSubpagePoolHead(subpage.elemSize);
        synchronized (head) {
            if (subpage.free(head, bitmapIdx & 0x3FFFFFFF)) {
                return;
            }
        }
    }
    freeBytes += runLength(memoryMapIdx);
    setValue(memoryMapIdx, depth(memoryMapIdx));
    updateParentsFree(memoryMapIdx);
}
```

if (bitmapIdx!= 0)判断当前缓冲区分配的级别是 Page 还是 SubPage，如果是 SubPage，则会找到相关的 SubPage 将其位图标记为 0，如果不是 SubPage，则通过分配内存的反向标记，将该内存标记为未使用。这段逻辑大家可以自行分析，如果之前分配相关的知识掌握扎实的话，这里的逻辑也不是很难。回到 PooledByteBuf 的 deallocate()方法。

```
protected final void deallocate() {
    if (handle >= 0) {
        final long handle = this.handle;
        this.handle = -1;
        memory = null;
        chunk.arena.free(chunk, handle, maxLength, cache);
        recycle();
    }
}
```

最后，通过 recycle()将释放的 ByteBuf 放入对象回收站，有关对象回收站的知识，会在以后的章节进行剖析。以上就是内存回收的大概逻辑。

11.4.8　SocketChannel 读取 ByteBuf 的过程

本节知识和之前分析过的很多知识有关联，我们不再重复介绍。因此，学习本节之前，可以回顾一下前面的几章内容，如客户端接入、客户端发送数据、服务端读取数据的流程等。首先看 NioEventLoop 的 processSelectedKey()方法。

```
private void processSelectedKey(SelectionKey k, AbstractNioChannel ch) {
    //获取 Channel 中的 Unsafe
    final AbstractNioChannel.NioUnsafe unsafe = ch.unsafe();
    //如果这个 Key 不是合法的，说明这个 Channel 可能有问题
    if (!k.isValid()) {
        //代码省略
    }
    try {
        //如果是合法的，则获得 Key 的 I/O 事件
        int readyOps = k.readyOps();
        //链接事件
        if ((readyOps & SelectionKey.OP_CONNECT) != 0) {
            int ops = k.interestOps();
            ops &= ~SelectionKey.OP_CONNECT;
            k.interestOps(ops);
            unsafe.finishConnect();
        }
        //写事件
        if ((readyOps & SelectionKey.OP_WRITE) != 0) {
            ch.unsafe().forceFlush();
        }
        //读事件和接受链接事件
        //如果当前 NioEventLoop 是 Worker 线程的话，这里就是 Op_Read 事件
        //如果当前 NioEventLoop 是 Boss 线程的话，这里就是 Op_Accept 事件
```

```
        if ((readyOps & (SelectionKey.OP_READ | SelectionKey.OP_ACCEPT)) != 0 || readyOps ==
0) {
            unsafe.read();
            if (!ch.isOpen()) {
                return;
            }
        }
    } catch (CancelledKeyException ignored) {
        unsafe.close(unsafe.voidPromise());
    }
}
```

if ((readyOps & (SelectionKey.OP_READ | SelectionKey.OP_ACCEPT)) != 0 || readyOps == 0)这里的判断表示轮询到大事件是 OP_READ 或者 OP_ACCEPT 事件。之前我们分析过，如果当前 NioEventLoop 是 Worker 线程的话，那么就是 OP_READ 事件，也就是读事件，表示客户端发来了数据流，会调用 Unsafe 的 read()方法进行读取。如果是 Worker 线程，那么这里的 Channel 是 NioServerSocketChannel，其绑定的 Unsafe 是 NioByteUnsafe，会调用 NioByteUnsafe 的 read()方法，代码如下。

```
public final void read() {
        final ChannelConfig config = config();
        final ChannelPipeline pipeline = pipeline();
        final ByteBufAllocator allocator = config.getAllocator();
        final RecvByteBufAllocator.Handle allocHandle = recvBufAllocHandle();
        allocHandle.reset(config);

        ByteBuf byteBuf = null;
        boolean close = false;
        try {
            do {
                byteBuf = allocHandle.allocate(allocator);
                allocHandle.lastBytesRead(doReadBytes(byteBuf));
                if (allocHandle.lastBytesRead() <= 0) {
                    byteBuf.release();
                    byteBuf = null;
                    close = allocHandle.lastBytesRead() < 0;
                    break;
                }

                allocHandle.incMessagesRead(1);
                readPending = false;
                pipeline.fireChannelRead(byteBuf);
                byteBuf = null;
            } while (allocHandle.continueReading());
```

```
            allocHandle.readComplete();
            pipeline.fireChannelReadComplete();

            if (close) {
                closeOnRead(pipeline);
            }
        } catch (Throwable t) {
            handleReadException(pipeline, byteBuf, t, close, allocHandle);
        } finally {
            if (!readPending && !config.isAutoRead()) {
                removeReadOp();
            }
        }
    }
}
```

首先获取 SocketChannel 的 config、pipeline 等相关属性，final ByteBufAllocator allocator = config.getAllocator()这一步是获取一个 ByteBuf 的内存分配器，用于分配 ByteBuf。这里会调用 DefaultChannelConfig 的 getAllocator()方法。

```
public ByteBufAllocator getAllocator() {
    return allocator;
}
```

这里返回 DefaultChannelConfig 的成员变量，我们看这个成员变量，代码如下。

```
private volatile ByteBufAllocator allocator = ByteBufAllocator.DEFAULT;
```

这里调用 ByteBufAllocator 的属性 DEFAULT，代码如下。

```
ByteBufAllocator DEFAULT = ByteBufUtil.DEFAULT_ALLOCATOR;
```

这里又调用了 ByteBufUtil 的静态属性 DEFAULT_ALLOCATOR，代码如下。

```
static final ByteBufAllocator DEFAULT_ALLOCATOR;
```

DEFAULT_ALLOCATOR 属性是在 static 块中初始化的，static 块的代码如下。

```
static {
    String allocType = SystemPropertyUtil.get(
            "io.netty.allocator.type", PlatformDependent.isAndroid() ? "unpooled" : "pooled");
    allocType = allocType.toLowerCase(Locale.US).trim();

    ByteBufAllocator alloc;
    if ("unpooled".equals(allocType)) {
        alloc = UnpooledByteBufAllocator.DEFAULT;
        logger.debug("-Dio.netty.allocator.type: {}", allocType);
    } else if ("pooled".equals(allocType)) {
```

```
        alloc = PooledByteBufAllocator.DEFAULT;
        logger.debug("-Dio.netty.allocator.type: {}", allocType);
    } else {
        alloc = PooledByteBufAllocator.DEFAULT;
        logger.debug("-Dio.netty.allocator.type: pooled (unknown: {})", allocType);
    }
    DEFAULT_ALLOCATOR = alloc;
    //代码省略
}
```

上述代码中，首先判断运行环境是不是 Android，如果不是 Android，返回“pooled”字符串保存在 AllocType 中，然后通过 if 判断局部变量 alloc = PooledByteBufAllocator.DEFAULT，最后将 alloc 赋值到成员变量 DEFAULT_ALLOCATOR，我们看 PooledByteBufAllocator 的 DEFAULT 属性，代码如下。

```
public static final PooledByteBufAllocator DEFAULT =
        new PooledByteBufAllocator(PlatformDependent.directBufferPreferred());
```

我们看到，这段代码通过 new 的方式创建了一个 PooledByteBufAllocator 对象，也就是基于申请一块连续内存进行缓冲区分配的缓冲区分配器。缓冲区分配器的知识，前面的章节已经进行过详细的剖析，这里不再赘述。回到 NioByteUnsafe 的 read()方法。

```
public final void read() {
        final ChannelConfig config = config();
        final ChannelPipeline pipeline = pipeline();
        final ByteBufAllocator allocator = config.getAllocator();
        final RecvByteBufAllocator.Handle allocHandle = recvBufAllocHandle();
        allocHandle.reset(config);

        ByteBuf byteBuf = null;
        boolean close = false;
        try {
            do {
                byteBuf = allocHandle.allocate(allocator);
                allocHandle.lastBytesRead(doReadBytes(byteBuf));
                if (allocHandle.lastBytesRead() <= 0) {
                    byteBuf.release();
                    byteBuf = null;
                    close = allocHandle.lastBytesRead() < 0;
                    break;
                }

                allocHandle.incMessagesRead(1);
                readPending = false;
                pipeline.fireChannelRead(byteBuf);
                byteBuf = null;
```

```
            } while (allocHandle.continueReading());

            allocHandle.readComplete();
            pipeline.fireChannelReadComplete();

            if (close) {
                closeOnRead(pipeline);
            }
        } catch (Throwable t) {
            handleReadException(pipeline, byteBuf, t, close, allocHandle);
        } finally {
            if (!readPending && !config.isAutoRead()) {
                removeReadOp();
            }
        }
    }
}
```

上面代码中，ByteBufAllocator allocator = config.getAllocator() 中的 allocator 就是 PooledByteBufAllocator。

final RecvByteBufAllocator.Handle allocHandle = recvBufAllocHandle()创建一个 Handle，之前章节讲过，Handle 是对 RecvByteBufAllocator 进行实际操作的对象，recvBufAllocHandle 的代码如下。

```
public RecvByteBufAllocator.Handle recvBufAllocHandle() {
    //如果不存在，则创建一个 Handle 的实例
    if (recvHandle == null) {
        recvHandle = config().getRecvByteBufAllocator().newHandle();
    }
    return recvHandle;
}
```

这是我们之前剖析过的逻辑，如果不存在，则创建 Handle 的实例，具体创建过程可以回顾前面的章节，这里就不再赘述。同样 allocHandle.reset(config)是将配置重置，前面章节也对其进行过剖析。重置完配置之后，进行 do-while 循环，有关循环终止条件 allocHandle.continueReading()，之前小节也有过详细剖析，这里也不再赘述。在 do-while 循环中，首先看 byteBuf = allocHandle.allocate(allocator) 这一步，这里传入了之前创建的 allocate 对象，也就是 PooledByteBufAllocator，这里会调用 DefaultMaxMessagesRecvByteBufAllocator 类的 allocate()方法。

```
public ByteBuf allocate(ByteBufAllocator alloc) {
    return alloc.ioBuffer(guess());
}
```

这里的 guess()方法，会调用 AdaptiveRecvByteBufAllocator 的 guess()方法，代码如下。

```
public int guess() {
    return nextReceiveBufferSize;
}
```

这里会返回 AdaptiveRecvByteBufAllocator 的成员变量 nextReceiveBufferSize，也就是下次所分配缓冲区的大小，根据之前学习的内容，第一次分配的时候会分配初始大小，也就是 1 024 字节。这样，alloc.ioBuffer(guess())就会分配一个 PooledByteBuf，我们看 AbstractByteBufAllocator 的 ioBuffer()方法的代码。

```
public ByteBuf ioBuffer(int initialCapacity) {
    if (PlatformDependent.hasUnsafe()) {
        return directBuffer(initialCapacity);
    }
    return heapBuffer(initialCapacity);
}
```

这里首先判断是否能获取 JDK 的 Unsafe 对象，默认为 true，所以会执行到 directBuffer(initialCapacity)中，最终会分配一个 PooledUnsafeDirectByteBuf 对象，具体分配流程我们在之前小节做过详细剖析。回到 NioByteUnsafe 的 read()方法，分配完 ByteBuf 之后，再看这一步 allocHandle.lastBytesRead(doReadBytes(byteBuf))。

首先看参数 doReadBytes(byteBuf)方法，这步是将 Channel 中的数据读取到刚分配的 ByteBuf 中，并返回读取的字节数，这里会调用 NioSocketChannel 的 doReadBytes()方法，代码如下。

```
protected int doReadBytes(ByteBuf byteBuf) throws Exception {
    final RecvByteBufAllocator.Handle allocHandle = unsafe().recvBufAllocHandle();
    allocHandle.attemptedBytesRead(byteBuf.writableBytes());
    return byteBuf.writeBytes(javaChannel(), allocHandle.attemptedBytesRead());
}
```

首先获得绑定在 Channel 中的 Handle，因为已经创建了 Handle，所以这里会直接获得。再看 allocHandle.attemptedBytesRead(byteBuf.writableBytes())这步，byteBuf.writableBytes()返回 ByteBuf 的可写字节数，也就是最多能从 Channel 中读取多少字节写入 ByteBuf，allocate 的 attemptedBytesRead 会把可写字节数设置到 DefaultMaxMessagesRecvByteBufAllocator 类的 attemptedBytesRead 属性中，跟到 DefaultMaxMessagesRecvByteBufAllocator 中的 attemptedBytesRead。

```
public void attemptedBytesRead(int bytes) {
    attemptedBytesRead = bytes;
}
```

继续看 doReadBytes()方法。往下看到最后，通过 byteBuf.writeBytes(javaChannel()， allocHandle.attemptedBytesRead())将 JDK 底层的 Channel 中的数据写入创建的 ByteBuf，并返回实际写入的字节数。回到 NioByteUnsafe 的 read() 方法中，继续看 allocHandle.lastBytesRead(doReadBytes(byteBuf))这步，前面剖析过 doReadBytes(byteBuf)返回的是实际写入 ByteBuf 的字节数，再看 lastBytesRead()方法，DefaultMaxMessagesRecvByteBufAllocator 的 lastBytesRead()方法的代码如下。

```
public final void lastBytesRead(int bytes) {
    lastBytesRead = bytes;
    totalBytesRead += bytes;
    if (totalBytesRead < 0) {
        totalBytesRead = Integer.MAX_VALUE;
    }
}
```

这里会赋值两个属性，lastBytesRead 代表最后读取的字节数，赋值为写入 ByteBuf 的字节数，totalBytesRead 表示总共读取的字节数，这里将写入的字节数追加。继续看 NioByteUnsafe 的 read()方法，如果最后一次读取数据为 0，说明已经将 Channel 中的数据全部读取完毕，将新创建的 ByteBuf 释放循环利用，并跳出循环。allocHandle.incMessagesRead(1)这步是增加消息的读取次数，因为最多循环 16 次，所以当消息次数增加到 16 时会结束循环。读取完毕之后，会通过 pipeline.fireChannelRead(byteBuf)传递 ChannelRead 事件。有关 ChannelRead 事件，我们在第 9 章也进行了详细的剖析。

至此，小伙伴们应该有个疑问，如果一次读取不完，就传递 ChannelRead 事件，那么 Server 接收到的数据有可能就是不完整的，其实关于这点，Netty 也做了相应的处理，我们会在之后的章节详细剖析 Netty 的半包处理机制。循环结束后，会执行到 allocHandle.readComplete()这一步。

我们知道第一次分配 ByteBuf 的初始容量是 1 024 字节，但是初始容量不一定满足所有的业务场景。Netty 中，将每次读取数据的字节数进行记录，然后之后次分配 ByteBuf 的时候，容量会尽可能地符合业务场景所需要的大小，具体实现方式就是在 readComplete()这一步体现的。跟进 AdaptiveRecvByteBufAllocator 的 readComplete()方法。

```
public void readComplete() {
    record(totalBytesRead());
}
```

这里调用了 record()方法，并且传入了这一次所读取的字节总数，代码如下。

```
private void record(int actualReadBytes) {
    if (actualReadBytes <= SIZE_TABLE[Math.max(0, index - INDEX_DECREMENT - 1)]) {
        if (decreaseNow) {
            index = Math.max(index - INDEX_DECREMENT, minIndex);
```

```
            nextReceiveBufferSize = SIZE_TABLE[index];
            decreaseNow = false;
        } else {
            decreaseNow = true;
        }
    } else if (actualReadBytes >= nextReceiveBufferSize) {
        index = Math.min(index + INDEX_INCREMENT, maxIndex);
        nextReceiveBufferSize = SIZE_TABLE[index];
        decreaseNow = false;
    }
}
```

首先看判断条件 if (actualReadBytes <= SIZE_TABLE[Math.max(0, index - INDEX_DECREMENT - 1)])。这里 index 是当前分配的缓冲区大小所在的 SIZE_TABLE 中的索引，将这个索引进行缩进，然后根据缩进后的索引找出 SIZE_TABLE 中所存储的内存值，判断是否大于等于这次读取的最大字节数，如果条件成立，说明分配的内存过大，需要缩容操作，我们看 if 块中与缩容相关的逻辑。首先 if (decreaseNow) 会判断是否立刻进行缩容操作，通常第一次不会进行缩容操作，然后会将 decreaseNow 设置为 true，代表下一次直接进行缩容操作。假设需要立刻进行缩容操作，我们看缩容操作的相关逻辑：index = Math.max(index - INDEX_DECREMENT, minIndex) 这一步将索引缩进一步，但不能小于最小索引值；然后通过 nextReceiveBufferSize = SIZE_TABLE[index]获取设置索引之后的内存，赋值给 nextReceiveBufferSize，也就是下次需要分配的大小，下次就会根据这个大小分配 ByteBuf 了，这样就实现了缩容操作。

再看 else if (actualReadBytes >= nextReceiveBufferSize)，可以判断这次读取字节的总量比上次分配的大小还要大，则进行扩容操作。扩容操作也很简单，索引步进，获得步进后的索引所对应的内存值，作为下次所需要分配的大小。

在 NioByteUnsafe 的 read() 方法中，经过缩容或者扩容操作之后，通过 pipeline.fireChannelReadComplete()传播 ChannelReadComplete()事件。以上就是读取客户端消息的相关流程。

12

第 12 章 Netty 编解码的艺术

上一章有一个遗留问题，就是 Server 在读取客户端数据的时候，如果一次读取不完整，就触发 channelRead 事件，那么 Netty 是如何处理这类问题的，本章会对此做详细剖析。

12.1 什么是拆包、粘包

12.1.1 TCP 拆包、粘包

TCP 是一个“流”协议。所谓流，就是没有界限的一长串二进制数据。TCP 作为传输层协议，并不了解上层业务数据的具体含义，它会根据 TCP 缓冲区的实际情况进行数据包的划分，所以在业务上认为是一个完整包的，可能会被 TCP 拆分成多个包进行发送，也有可能把多个小的包封装成一个大的数据包发送，这就是所谓的 TCP 拆包和粘包问题。

同样，在 Netty 的编码器中，也会对半包和粘包问题做相应的处理。什么是半包，顾名思义，就是不完整的数据包，因为 Netty 在轮询读事件的时候，每次从 Channel 中读取的数据，不一定是一个完整的数据包，这种情况就叫作半包。粘包同样也不难理解，Client 往 Server 发送数据包时，如果发送频繁很有可能会将多个数据包的数据都发送到通道中，Server 在读取的时候可能会读取

到超过一个完整数据包的长度，这种情况叫作粘包。有关半包和粘包，如下图所示。

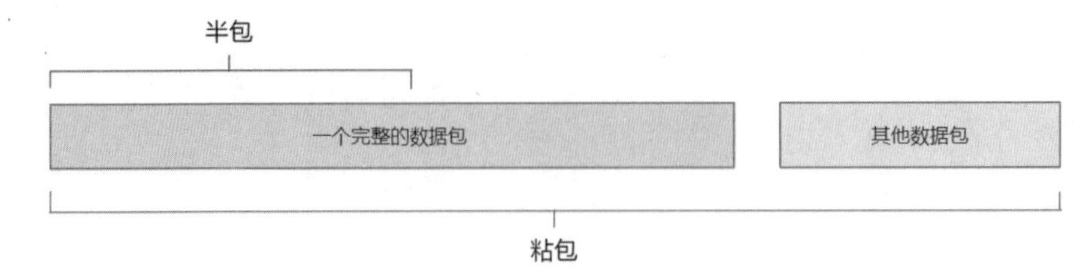

12.1.2 粘包问题的解决策略

由于底层的 TCP 无法理解上层的业务数据，所以在底层是无法保证数据包不被拆分和重组的，这个问题只能通过上层的应用协议栈设计来解决。目前业界主流协议的解决方案可以归纳如下。

（1）消息定长，报文长度固定，例如每个报文的长度固定为 200 字节，如果不够空位补空格。

（2）报尾添加特殊分隔符，例如每条报文结束都添加回车换行符（如 FTP）或者指定特殊字符作为报文分隔符，接收方通过特殊分隔符区分报文。

（3）将消息分为消息头和消息体，消息头包含表示信息的总长度（或者消息体长度）的属性。

（4）更复杂的自定义应用层协议。

Netty 对半包或者粘包的处理其实也很简单。通过之前的学习知道，每个 Handler 都是和 Channel 唯一绑定的，一个 Handler 只对应一个 Channel，所以 Channel 中的数据读取的时候经过解析，如果不是一个完整的数据包，则解析失败，将这个数据包进行保存，等下次解析时再和这个数据包进行组装解析，直到解析到完整的数据包，才会将数据包向下传递。

12.2 什么是编解码

12.2.1 编解码技术

通常我们习惯将编码（Encode）称为序列化（Serialization），它将对象序列化为字节数组，用于网络传输、数据持久化或者其他用途。反之，解码（Decode）/反序列化（Deserialization）把从网络、磁盘等读取的字节数组还原成原始对象（通常是原始对象的拷贝），以便后续的业务逻辑操作。进行远程跨进程服务调用时（例如 RPC），需要使用特定的编解码技术，对需要进行网络传输的对象做编码或者解码，以便完成远程调用。

12.2.2　Netty 为什么要提供编解码框架

作为一个高性能的 NIO 通信框架，编解码框架是 Netty 的重要组成部分。尽管站在微内核的角度看，编解码框架并不是 Netty 微内核的组成部分，但是通过 ChannelHandler 定制扩展出的编解码框架却是不可或缺的。

然而，在 Netty 中，从网络读取的 Inbound 消息，需要经过解码，将二进制数据报转换成应用层协议消息或者业务消息，才能够被上层的应用逻辑识别和处理；同理，用户发送到网络的 Outbound 业务消息，需要经过编码转换成二进制字节数组（对于 Netty 就是 ByteBuf）才能够发送到网络对端。编码和解码功能是 NIO 框架的有机组成部分，无论是由业务定制扩展实现，还是 NIO 框架内置编解码能力，该功能是必不可少的。

为了降低用户的开发难度，Netty 对常用的功能和 API 做了装饰，以屏蔽底层的实现细节。编解码功能的定制，对于熟悉 Netty 底层实现的开发者而言，直接基于 ChannelHandler 扩展开发，难度并不是很大。但是对于大多数初学者或者不愿意去了解底层实现细节的用户，需要给他们提供更简单的类库和 API，而不是 ChannelHandler。

Netty 在这方面做得非常出色，针对编解码功能，它既提供了通用的编解码框架供用户扩展，也提供了常用的编解码类库供用户直接使用。在保证定制扩展性的基础上，尽量降低用户的开发工作量和开发门槛，提升开发效率。

Netty 预置的编解码功能包括 Base64、Bytes、Compression、JSON、Marshalling、Protobuf、Serialization、XML 等，如下图所示。

- handler
  - codec
    - base64
    - bytes
    - compression
    - json
    - marshalling
    - protobuf
    - serialization
    - string
    - xml

12.3 Netty 中常用的解码器

Netty 默认提供了多个解码器，可以进行分包操作，满足 99%的编码需求。

12.3.1 ByteToMessageDecoder 抽象解码器

使用 NIO 进行网络编程时，往往需要将读取到的字节数组或者字节缓冲区解码为业务可以使用的 POJO 对象（Plain Ordinary Java Object，普通的 Java 对象）。因此 Netty 提供了 ByteToMessageDecoder 抽象解码类工具。

用户自定义解码器继承 ByteToMessageDecoder，只需要实现 void decode(ChannelHandler Context ctx, ByteBuf in, List<Object> out)抽象方法即可完成 ByteBuf 到 POJO 对象的解码。

由于 ByteToMessageDecoder 并没有考虑 TCP 粘包和拆包等场景，用户自定义解码器需要自己处理"读半包"问题。正因为如此，大多数场景不会直接继承 ByteToMessageDecoder，而是继承另外一些更高级的解码器来屏蔽半包的处理。

实际项目中，通常将 LengthFieldBasedFrameDecoder 和 ByteToMessageDecoder 组合使用，前者负责将网络读取的数据报解码为整包消息，后者负责将整包消息解码为最终的业务对象。除了和其他解码器组合形成新的解码器，ByteToMessageDecoder 也是很多基础解码器的父类，它的继承关系如下图所示。

- ByteToMessageDecoder (io.netty.handler.codec)
  - AbstractMemcacheObjectDecoder (io.netty.handler.codec.memcache)
    - AbstractBinaryMemcacheDecoder (io.netty.handler.codec.memcache.binary)
      - BinaryMemcacheRequestDecoder (io.netty.handler.codec.memcache.binary)
      - BinaryMemcacheResponseDecoder (io.netty.handler.codec.memcache.binary)
  - AbstractSniHandler (io.netty.handler.ssl)
  - Anonymous in decoder in ByteToMessageCodec (io.netty.handler.codec)
  - BigIntegerDecoder (io.netty.example.factorial)
  - Bzip2Decoder (io.netty.handler.codec.compression)
  - DelimiterBasedFrameDecoder (io.netty.handler.codec)
  - FastLzFrameDecoder (io.netty.handler.codec.compression)
  - FixedLengthFrameDecoder (io.netty.handler.codec)
  - HAProxyMessageDecoder (io.netty.handler.codec.haproxy)
  - Http2ConnectionHandler (io.netty.handler.codec.http2)
  - HttpObjectDecoder (io.netty.handler.codec.http)
  - JsonObjectDecoder (io.netty.handler.codec.json)
  - LengthFieldBasedFrameDecoder (io.netty.handler.codec)

下面来看 ByteToMessageDecoder 类的定义代码。

```
public abstract class ByteToMessageDecoder extends ChannelInboundHandlerAdapter{
    //类体省略
}
```

从上面代码可以看出，ByteToMessageDecoder 继承了 ChannelInboundHandlerAdapter，根据之前的学习，我们知道这是个 Inbound 类型的 Handler，也就是处理流向自身事件的 Handler。其次，该类通过 abstract 关键字修饰，说明是个抽象类，在实际使用的时候，并不是直接使用这个类，而是使用其子类，类定义了解码器的骨架方法，具体实现逻辑交给子类，同样，在半包处理中也是由该类实现的。Netty 中很多解码器都实现了这个类，也可以通过实现该类自定义解码器。

我们重点关注该类的 cumulation 属性，它是有关半包处理的关键属性，从 12.1.2 节中我们知道，Netty 会将不完整的数据包进行保存，就保存在这个属性中。我们知道，ByteBuf 读取完数据会传递 ChannelRead 事件，传播过程中会调用 Handler 的 channelRead 方法。ByteToMessageDecoder 的 channelRead()方法是编码的关键部分。来看 channelRead()方法的代码。

```
public void channelRead(ChannelHandlerContext ctx, Object msg) throws Exception {
    //如果 message 是 ByteBuf 类型
    if (msg instanceof ByteBuf) {
        //简单当成一个 ArrayList，用于盛放解析的对象
        CodecOutputList out = CodecOutputList.newInstance();
        try {
            ByteBuf data = (ByteBuf) msg;
            //当前累加器为空，说明这是第一次从 I/O 流里面读取数据
            first = cumulation == null;
            if (first) {
                //如果是第一次，则将累加器赋值为刚读进来的对象
                cumulation = data;
            } else {
                //如果不是第一次，则把当前累加的数据和读进来的数据进行累加
                cumulation = cumulator.cumulate(ctx.alloc(), cumulation, data);
            }
            //调用子类的方法进行解析
            callDecode(ctx, cumulation, out);
        } catch (DecoderException e) {
            throw e;
        } catch (Throwable t) {
            throw new DecoderException(t);
        } finally {
            if (cumulation != null && !cumulation.isReadable()) {
                numReads = 0;
                cumulation.release();
                cumulation = null;
            } else if (++ numReads >= discardAfterReads) {
                numReads = 0;
                discardSomeReadBytes();
```

```
            }
            //记录 List 长度
            int size = out.size();
            decodeWasNull = !out.insertSinceRecycled();
            //向下传播
            fireChannelRead(ctx, out, size);
            out.recycle();
        }
    } else {
        //不是 ByteBuf 类型则向下传播
        ctx.fireChannelRead(msg);
    }
}
```

这个方法比较长，现在一步步剖析。首先判断如果传来的数据是 ByteBuf，则进入 if 块中，可以把 CodecOutputList out = CodecOutputList.newInstance()当成一个 ArrayList，用于保存解码完成的数据；ByteBuf data = (ByteBuf) msg 这步将数据转化成 ByteBuf；first = cumulation == null 表示如果 cumulation == null，说明没有存储半包数据，则将当前的数据保存在属性 Cumulation 中；如果 cumulation != null，说明存储了半包数据，则通过 cumulator.cumulate(ctx.alloc(), cumulation, data)将读取的数据和原来的数据进行累加，保存在 Cumulation 属性中。我们看 cumulator 属性的定义，代码如下。

```
private Cumulator cumulator = MERGE_CUMULATOR;
```

这里调用了其静态属性 MERGE_CUMULATOR，代码如下。

```
public static final Cumulator MERGE_CUMULATOR = new Cumulator() {
    @Override
    public ByteBuf cumulate(ByteBufAllocator alloc, ByteBuf cumulation, ByteBuf in) {
        ByteBuf buffer;
        //不能超过最大内存
        if (cumulation.writerIndex() > cumulation.maxCapacity() - in.readableBytes()
                || cumulation.refCnt() > 1) {
            buffer = expandCumulation(alloc, cumulation, in.readableBytes());
        } else {
            buffer = cumulation;
        }
        //将当前数据写入缓冲区
        buffer.writeBytes(in);
        in.release();
        return buffer;
    }
};
```

这里创建了 Cumulator 类型的静态对象，并重写了 cumulate()方法。这个 cumulate()方法就是用于将 ByteBuf 进行拼接的方法。在方法中，首先判断 Cumulation 的写指针+In 的可读字节数是否超过了 Cumulation 的最大长度，如果超过了，将对 Cumulation 进行扩容；如果没超过，则将其赋值到局部变量 Buffer 中。然后将 In 的数据写入 Buffer，将 In 进行释放，返回写入数据后的 ByteBuf。回到 channelRead()方法，最后调用 callDecode(ctx, cumulation, out)方法进行解码，这里传入了 Context 对象、缓冲区 Cumulation 和集合 Out。我们跟进 callDecode(ctx, cumulation, out)方法。

```
protected void callDecode(ChannelHandlerContext ctx, ByteBuf in, List<Object> out) {
    try {
        //只要累加器里面有数据
        while (in.isReadable()) {
            int outSize = out.size();
            //判断当前 List 是否有对象
            if (outSize > 0) {
                //如果有对象，则向下传播事件
                fireChannelRead(ctx, out, outSize);
                //清空当前 List
                out.clear();
                //解码过程中判断如果 ctx 被删除就跳出循环
                if (ctx.isRemoved()) {
                    break;
                }
                outSize = 0;
            }
            //当前可读数据长度
            int oldInputLength = in.readableBytes();
            //子类实现
            //子类解析，解析完对象放到 Out 里面
            decode(ctx, in, out);
            if (ctx.isRemoved()) {
                break;
            }
            //List 解析前大小和解析后长度一样（什么都没有解析出来）
            if (outSize == out.size()) {
                //原来可读的长度==解析后可读长度
                //说明没有读取数据（当前累加的数据并没有拼成一个完整的数据包）
                if (oldInputLength == in.readableBytes()) {
                    //跳出循环（下次再读取数据才能进行后续的解析）
                    break;
                } else {
                    //没有解析到数据，但是进行读取了
                    continue;
                }
            }
            //Out 里面有数据，但是没有从累加器读取数据
```

```
            if (oldInputLength == in.readableBytes()) {
                throw new DecoderException(
                        StringUtil.simpleClassName(getClass()) +
                        ".decode() did not read anything but decoded a message.");
            }

            if (isSingleDecode()) {
                break;
            }
        }
    } catch (DecoderException e) {
        throw e;
    } catch (Throwable cause) {
        throw new DecoderException(cause);
    }
}
```

分析下上面的源码，首先循环判断传入的 ByteBuf 是否还有可读字节，如果有可读字节说明没有解码完成，则继续循环解码。然后判断集合 Out 的 size 值，如果 size 大于 1，说明 Out 中装入了解码完成之后的数据，接下来将事件向下传播并清空 Out。

因为第一次解码 Out 是空的，所以不会进入 if 块，这部分稍后分析。继续往下看，通过 int oldInputLength = in.readableBytes()获取当前 ByteBuf，其实就是属性 Cumulation 的可读字节数，这里就是一个用于比较的备份。继续往下看，decode(ctx, in, out)方法是最终的解码操作，这步会读取 Cumulation 并且将解码后的数据放入集合 Out 中，在 ByteToMessageDecoder 中的该方法是一个抽象方法，让子类进行实现，Netty 很多解码都是继承了 ByteToMessageDecoder 并实现了 decode 方法从而完成了解码操作，同样我们也可以遵循相应的规则进行自定义解码器，在之后的小节中会讲解 Netty 自定义的解码器，并剖析相关的实现细节。

继续往下看，if (outSize == out.size()) 这个判断表示将解析之前的 Out 大小和解析之后的 Out 大小进行比较，如果相同，说明并没有解析出数据，进入 if 块中。if (oldInputLength == in.readableBytes()) 表示 Cumulation 的可读字节数在解析之前和解析之后是相同的，说明解码方法中并没有解析数据，也就是当前的数据并不是一个完整的数据包，则跳出循环，留给下次解析；否则，说明没有解析到数据，但是读取了，所以跳过该次循环进入下次循环。最后判断 if (oldInputLength == in.readableBytes())，这里代表 Out 中有数据，但是并没有从 Cumulation 读数据，说明这个 Out 的内容是非法的，直接抛出异常。现在回到 channelRead()方法，关注 finally 代码块中的内容。

```
finally {
    if (cumulation != null && !cumulation.isReadable()) {
        numReads = 0;
```

```
        cumulation.release();
        cumulation = null;
    } else if (++ numReads >= discardAfterReads) {
        numReads = 0;
        discardSomeReadBytes();
    }
    //记录 List 长度
    int size = out.size();
    decodeWasNull = !out.insertSinceRecycled();
    //向下传播
    fireChannelRead(ctx, out, size);
    out.recycle();
}
```

首先判断 Cumulation 不为 null，并且没有可读字节，则将累加器进行释放，并设置为 null，之后记录 Out 的长度，通过 fireChannelRead(ctx, out, size)将 ChannelRead 事件进行向下传播，并回收 Out 对象。跟进 fireChannelRead(ctx, out, size)方法看代码。

```
static void fireChannelRead(ChannelHandlerContext ctx, CodecOutputList msgs, int numElements)
{
    //遍历 List
    for (int i = 0; i < numElements; i ++) {
        //逐个向下传递
        ctx.fireChannelRead(msgs.getUnsafe(i));
    }
}
```

上面代码中遍历 Out 集合，并将里面的元素逐个向下传递。以上就是有关解码的骨架逻辑。

12.3.2 LineBasedFrameDecoder 行解码器

LineBasedFrameDecoder 是回车换行解码器，如果用户发送的消息以回车换行符（以\r\n 或者直接以\n 结尾）作为消息结束的标识，则可以直接使用 Netty 的 LineBasedFrameDecoder 对消息进行解码，只需要在初始化 Netty 服务端或者客户端时将 LineBasedFrameDecoder 正确地添加到 ChannelPipeline 中即可，不需要自己重新实现一套换行解码器。

LineBasedFrameDecoder 的工作原理是它依次遍历 ByteBuf 中的可读字节，判断是否有“\n”或者“\r\n”，如果有，就以此位置为结束位置，从可读索引到结束位置区间的字节就组成了一行。它是以换行符为结束标志的解码器，支持携带结束符或者不携带结束符两种解码方式，同时支持配置单行的最大长度。如果连续读取到最大长度后仍然没有发现换行符，就会抛出异常，同时忽略之前读到的异常码流，防止由于数据报没有携带换行符导致接收到的 ByteBuf 无限制积压，引起系统内存溢出。它的使用效果如下。

```
解码之前：
+------------------------------------------------------------------+
                          接收的数据报
"This is a netty example for using the nio framework.\r\n When you"
+------------------------------------------------------------------+
解码之后的 ChannelHandler 接收到的 Object 如下：
+------------------------------------------------------------------+
                          解码之后的文本消息
"This is a netty example for using the nio framework."
+------------------------------------------------------------------+
```

通常情况下，LineBasedFrameDecoder 会和 StringDecoder 配合使用，组合成按行切换的文本解码器，对于文本类协议的解析，文本换行解码器非常实用，例如对 HTTP 消息头的解析、FTP 消息的解析等。

下面简单给出文本换行解码器的使用示例。

```
pipeline.addLast(new LineBasedFrameDecoder(1024));
pipeline.addLast(new StringDecoder());
```

初始化 Channel 的时候，首先将 LineBasedFrameDecoder 添加到 ChannelPipcline 中，然后依次添加字符串解码器 StringDecoder、业务 Handler。

接下来看 LineBasedFrameDecoder 的源码，LineBasedFrameDecoder 也继承了 ByteToMessageDecoder，其参数定义如下。

```
//数据包的最大长度，超过该长度会进行丢弃模式
private final int maxLength;
//超出最大长度是否要抛出异常
private final boolean failFast;
//最终解析的数据包是否带有换行符
private final boolean stripDelimiter;
//为 true 说明当前解码过程为丢弃模式
private boolean discarding;
//丢弃了多少字节
private int discardedBytes;
```

其中的丢弃模式，我们会在源码中看到其含义。我们看 decode()方法的代码。

```
protected final void decode(ChannelHandlerContext ctx, ByteBuf in, List<Object> out) throws
Exception {
    Object decoded = decode(ctx, in);
    if (decoded != null) {
        out.add(decoded);
    }
}
```

这里 decode()方法调用重载的 decode()方法，并将解码后的内容放到 Out 集合中。我们看重载的 decode()方法，代码如下。

```
protected Object decode(ChannelHandlerContext ctx, ByteBuf buffer) throws Exception {
    //找这行的结尾
    final int eol = findEndOfLine(buffer);
    if (!discarding) {
        if (eol >= 0) {
            final ByteBuf frame;
            //计算从换行符到可读字节之间的长度
            final int length = eol - buffer.readerIndex();
            //获得分隔符长度，如果是\r\n 结尾，分隔符长度为 2
            final int delimLength = buffer.getByte(eol) == '\r'? 2 : 1;

            //如果长度大于最大长度
            if (length > maxLength) {
                //指向换行符之后的可读字节（这段数据完全丢弃）
                buffer.readerIndex(eol + delimLength);
                //传播异常事件
                fail(ctx, length);
                return null;
            }
            //如果这次解析的数据是有效的
            //分隔符是否算在完整数据包里
            //true 为丢弃分隔符
            if (stripDelimiter) {
                //截取有效长度
                frame = buffer.readRetainedSlice(length);
                //跳过分隔符的字节
                buffer.skipBytes(delimLength);
            } else {
                //包含分隔符
                frame = buffer.readRetainedSlice(length + delimLength);
            }

            return frame;
        } else {
            //如果没找到分隔符（非丢弃模式）
            //可读字节长度
            final int length = buffer.readableBytes();
            //如果超过能解析的最大长度
            if (length > maxLength) {
                //将当前长度标记为可丢弃的
                discardedBytes = length;
                //直接将读指针移动到写指针
                buffer.readerIndex(buffer.writerIndex());
                //标记为丢弃模式
```

```
                discarding = true;
                //超过最大长度抛出异常
                if (failFast) {
                    fail(ctx, "over " + discardedBytes);
                }
            }
            //没有超过，则直接返回
            return null;
        }
    } else {
        //丢弃模式
        if (eol >= 0) {
            //找到分隔符
            //当前丢弃的字节（前面已经丢弃的+现在丢弃的位置-写指针）
            final int length = discardedBytes + eol - buffer.readerIndex();
            //当前换行符长度为多少
            final int delimLength = buffer.getByte(eol) == '\r'? 2 : 1;
            //读指针直接移到换行符+换行符的长度
            buffer.readerIndex(eol + delimLength);
            //当前丢弃的字节为0
            discardedBytes = 0;
            //设置为未丢弃模式
            discarding = false;
            //丢弃完字节之后触发异常
            if (!failFast) {
                fail(ctx, length);
            }
        } else {
            //累计已丢弃的字节个数+当前可读的长度
            discardedBytes += buffer.readableBytes();
            //移动
            buffer.readerIndex(buffer.writerIndex());
        }
        return null;
    }
}
```

final int eol=findEndOfLine(buffer)是寻找当前行的结尾的索引值，也就是\r\n 或者是\n，如下图所示。

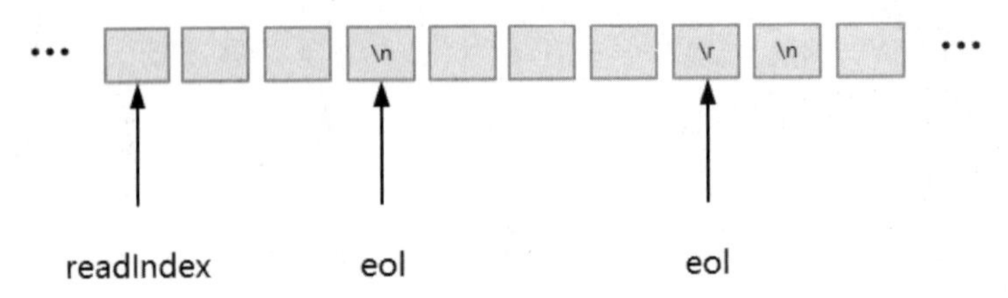

从上图中不难看出，如果是以\n 结尾的，返回的索引值是\n 的索引值，如果是\r\n 结尾的，返回的索引值是\r 的索引值。

我们看 findEndOfLine(buffer)方法的代码。

```
private static int findEndOfLine(final ByteBuf buffer) {
    //找到\n 这个字节
    int i = buffer.forEachByte(ByteProcessor.FIND_LF);
    //如果找到了，并且前面的字符是\r，则指向\r 字节
    if (i > 0 && buffer.getByte(i - 1) == '\r') {
        i--;
    }
    return i;
}
```

从上面代码看到，通过一个 forEachByte()方法找\n 这个字节，如果找到了，并且前面是\r，则返回\r 的索引，否则返回\n 的索引。回到重载的 decode()方法，if (!discarding)判断是否为非丢弃模式，默认是非丢弃模式，所以进入 if 块中。根据 if (eol >= 0) 判断是否找到了换行符，来看非丢弃模式下找到换行符的相关逻辑。

```
final ByteBuf frame;
final int length = eol - buffer.readerIndex();
final int delimLength = buffer.getByte(eol) == '\r'? 2 : 1;
if (length > maxLength) {
    buffer.readerIndex(eol + delimLength);
    fail(ctx, length);
    return null;
}
if (stripDelimiter) {
    frame = buffer.readRetainedSlice(length);
    buffer.skipBytes(delimLength);
} else {
    frame = buffer.readRetainedSlice(length + delimLength);
}

return frame;
```

首先获得换行符到可读字节之间的长度，然后获取换行符的长度，如果是\n 结尾，则长度为 1；如果是\r 结尾，则长度为 2。if (length > maxLength)代表如果长度超过最大长度，则直接通过 readerIndex(eol + delimLength)这种方式将读指针指向换行符之后的字节，说明换行符之前的字节需要完全丢弃，如下图所示。

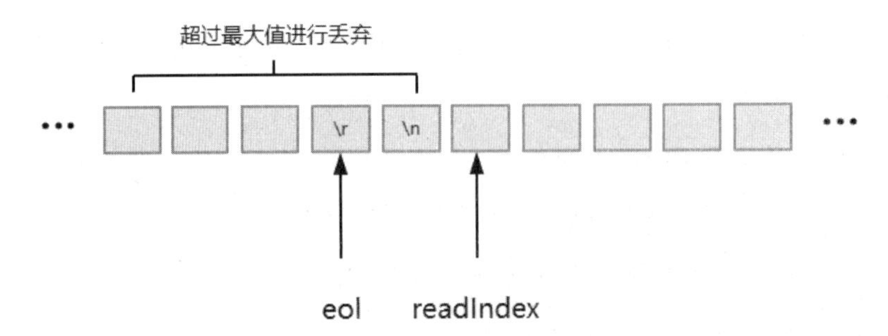

丢弃之后通过 fail 方法传播异常，并返回 null。继续往下看，执行到下一步，说明解析出来的数据长度没有超过最大长度，说明是有效数据包。if(stripDelimiter)表示是否要将分隔符放在完整数据包里面，如果是 true，则说明要丢弃分隔符，然后截取有效长度，并跳过分隔符长度，将包含的分隔符进行截取。

以上就是非丢弃模式下找到换行符的相关逻辑。再看非丢弃模式下没有找到换行符的相关逻辑，也就是非丢弃模式下 if(eol >= 0)中的 else 块，代码如下。

```
final int length = buffer.readableBytes();
if (length > maxLength) {
    discardedBytes = length;
    buffer.readerIndex(buffer.writerIndex());
    discarding = true;
    if (failFast) {
        fail(ctx, "over " + discardedBytes);
    }
}
return null;
```

首先通过 final int length = buffer.readableBytes()获取所有的可读字节数，然后判断可读字节数是否超过最大值，如果超过最大值，则属性 discardedBytes 被标记为这个长度，代表这段内容要进行丢弃，如下图所示。

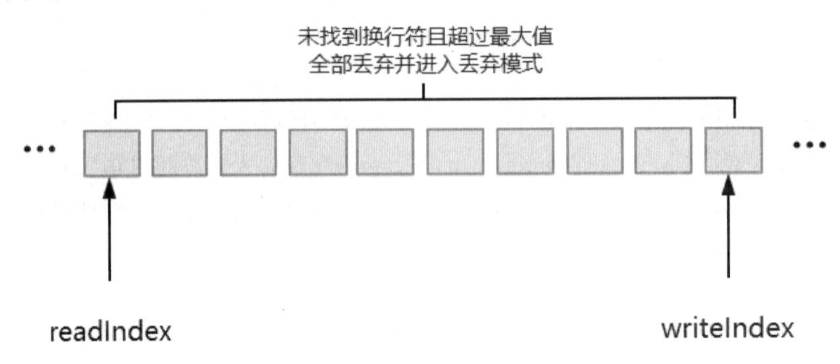

这里 buffer.readerIndex(buffer.writerIndex())直接将读指针移动到写指针，并且将 discarding 设置为 true，就是丢弃模式。如果可读字节没有超过最大长度，则返回 null，表示什么都没解析出来，等着下次解析。再看丢弃模式的处理逻辑，也就是 if (!discarding)中的 else 块，这里也分两种情况，根据 if (eol >= 0) 判断是否找到了分隔符。首先看找到分隔符的解码逻辑。

```
final int length = discardedBytes + eol - buffer.readerIndex();
final int delimLength = buffer.getByte(eol) == '\r'? 2 : 1;
buffer.readerIndex(eol + delimLength);
discardedBytes = 0;
discarding = false;
if (!failFast) {
    fail(ctx, length);
}
```

如果找到换行符，则需要将换行符之前的数据全部丢弃掉，如下图所示。

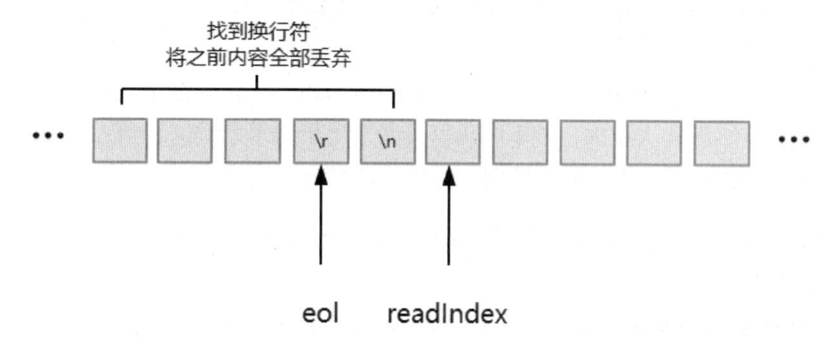

final int length = discardedBytes + eol - buffer.readerIndex()获得丢弃的字节总数，也就是之前丢弃的字节数+现在需要丢弃的字节数。然后计算换行符的长度，如果是\n 则是 1，如果是\r\n 就是 2。buffer.readerIndex(eol + delimLength)将读指针移动到换行符之后的位置，然后将 discarding 设置为 false，表示当前是非丢弃状态。再看丢弃模式下未找到换行符的情况，也就是丢弃模式下 if (eol >= 0) 中的 else 块，代码如下。

```
discardedBytes += buffer.readableBytes();
buffer.readerIndex(buffer.writerIndex());
```

这里做的事情非常简单，就是累计丢弃的字节数，并将读指针移动到写指针，即把数据全部丢弃。最后在丢弃模式下，decode()方法返回 null，代表本次没有解析出任何数据。以上就是行解码器的相关逻辑。

12.3.3 DelimiterBasedFrameDecoder 分隔符解码器

DelimiterBasedFrameDecoder 分隔符解码器是按照指定分隔符进行解码的解码器，通过分隔符可以将二进制流拆分成完整的数据包。回车换行解码器实际上是一种特殊的 DelimiterBasedFrameDecoder 解码器。

分隔符解码器在实际工作中有很广泛的应用，笔者所从事的电信行业，很多简单的文本私有协议都以特殊的分隔符作为消息结束的标识，特别是那些使用长连接的基于文本的私有协议。

关于分隔符的指定，与大家的习惯不同的是，分隔符并非以 Char 或者 String 作为构造参数，而是以 ByteBuf。下面就结合实际例子给出它的用法。假如消息以“$_”作为分隔符，服务端或者客户端初始化 ChannelPipeline 的代码实例如下。

```
ByteBuf delimiter = Unpooled.copiedBuffer("$_".getBytes());
pipeline.addLast(new DelimiterBasedFrameDecoder(1024,delimiter));
pipeline.addLast(new StringDecoder());
```

首先将“$_”转换成 ByteBuf 对象，作为参数构造 DelimiterBasedFrameDecoder，将其添加到 ChannelPipeline 中，然后依次添加字符串解码器（通常用于文本解码）和用户 Handler，请注意解码器和 Handler 的添加顺序，如果顺序颠倒，会导致消息解码失败。

DelimiterBasedFrameDecoder 同样继承了 ByteToMessageDecoder 并重写了 decode()方法，来看其中一个构造方法的代码。

```
public DelimiterBasedFrameDecoder(int maxFrameLength, ByteBuf... delimiters) {
    this(maxFrameLength, true, delimiters);
}
```

其中，参数 maxFrameLength 代表最大长度，delimiters 是个可变参数，可以支持多个分隔符进行解码。跟进 decode()方法，代码如下。

```
protected final void decode(ChannelHandlerContext ctx, ByteBuf in, List<Object> out) throws
Exception {
    Object decoded = decode(ctx, in);
    if (decoded != null) {
        out.add(decoded);
    }
}
```

这里同样调用了重载的 decode()方法并将解析好的数据添加到集合 List 中，其父类就可以遍历 Out，并将内容传播。重载的 decode()方法的代码如下。

```
protected Object decode(ChannelHandlerContext ctx, ByteBuf buffer) throws Exception {
    //行处理器(1)
```

```
if (lineBasedDecoder != null) {
    return lineBasedDecoder.decode(ctx, buffer);
}
int minFrameLength = Integer.MAX_VALUE;
ByteBuf minDelim = null;

//找到最小长度的分隔符(2)
for (ByteBuf delim: delimiters) {
    //每个分隔符分隔的数据包长度
    int frameLength = indexOf(buffer, delim);
    if (frameLength >= 0 && frameLength < minFrameLength) {
        minFrameLength = frameLength;
        minDelim = delim;
    }
}
//解码(3)
//已经找到分隔符
if (minDelim != null) {
    int minDelimLength = minDelim.capacity();
    ByteBuf frame;

    //当前分隔符是否处于丢弃模式
    if (discardingTooLongFrame) {
        //首先设置为非丢弃模式
        discardingTooLongFrame = false;
        //丢弃
        buffer.skipBytes(minFrameLength + minDelimLength);

        int tooLongFrameLength = this.tooLongFrameLength;
        this.tooLongFrameLength = 0;
        if (!failFast) {
            fail(tooLongFrameLength);
        }
        return null;
    }
    //处于非丢弃模式
    //当前找到的数据包，大于允许的数据包
    if (minFrameLength > maxFrameLength) {
        //当前数据包+最小分隔符长度 全部丢弃
        buffer.skipBytes(minFrameLength + minDelimLength);
        //传递异常事件
        fail(minFrameLength);
        return null;
    }
    //如果是正常的长度
    //解析出来的数据包是否忽略分隔符
    if (stripDelimiter) {
```

```
            //如果不包含分隔符
            //截取
            frame = buffer.readRetainedSlice(minFrameLength);
            //跳过分隔符
            buffer.skipBytes(minDelimLength);
        } else {
            //截取包含分隔符的长度
            frame = buffer.readRetainedSlice(minFrameLength + minDelimLength);
        }

        return frame;
    } else {
        //如果没有找到分隔符
        //非丢弃模式
        if (!discardingTooLongFrame) {
            //可读字节大于允许的解析出来的长度
            if (buffer.readableBytes() > maxFrameLength) {
                //将这个长度记录下
                tooLongFrameLength = buffer.readableBytes();
                //跳过这段长度
                buffer.skipBytes(buffer.readableBytes());
                //标记当前处于丢弃状态
                discardingTooLongFrame = true;
                if (failFast) {
                    fail(tooLongFrameLength);
                }
            }
        } else {
            tooLongFrameLength += buffer.readableBytes();
            buffer.skipBytes(buffer.readableBytes());
        }
        return null;
    }
}
```

这里的方法也比较长，通过拆分进行剖析：①行处理器；②找到最小长度分隔符；③解码。首先看第一步行处理器的代码。

```
if (lineBasedDecoder != null) {
    return lineBasedDecoder.decode(ctx, buffer);
}
```

这里首先判断成员变量 lineBasedDecoder 是否为空，如果不为空则直接调用 lineBasedDecoder 的 decode()方法进行解码，lineBasedDecoder 实际上就是上一小节剖析的 LineBasedFrameDecoder 解码器。这个成员变量会在分隔符是\n 和\r\n 的时候进行初始化。我们看初始化该属性的构造方法。

```
public DelimiterBasedFrameDecoder(
        int maxFrameLength, boolean stripDelimiter, boolean failFast, ByteBuf... delimiters)
{
    //代码省略
    //如果是基于行的分隔
    if (isLineBased(delimiters) && !isSubclass()) {
        //初始化行处理器
        lineBasedDecoder = new LineBasedFrameDecoder(maxFrameLength, stripDelimiter,
failFast);
        this.delimiters = null;
    } else {
        //代码省略
    }
    //代码省略
}
```

这里 isLineBased(delimiters)会判断是否是基于行的分隔，代码如下。

```
private static boolean isLineBased(final ByteBuf[] delimiters) {
    //分隔符长度不为 2
    if (delimiters.length != 2) {
        return false;
    }
    //获取第一个分隔符
    ByteBuf a = delimiters[0];
    //获取第二个分隔符
    ByteBuf b = delimiters[1];
    if (a.capacity() < b.capacity()) {
        a = delimiters[1];
        b = delimiters[0];
    }
    //确保 a 是\r\n 分隔符，确保 b 是\n 分隔符
    return a.capacity() == 2 && b.capacity() == 1
            && a.getByte(0) == '\r' && a.getByte(1) == '\n'
            && b.getByte(0) == '\n';
}
```

首先判断长度等于 2，直接返回 false。然后获取第一个 ByteBuf a 和第二个 ByteBuf b，判断 a 的第一个分隔符是不是\r，a 的第二个分隔符是不是\n，b 的第一个分隔符是不是\n，如果都为 true，则条件成立。回到 decode()方法中，看第 2 步，找到最小长度的分隔符。这里最小长度的分隔符，就是从读指针开始找到最近的分隔符，代码如下。

```
for (ByteBuf delim: delimiters) {
    //每个分隔符分隔的数据包长度
    int frameLength = indexOf(buffer, delim);
    if (frameLength >= 0 && frameLength < minFrameLength) {
```

```
            minFrameLength = frameLength;
            minDelim = delim;
        }
    }
```

这里会遍历所有的分隔符，找到每个分隔符到读指针的数据包长度。再通过 if 判断，找到长度最小的数据包的长度，然后保存当前数据包的分隔符，如下图所示。

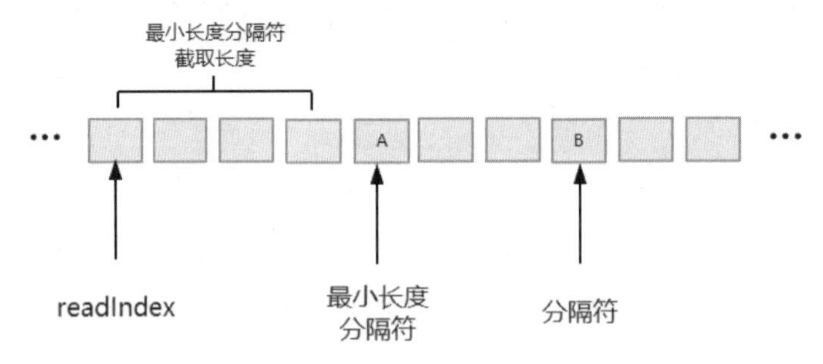

假设 A 和 B 同为分隔符，因为 A 分隔符到读指针的长度小于 B 分隔符到读指针的长度，所以会找到最小的分隔符 A，分隔符的最小长度就是 readIndex 到 A 的长度。继续看第 3 步，解码。if (minDelim != null) 表示已经找到最小长度分隔符，继续看 if 块中的逻辑，代码如下。

```
int minDelimLength = minDelim.capacity();
ByteBuf frame;
if (discardingTooLongFrame) {
    discardingTooLongFrame = false;
    buffer.skipBytes(minFrameLength + minDelimLength);
    int tooLongFrameLength = this.tooLongFrameLength;
    this.tooLongFrameLength = 0;
    if (!failFast) {
        fail(tooLongFrameLength);
    }
    return null;
}
if (minFrameLength > maxFrameLength) {
    buffer.skipBytes(minFrameLength + minDelimLength);
    fail(minFrameLength);
    return null;
}
if (stripDelimiter) {
    frame = buffer.readRetainedSlice(minFrameLength);
    buffer.skipBytes(minDelimLength);
} else {
    frame = buffer.readRetainedSlice(minFrameLength + minDelimLength);
}
return frame;
```

if (discardingTooLongFrame) 表示当前是否处于非丢弃模式，如果是丢弃模式，则进入 if 块。因为第一个不是丢弃模式，所以先分析 if 块后面的逻辑。if (minFrameLength > maxFrameLength) 判断当前找到的数据包长度大于最大长度，这个最大长度是创建解码器的时候设置的，如果超过了最大长度，就通过 buffer.skipBytes(minFrameLength + minDelimLength) 方式跳过数据包+分隔符的长度，也就是将这部分数据进行完全丢弃。继续往下看，如果长度不大于最大允许长度，则通过 if (stripDelimiter) 判断解析出来的数据包是否包含分隔符，如果不包含分隔符，则截取数据包的长度后，跳过分隔符。

再回去看 if (discardingTooLongFrame) 中 if 块的逻辑，也就是丢弃模式。首先将 discardingTooLongFrame 设置为 false，标记为非丢弃模式，然后通过 buffer.skipBytes (minFrameLength + minDelimLength) 将数据包+分隔符长度的字节数跳过，也就是进行丢弃，之后抛出异常。分析完找到分隔符之后的丢弃模式和非丢弃模式的逻辑处理，我们再分析没找到分隔符的逻辑处理，也就是 if (minDelim != null) 中的 else 块，代码如下。

```
if (!discardingTooLongFrame) {
    if (buffer.readableBytes() > maxFrameLength) {
        tooLongFrameLength = buffer.readableBytes();
        buffer.skipBytes(buffer.readableBytes());
        discardingTooLongFrame = true;
        if (failFast) {
            fail(tooLongFrameLength);
        }
    }
} else {
    tooLongFrameLength += buffer.readableBytes();
    buffer.skipBytes(buffer.readableBytes());
}
return null;
```

首先通过 if (!discardingTooLongFrame)判断是否为非丢弃模式，如果是非丢弃模式，则进入 if 块。在 if 块中，通过 if (buffer.readableBytes() > maxFrameLength) 判断当前可读字节数是否大于最大允许的长度，如果大于最大允许的长度，则将可读字节数设置到 tooLongFrameLength 的属性中，代表丢弃的字节数，然后通过 buffer.skipBytes(buffer.readableBytes()) 将累计器中所有的可读字节进行丢弃，最后将 discardingTooLongFrame 设置为 true，也就是丢弃模式，之后抛出异常。如果 if (!discardingTooLongFrame) 为 false，也就是当前处于丢弃模式，则追加 tooLongFrameLength 也就是丢弃的字节数的长度，并通过 buffer.skipBytes(buffer.readableBytes()) 将所有的字节继续进行丢弃。以上就是分隔符解码器的相关逻辑。

12.3.4 FixedLengthFrameDecoder 固定长度解码器

FixedLengthFrameDecoder 固定长度解码器能够按照指定的长度对消息进行自动解码，开发者不需要考虑 TCP 的粘包和拆包等问题，非常实用。

对于定长消息，如果消息实际长度小于定长，则往往会进行补位操作，它在一定程度上导致了空间和资源的浪费。但是它的优点也是非常明显的，编解码比较简单，因此在实际项目中仍然有一定的应用场景。

利用 FixedLengthFrameDecoder 解码器，无论一次接收到多少数据报，它都会按照构造函数中设置的固定长度进行解码，如果是半包消息，FixedLengthFrameDecoder 会缓存半包消息并等待下个包到达后进行拼包，直到读取到一个完整的包。假如单条消息的长度是 20 字节，使用 FixedLengthFrameDecoder 解码器的效果如下。

```
解码前：
+--------------------------------------------------------------+
                        接收到的数据报
"HELLO NETTY FOR USER DEVELOPER"
+--------------------------------------------------------------+
解码后：
+--------------------------------------------------------------+
                        解码后的数据报
"HELLO NETTY FOR USER"
+--------------------------------------------------------------+
```

其类的定义代码如下。

```
public class FixedLengthFrameDecoder extends ByteToMessageDecoder {
    //长度大小
    private final int frameLength;
    public FixedLengthFrameDecoder(int frameLength) {
        if (frameLength <= 0) {
            throw new IllegalArgumentException(
                    "frameLength must be a positive integer: " + frameLength);
        }
        //保存当前 frameLength
        this.frameLength = frameLength;
    }
    @Override
    protected final void decode(ChannelHandlerContext ctx, ByteBuf in, List<Object> out) throws
Exception {
        //通过 ByteBuf 去解码，解码到对象之后添加到 out 上
        Object decoded = decode(ctx, in);
        if (decoded != null) {
            //将解析的 ByteBuf 添加到对象里面
```

```
            out.add(decoded);
        }
    }
    protected Object decode(
            @SuppressWarnings("UnusedParameters") ChannelHandlerContext ctx, ByteBuf in)
throws Exception {
        //字节是否小于这个固定长度
        if (in.readableBytes() < frameLength) {
            return null;
        } else {
            //当前累加器中截取这个长度的数值
            return in.readRetainedSlice(frameLength);
        }
    }
}
```

我们看到 FixedLengthFrameDecoder 类继承了 ByteToMessageDecoder，重写了 decode()方法，这个类只有一个叫作 frameLength 的属性，并在构造方法中初始化了该属性。再看 decode()方法，在 decode()方法中又调用了自身另一个重载的 decode()方法进行解析，解析出来之后将解析后的数据放在集合 Out 中。再看重载的 decode()方法，重载的 decode()方法中首先判断累加器的字节数是否小于固定长度，如果小于固定长度则返回 null，代表不是一个完整的数据包，直接返回 null。如果大于等于固定长度，则直接从累加器中截取这个长度的数值，in.readRetainedSlice(frameLength)会返回一个新的截取后的 ByteBuf，并将原来的累加器读指针后移 frameLength 字节。如果累计器中还有数据，则会通过 ByteToMessageDecoder 中 callDecode()方法里的 while 循环方式，继续进行解码。这样，就实现了固定长度的解码工作。

12.3.5　LengthFieldBasedFrameDecoder 通用解码器

了解 TCP 通信机制的小伙伴应该都知道 TCP 底层的粘包和拆包，当我们在接收消息的时候，显示不能认为读取到的报文就是个整包消息，特别是对于采用非阻塞 I/O 和长连接通信的程序。

如何区分一个整包消息，通常有四种做法。

（1）固定长度，例如每 120 字节代表一个整包消息，不足的前面补位。解码器在处理这类定常消息的时候比较简单，每次读到指定长度的字节后再进行解码。

（2）通过回车换行符区分消息，例如 HTTP。这类区分消息的方式多用于文本协议。

（3）通过特定的分隔符区分整包消息。

（4）通过在协议头/消息头中设置长度属性来标识整包消息。

前三种解码器前面章节已经做了详细介绍，下面介绍最后一种通用解码器——LengthFieldBasedFrameDecoder。

大多数的协议（私有或者公有）的协议头中会携带长度属性，用于标识消息体或者整包消息的长度，例如 SMPP（Short Message Peer to Peer，短消息对等协议）、HTTP 等。由于基于长度解码需求的通用性，以及为了降低用户的协议开发难度，Netty 提供了 LengthFieldBasedFrameDecoder，自动屏蔽 TCP 底层的拆包和粘包问题，只需要传入正确的参数，即可轻松解决“读半包”问题。

下面看如何通过不同的参数组合来实现不同的“半包”读取策略。第一种常用的方式是消息的第一个属性是长度属性，后面是消息体，消息头中只包含一个长度属性。它的消息结构定义如下所示。

```
+---------+----------------+
| Length   |  Actual Content  |
| 0x000C   |  "HELLO,WORLD"   |
+---------+----------------+
```

使用以下参数组合进行解码。

（1）lengthFieldOffset = 0。

（2）lengthFieldLength = 2。

（3）lengthAdjustment = 0。

（4）initialBytesToStrip = 0。

解码后的字节缓冲区内容如下所示。

```
+---------+----------------+
| Length   |  Actual Content  |
| 0x000C   |  "HELLO,WORLD"   |
+---------+----------------+
```

通过 ByteBuf.readableBytes()方法可以获取当前消息的长度，解码后的字节缓冲区可以不携带长度属性，由于长度属性在起始位置并且长度为 2，所以将 initialBytesToStrip 设置为 2，参数组合修改如下。

（1）lengthFieldOffset = 0。

（2）lengthFieldLength = 2。

（3）lengthAdjustment = 0。

（4）initialBytesToStrip = 2。

解码后的字节缓冲区内容如下所示。

```
+------------------+
|  Actual Content  |
|  "HELLO,WORLD"   |
+------------------+
```

解码后的字节缓冲区丢弃了长度属性，仅仅包含消息体，对于大多数的协议，解码之后消息长度没有用处，因此可以丢弃。在大多数应用场景中，长度属性仅用来标识消息体的长度，这类协议通常由消息长度属性+消息体组成，如上所示的几个例子。但是，对于某些协议，长度属性还包含了消息头的长度。在这种应用场景中，往往需要使用 lengthAdjustment 进行修正。由于整个消息（包含消息头）的长度往往大于消息体的长度，所以，lengthAdjustment 为负数。下面的代码展示了通过指定 lengthAdjustment 属性来包含消息头的长度。

```
(1) lengthFieldOffset = 0。
(2) lengthFieldLength = 2。
(3) lengthAdjustment = -2。
(4) initialBytesToStrip = 0。
```

解码之前的码流如下。

```
+---------+----------------+
| Length  | Actual Content  |
| 0x000E  | "HELLO,WORLD"   |
+---------+----------------+
```

解码之后的码流如下。

```
+---------+----------------+
| Length  | Actual Content  |
| 0x000E  | "HELLO,WORLD"   |
+---------+----------------+
```

由于协议种类繁多，并不是所有的协议都将长度属性放在消息头的首位，当标识消息长度的属性位于消息头的中间或者尾部时，需要使用 lengthFieldOffset 属性进行标识，下面的参数组合给出了如何解决消息长度属性不在首位的问题。

（1）lengthFieldOffset = 2。

（2）lengthFieldLength = 3。

（3）lengthAdjustment = 0。

（4）initialBytesToStrip = 0。

其中 lengthFieldOffset 表示长度属性在消息头中偏移的字节数，lengthFieldLength 表示长度属性自身的长度，下面来看解码效果。

解码之前的码流如下。

```
+-----------+--------+----------------+
| Header 1  | Length | Actual Content |
|  0xCAFE   |0x00000C| "HELLO,WORLD"  |
+-----------+--------+----------------+
```

解码之后的码流如下。

```
+-----------+--------+----------------+
| Header 1  | Length | Actual Content |
|  0xCAFE   |0x00000C| "HELLO,WORLD"  |
+-----------+--------+----------------+
```

由于消息头 1 的长度为 2，所以长度属性的偏移量为 2；消息长度属性 Length 为 3，所以 lengthFieldLength 值为 3。由于长度属性仅仅标识消息体的长度，所以 lengthAdjustment 和 initialBytesToStrip 都为 0。

最后一种场景是长度属性夹在两个消息头之间或者长度属性位于消息头的中间，前后都有其他消息头字段。在这种场景下如果想忽略长度属性以及其前面的其他消息头属性，则可以通过 initialBytesToStrip 参数来跳过要忽略的字节长度，它的组合配置如下。

（1）lengthFieldOffset = 1。

（2）lengthFieldLength = 2。

（3）lengthAdjustment = 1。

（4）initialBytesToStrip = 3。

解码之前的码流（16 字节）如下。

```
+------+--------+------+----------------+
| HDR1 | Length | HDR2 | Actual Content |
| 0xCA | 0x000C | 0xFE | "HELLO,WORLD"  |
+------+--------+------+----------------+
```

解码之后的码流（13 字节）如下。

```
+------+----------------+
| HDR2 | Actual Content |
| 0xFE | "HELLO,WORLD"  |
+------+----------------+
```

由于 HDR1 的长度为 1，所以长度属性的偏移量 lengthFieldOffset 为 1；长度属性为 2 字节，所以 lengthFieldLength 为 2。由于长度属性是消息体的长度，解码后如果携带消息头中的属性，则需要使用 lengthAdjustment 进行调整，此处它的值为 1，代表的是 HDR2 的长度，最后由于解码后的缓冲区要忽略长度属性和 HDR1 部分，所以 lengthAdjustment 为 3。解码后的结果为 13 字节，HDR1 和 Length 属性被忽略。

事实上，通过 4 个参数的不同组合，可以达到不同的解码效果，用户在使用过程中可以根据业务的实际情况进行灵活调整。

由于 TCP 存在粘包和组包问题，所以通常情况下用户需要自己处理半包消息。利用 LengthFieldBasedFrameDecoder 解码器可以自动解决半包问题，它的习惯用法如下。

```
pipeline.addLast("frameDecoder", new LengthFieldBasedFrameDecoder(65536,0,2));
```

在 Pipeline 中增加 LengthFieldBasedFrameDecoder 解码器，指定正确的参数组合，可以将 Netty 的 ByteBuf 解码成整包消息，后面的用户解码器获取的就是个完整的数据报，按照逻辑正常进行解码即可，不再需要额外考虑“读半包”问题，降低了用户的开发难度。

12.4　Netty 编码器原理和数据输出

Netty 默认提供了丰富的编解码框架供用户集成使用，我们只对较常用的 Java 序列化编码器进行讲解。其他的编码器实现方式大同小异。其实编码器和解码器比较类似，编码器也是一个 Handler，并且属于 OutboundHandler，就是将准备发出去的数据进行拦截，拦截之后进行相应的处理后再次进行发送，如果理解了解码器，那么编码器的相关内容理解起来也比较容易。

12.4.1　WriteAndFlush 事件传播

我们在前面的章节学习 Pipeline 的时候，讲解了 write 事件的传播过程，但在实际使用的时候，通常不会调用 Channel 的 write()方法，因为该方法只会写入发送数据的缓存，所以并不会直接写入 Channel。如果想写入 Channel，还需要调用 flush()方法。在实际使用过程中，我们用的更多的是 writeAndFlush()方法，这个方法既能将数据写入发送数据的缓存，也能刷新到 Channel。我们看一个比较简单的使用场景。

```
public void channelRead(ChannelHandlerContext ctx, Object msg) throws Exception {
    ctx.channel().writeAndFlush("test data");
}
```

通过上面代码的方式，可以将数据发送到 Channel 中，对方可以收到响应。简单回顾一下，AbstractChannel 的 writeAndFlush()方法的代码如下。

```
public ChannelFuture writeAndFlush(Object msg) {
    return pipeline.writeAndFlush(msg);
}
```

继续跟进 DefaultChannelPipeline 中的 writeAndFlush()方法。

```
public final ChannelFuture writeAndFlush(Object msg) {
    return tail.writeAndFlush(msg);
}
```

可以看到，writeAndFlush 从 Tail 节点进行传播。有关事件传播，我们在 Pipeline 中进行过剖析。继续跟进 AbstractChannelHandlerContext 中的 writeAndFlush()方法，代码如下。

```
public ChannelFuture writeAndFlush(Object msg) {
    return writeAndFlush(msg, newPromise());
}
```

继续看代码。

```
public ChannelFuture writeAndFlush(Object msg, ChannelPromise promise) {
    if (msg == null) {
        throw new NullPointerException("msg");
    }
    if (!validatePromise(promise, true)) {
        ReferenceCountUtil.release(msg);
        // cancelled
        return promise;
    }
    write(msg, true, promise);
    return promise;
}
```

继续看 write()方法。

```
private void write(Object msg, boolean flush, ChannelPromise promise) {
    //findContextOutbound()寻找前一个 Outbound 节点
    //最后到 Head 节点结束
    AbstractChannelHandlerContext next = findContextOutbound();
    final Object m = pipeline.touch(msg, next);
    EventExecutor executor = next.executor();
    if (executor.inEventLoop()) {
```

```
        if (flush) {
            next.invokeWriteAndFlush(m, promise);
        } else {
            //没有调 flush
            next.invokeWrite(m, promise);
        }
    } else {
        AbstractWriteTask task;
        if (flush) {
            task = WriteAndFlushTask.newInstance(next, m, promise);
        }  else {
            task = WriteTask.newInstance(next, m, promise);
        }
        safeExecute(executor, task, promise, m);
    }
}
```

上面代码的逻辑我们也不陌生，找到下一个节点，因为 writeAndFlush()方法是从 Tail 节点开始的，并且是 Outbound 事件，所以会找到 Tail 节点的上一个 OutBoundHandler，有可能是编码器，也有可能是业务处理的 Handler。if (executor.inEventLoop()) 判断是否是 EventLoop 线程，如果不是，则封装成 Task 通过 NioEventLoop 异步执行，这里先按照是 EventLoop 线程分析。首先，通过 flush 判断是否调用了 flush，这里显然是 true，我们调用的方法是 writeAndFlush()方法，代码如下。

```
private void invokeWriteAndFlush(Object msg, ChannelPromise promise) {
    if (invokeHandler()) {
        //写入
        invokeWrite0(msg, promise);
        //刷新
        invokeFlush0();
    } else {
        writeAndFlush(msg, promise);
    }
}
```

这里就真相大白了，其实在 writeAndFlush()方法中，首先调用 write()方法，完成之后再调用 flush()方法进行刷新。跟进 invokeWrite0()方法。

```
private void invokeWrite0(Object msg, ChannelPromise promise) {
    try {
        //调用当前 Handler 的 write()方法
        ((ChannelOutboundHandler) handler()).write(this, msg, promise);
    } catch (Throwable t) {
        notifyOutboundHandlerException(t, promise);
```

```
    }
}
```

该方法我们在 Pipeline 中已经进行过分析，就是调用当前 Handler 的 write()方法，如果当前 Handler 中 write()方法是继续往下传播，则会继续传播写事件，直到传播到 Head 节点，最后会执行到 HeadContext 的 write()方法中，代码如下。

```
public void write(ChannelHandlerContext ctx, Object msg, ChannelPromise promise) throws
Exception {
    unsafe.write(msg, promise);
}
```

通过当前 Channel 的 Unsafe 对象将当前消息写入缓存，回到 invokeWriteAndFlush()方法。

```
private void invokeWriteAndFlush(Object msg, ChannelPromise promise) {
    if (invokeHandler()) {
        //写入
        invokeWrite0(msg, promise);
        //刷新
        invokeFlush0();
    } else {
        writeAndFlush(msg, promise);
    }
}
```

再看 invokeFlush0()方法的代码。

```
private void invokeFlush0() {
    try {
        ((ChannelOutboundHandler) handler()).flush(this);
    } catch (Throwable t) {
        notifyHandlerException(t);
    }
}
```

同样，这里会调用当前 Handler 的 flush()方法，如果当前 Handler 的 flush()方法是继续传播 flush 事件，则 flush 事件会继续往下传播，直到最后会调用 Head 节点的 flush()方法。HeadContext 的 flush()方法的代码如下。

```
public void flush(ChannelHandlerContext ctx) throws Exception {
    unsafe.flush();
}
```

这里，当前 Channel 的 Unsafe 对象通过调用 flush()方法将缓存的数据刷新到 Channel 中。以上就是 writeAndFlush()方法的相关逻辑，整体上比较简单，掌握了 Pipeline 的小伙伴应该很容易理解。

12.4.2　MessageToByteEncoder 抽象编码器

同解码器一样，编码器中也有一个抽象类叫 MessageToByteEncoder，定义了编码器的骨架方法，具体编码逻辑交给子类实现。解码器同样也是个 Handler，将写出的数据进行截取处理。我们知道，写数据的时候会传递 write 事件，传递过程中会调用 Handler 的 write()方法，所以编码器可以重写 write()方法，将数据编码成二进制字节流再传递 write 事件。首先来看 MessageToByteEncoder 的类声明：MessageToByteEncoder 负责将 POJO 对象编码成 ByteBuf，用户的编码器继承 MessageToByteEncoder，实现 void encode(ChannelHandlerContext ctx, I msg, ByteBuf out)接口，示例代码如下。

```
public class IntegerEncoder extends MessageToByteEncoder<Integer> {
    @Override
    public void encode(ChannelHandlerContext ctx, Integer msg,ByteBuf out)
       throws Exception {
           out.writeInt(msg);
    }
}
```

它的实现原理如下：调用 write()方法时，首先判断当前编码器是否支持需要发送的消息，如果不支持则直接透传；如果支持，则判断缓冲区的类型，对于直接内存分配 ioBuffer（堆外内存），对于堆内内存通过 heapBuffer()方法分配，源码如下。

```
public void write(ChannelHandlerContext ctx, Object msg, ChannelPromise promise) throws
Exception {
        ByteBuf buf = null;
        try {
            if (acceptOutboundMessage(msg)) {
                @SuppressWarnings("unchecked")
                I cast = (I) msg;
                buf = allocateBuffer(ctx, cast, preferDirect);
                try {
                    encode(ctx, cast, buf);
                } finally {
                    ReferenceCountUtil.release(cast);
                }

                if (buf.isReadable()) {
                    ctx.write(buf, promise);
                } else {
                    buf.release();
                    ctx.write(Unpooled.EMPTY_BUFFER, promise);
                }
                buf = null;
            } else {
```

```
            ctx.write(msg, promise);
        }
    } catch (EncoderException e) {
        throw e;
    } catch (Throwable e) {
        throw new EncoderException(e);
    } finally {
        if (buf != null) {
            buf.release();
        }
    }
}
```

编码使用的缓冲区分配完成之后，调用 encode()抽象方法进行编码，它由子类负责具体实现，方法定义如下。

```
protected abstract void encode(ChannelHandlerContext ctx, I msg, ByteBuf out) throws Exception;
```

编码完成之后，调用 ReferenceCountUtil 的 release()方法释放编码对象 msg。对编码后的 ByteBuf 进行以下判断。

（1）如果缓冲区包含可发送的字节，则调用 ChannelHandlerContext 的 write()方法发送 ByteBuf。

（2）如果缓冲区没有包含可写的字节，则需要释放编码后的 ByteBuf，将一个空的 ByteBuf 写入 ChannelHandlerContext。

发送操作完成之后，在方法退出之前释放编码缓冲区的 ByteBuf 对象。

12.4.3 写入 Buffer 队列

12.4.1 节我们介绍过，writeAndFlush()方法其实最终会调用 write()和 flush()方法，write()方法最终会传递到 Head 节点，调用 HeadContext 的 write()方法，代码如下。

```
public void write(ChannelHandlerContext ctx, Object msg, ChannelPromise promise) throws
Exception {
    unsafe.write(msg, promise);
}
```

上面代码中通过 Unsafe 对象的 write()方法，将消息写入缓存。AbstractUnsafe 的 write()方法的代码如下。

```
public final void write(Object msg, ChannelPromise promise) {
    assertEventLoop();
    //负责缓冲写进来的 ByteBuf
```

```
    ChannelOutboundBuffer outboundBuffer = this.outboundBuffer;
    if (outboundBuffer == null) {
        safeSetFailure(promise, WRITE_CLOSED_CHANNEL_EXCEPTION);
        ReferenceCountUtil.release(msg);
        return;
    }
    int size;
    try {
        //非堆外内存转化为堆外内存
        msg = filterOutboundMessage(msg);
        size = pipeline.estimatorHandle().size(msg);
        if (size < 0) {
            size = 0;
        }
    } catch (Throwable t) {
        safeSetFailure(promise, t);
        ReferenceCountUtil.release(msg);
        return;
    }
    //插入写队列
    outboundBuffer.addMessage(msg, size, promise);
}
```

首先看 ChannelOutboundBuffer outboundBuffer = this.outboundBuffer，ChannelOutboundBuffer 的功能就是缓存写入的 ByteBuf。继续看 try 块中的 msg = filterOutboundMessage(msg)，这步的意义就是将非堆外内存转化为堆外内存，filterOutboundMessage() 方法最终会调用 AbstractNioByteChannel 中的 filterOutboundMessage()方法。

```
protected final Object filterOutboundMessage(Object msg) {
    if (msg instanceof ByteBuf) {
        ByteBuf buf = (ByteBuf) msg;
        //是堆外内存，直接返回
        if (buf.isDirect()) {
            return msg;
        }
        return newDirectBuffer(buf);
    }
    if (msg instanceof FileRegion) {
        return msg;
    }
    throw new UnsupportedOperationException(
            "unsupported message type: " + StringUtil.simpleClassName(msg) + EXPECTED_TYPES);
}
```

首先判断 msg 是否是 ByteBuf 对象，如果是，则判断是否是堆外内存。如果是堆外内存，则直接返回，否则，通过 newDirectBuffer(buf)方式转化为堆外内存。回到 write()方法，outboundBuffer.

addMessage(msg, size, promise)将已经转化为堆外内存的 msg 插入写队列。跟到 addMessage()方法中，这是 ChannelOutboundBuffer 中的方法，代码如下。

```
public void addMessage(Object msg, int size, ChannelPromise promise) {
    Entry entry = Entry.newInstance(msg, size, total(msg), promise);
    if (tailEntry == null) {
        flushedEntry = null;
        tailEntry = entry;
    } else {
        Entry tail = tailEntry;
        tail.next = entry;
        tailEntry = entry;
    }
    if (unflushedEntry == null) {
        unflushedEntry = entry;
    }
    incrementPendingOutboundBytes(size, false);
}
```

首先通过 Entry.newInstance(msg, size, total(msg), promise) 的方式将 msg 封装成 Entry，然后通过调整 tailEntry、flushedEntry、unflushedEntry 三个指针，完成 Entry 的添加。这三个指针均是 ChannelOutboundBuffer 的成员变量。

- flushedEntry 指向第一个被 flush 的 Entry。
- unflushedEntry 指向第一个未被 flush 的 Entry。
- tailEntry 指向最后一个 Entry。

也就是说，从 flushedEntry 到 unflushedEntry 的 Entry，都是已经被 flush 的 Entry。从 unflushedEntry 到 tailEntry 的 Entry 都是没被 flush 的 Entry。回到代码中，创建 Entry 之后首先判断尾指针是否为空，在第一次添加的时候，均是空，所以会将 flushedEntry 设置为 null，并且将尾指针设置为当前创建的 Entry，最后判断 unflushedEntry 是否为空，如果第一次添加这里也是空，可以将 unflushedEntry 设置为新创建的 Entry。第一次调用 write()方法的结果如下图所示。

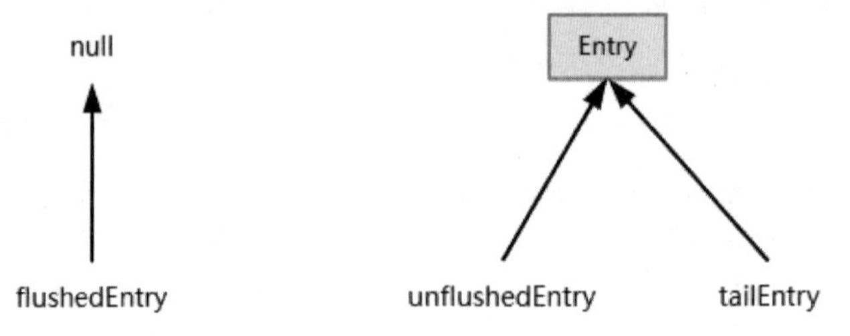

如果不是第一次调用 write()方法，则会进入 if (tailEntry == null) 中的 else 块。

- Entry tail = tailEntry，这里 Tail 就是当前尾节点。
- tail.next = entry，代表尾节点的下一个节点指向新创建的 Entry。
- tailEntry = entry，将尾节点也指向 Entry。

这样就完成了添加操作，其实就是将新创建的节点追加到原来尾节点之后，第二次添加时，if (unflushedEntry == null) 会返回 false，所以不会进入 if 块。第二次调用 write()方法之后指针的指向情况如下图所示。

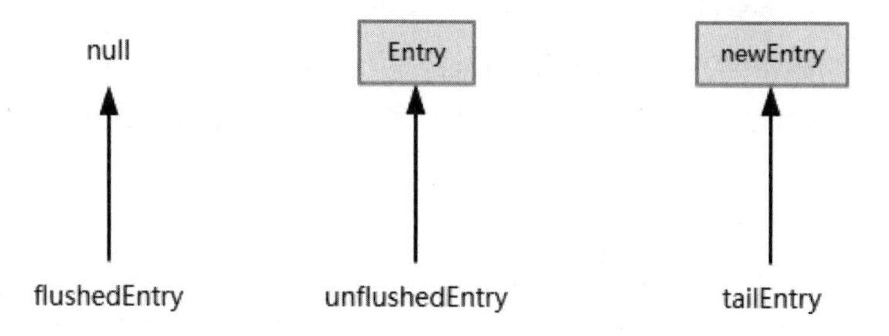

以后每次调用 write()方法，如果没有调用 flush 的话都会在尾节点之后进行追加。回到代码中，看这一步 incrementPendingOutboundBytes(size, false)，这步统计当前有多少字节需要被写出，代码如下。

```
private void incrementPendingOutboundBytes(long size, boolean invokeLater) {
    if (size == 0) {
        return;
    }
    //TOTAL_PENDING_SIZE_UPDATER 当前缓冲区里面有多少待写的字节
    long newWriteBufferSize = TOTAL_PENDING_SIZE_UPDATER.addAndGet(this, size);
    //getWriteBufferHighWaterMark() 最高不能超过 64KB
    if (newWriteBufferSize > channel.config().getWriteBufferHighWaterMark()) {
        setUnwritable(invokeLater);
    }
}
```

看这一步：long newWriteBufferSize = TOTAL_PENDING_SIZE_UPDATER.addAndGet(this, size)。TOTAL_PENDING_SIZE_UPDATER 表示当前缓冲区还有多少待写的字节，addAndGet 将当前 ByteBuf 的长度进行累加，累加到 newWriteBufferSize 中。继续判断 if (newWriteBufferSize > channel.config().getWriteBufferHighWaterMark())。channel.config().getWriteBufferHighWaterMark() 表示写 Buffer 的高水位值，默认是 64KB，也就是说写 Buffer 的最大长度不能超过 64KB。如果超过 64KB，则会调用 setUnwritable(invokeLater)方法设置写状态。

```
private void setUnwritable(boolean invokeLater) {
    for (;;) {
```

```
        final int oldValue = unwritable;
        final int newValue = oldValue | 1;
        if (UNWRITABLE_UPDATER.compareAndSet(this, oldValue, newValue)) {
            if (oldValue == 0 && newValue != 0) {
                fireChannelWritabilityChanged(invokeLater);
            }
            break;
        }
    }
}
```

这里通过自旋和 CAS 操作，传播一个 ChannelWritabilityChanged 事件，最终会调用 Handler 的 channelWritabilityChanged()方法进行处理。以上就是写 Buffer 的相关逻辑。

12.4.4 刷新 Buffer 队列

通过上一节的学习，我们知道 flush()方法通过事件传递，会传递到 HeadContext 的 flush()方法。

```
public void flush(ChannelHandlerContext ctx) throws Exception {
    unsafe.flush();
}
```

最终会调用 AbstractUnsafe 的 flush()方法。

```
public final void flush() {
    assertEventLoop();
    ChannelOutboundBuffer outboundBuffer = this.outboundBuffer;
    if (outboundBuffer == null) {
        return;
    }
    outboundBuffer.addFlush();
    flush0();
}
```

首先也是获取 ChannelOutboundBuffer 对象，然后看下面这一步。

```
outboundBuffer.addFlush();
```

这一步同样也是调整 ChannelOutboundBuffer 的指针，跟进 addFlush()方法。

```
public void addFlush() {
    Entry entry = unflushedEntry;
    if (entry != null) {
        if (flushedEntry == null) {
            flushedEntry = entry;
        }
        do {
            flushed ++;
```

```
            if (!entry.promise.setUncancellable()) {
                int pending = entry.cancel();
                decrementPendingOutboundBytes(pending, false, true);
            }
            entry = entry.next;
        } while (entry != null);
        unflushedEntry = null;
    }
}
```

上述代码中，首先声明一个 Entry 指向 unflushedEntry，也就是第一个未被 flush 的 Entry。通常情况下，因为 unflushedEntry 不为空，所以进入 if。在未刷新前，因为 flushedEntry 通常为空，所以会执行到 flushedEntry = entry，也就是 flushedEntry 指向 Entry。经过上述操作，缓冲区的指针情况如下图所示。

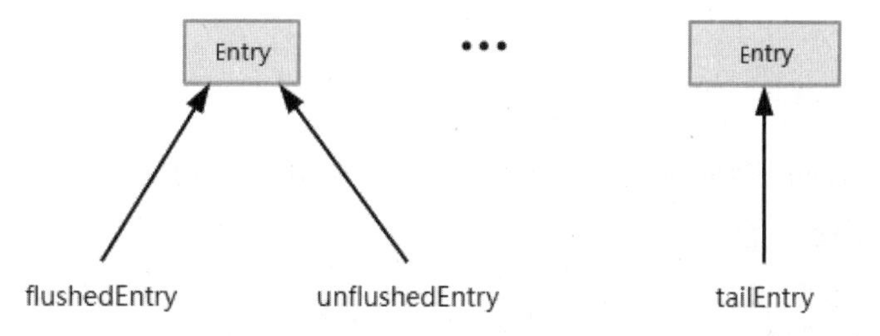

然后通过 do-while 不断寻找 unflushedEntry 后面的节点，直到没有节点为止，flushed 自增代表需要刷新多少个节点。循环中我们关注这一步。

```
decrementPendingOutboundBytes(pending, false, true);
```

这一步是统计缓冲区中的字节数的，但是和上一小节的 incrementPendingOutboundBytes 正好相反，因为这里是刷新，所以要减掉刷新后的字节数，代码如下。

```
private void decrementPendingOutboundBytes(long size, boolean invokeLater, boolean
notifyWritability) {
    if (size == 0) {
        return;
    }
    //从总的大小减去
    long newWriteBufferSize = TOTAL_PENDING_SIZE_UPDATER.addAndGet(this, -size);
    //直到减至某一个小于 32 字节的阈值
    if (notifyWritability && newWriteBufferSize <
channel.config().getWriteBufferLowWaterMark()) {
        //设置写状态
        setWritable(invokeLater);
    }
}
```

同样，TOTAL_PENDING_SIZE_UPDATER 代表缓冲区的字节数，这里 addAndGet 中的参数是 -size，也就是减掉 size 的长度。再看 if (notifyWritability && newWriteBufferSize < channel.config().getWriteBufferLowWaterMark())，getWriteBufferLowWaterMark()代表写 Buffer 的低水位值，也就是 32KB，如果写 Buffer 的长度小于这个数，就通过 setWritable()方法设置写状态，即通道由原来的不可写改成可写。回到 addFlush()方法，遍历 do-while 循环结束之后，将 unflushedEntry 指为空，代表所有的 Entry 都是可写的。经过上述操作，缓冲区的指针情况如下图所示。

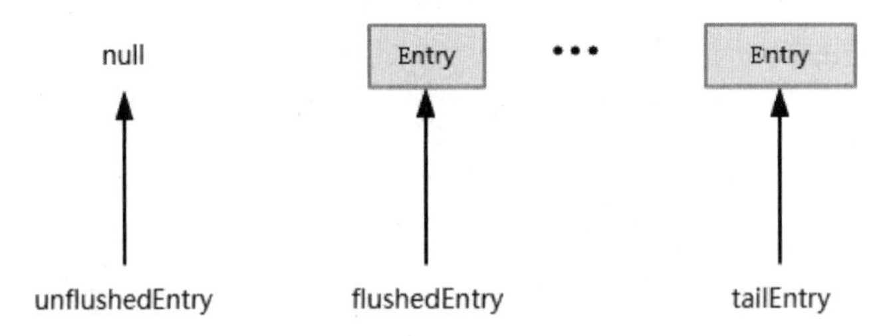

回到 AbstractUnsafe 的 flush()方法，指针调整完后，flush0()方法的代码如下。

```
protected void flush0() {
    if (inFlush0) {
        return;
    }
    final ChannelOutboundBuffer outboundBuffer = this.outboundBuffer;
    if (outboundBuffer == null || outboundBuffer.isEmpty()) {
        return;
    }
    inFlush0 = true;
    if (!isActive()) {
        try {
            if (isOpen()) {
                outboundBuffer.failFlushed(FLUSH0_NOT_YET_CONNECTED_EXCEPTION, true);
            } else {
                outboundBuffer.failFlushed(FLUSH0_CLOSED_CHANNEL_EXCEPTION, false);
            }
        } finally {
            inFlush0 = false;
        }
        return;
    }
    try {
        doWrite(outboundBuffer);
    } catch (Throwable t) {
        if (t instanceof IOException && config().isAutoClose()) {
            close(voidPromise(), t, FLUSH0_CLOSED_CHANNEL_EXCEPTION, false);
```

```
        } else {
            outboundBuffer.failFlushed(t, true);
        }
    } finally {
        inFlush0 = false;
    }
}
```

if (inFlush0) 表示判断当前 Flush 是否在进行中，如果在进行中，则返回，避免重复进入。我们重点关注 doWrite()方法，AbstractNioByteChannel 的 doWrite()方法的代码如下。

```
protected void doWrite(ChannelOutboundBuffer in) throws Exception {
    int writeSpinCount = -1;
    boolean setOpWrite = false;
    for (;;) {
        //每次获取当前节点
        Object msg = in.current();
        if (msg == null) {
            clearOpWrite();
            return;
        }
        if (msg instanceof ByteBuf) {
            //转化成 ByteBuf
            ByteBuf buf = (ByteBuf) msg;
            //如果没有可写的值
            int readableBytes = buf.readableBytes();
            if (readableBytes == 0) {
                //移除
                in.remove();
                continue;
            }
            boolean done = false;
            long flushedAmount = 0;
            if (writeSpinCount == -1) {
                writeSpinCount = config().getWriteSpinCount();
            }
            for (int i = writeSpinCount - 1; i >= 0; i --) {
                //将 Buffer 写入 Socket
                //localFlushedAmount 代表向 JDK 底层写了多少字节
                int localFlushedAmount = doWriteBytes(buf);
                //如果一个字节没写则直接 Break
                if (localFlushedAmount == 0) {
                    setOpWrite = true;
                    break;
                }
                //统计总共写了多少字节
                flushedAmount += localFlushedAmount;
```

```
                //如果 Buffer 全部写入 JDK 底层
                if (!buf.isReadable()) {
                    //标记全部写入
                    done = true;
                    break;
                }
            }
            in.progress(flushedAmount);
            if (done) {
                //移除当前对象
                in.remove();
            } else {
                break;
            }
        } else if (msg instanceof FileRegion) {
            //代码省略
        } else {
            throw new Error();
        }
    }
    incompleteWrite(setOpWrite);
}
```

上面代码中，首先是一个无限 for 循环，Object msg = in.current() 这一步是获取 flushedEntry 指向的 Entry 中的 msg，current()方法的代码如下。

```
public Object current() {
    Entry entry = flushedEntry;
    if (entry == null) {
        return null;
    }
    return entry.msg;
}
```

这里直接获取 flushedEntry 指向的 Entry 中关联的 msg，也就是一个 ByteBuf。回到 doWrite()方法：

- 如果 msg 为 null，说明没有可以刷新的 Entry，则调用 clearOpWrite()方法清除写标识。
- 如果 msg 不为 null，则判断是否是 ByteBuf 类型，如果是 ByteBuf，就进入 if 块中的逻辑。

在 if 块中，首先将 msg 转化为 ByteBuf，然后判断 ByteBuf 是否可读，如果不可读，则通过 in.remove()将当前的 ByteBuf 所关联的 Entry 移除，然后跳过这次循环进入下次循环。remove()方法稍后分析，这里先继续往下看。boolean done = false 这里设置一个标识，标识刷新操作是否执行完成，默认值 false 代表执行到这里没有执行完。writeSpinCount = config().getWriteSpinCount()

获得一个写操作的循环次数，默认值是 16，然后根据这个循环次数进行循环写操作。在循环中，关注下面这一步。

```
int localFlushedAmount = doWriteBytes(buf);
```

这一步就是将 Buffer 的内容写入 Channel，并返回写的字节数，这里会调用 NioSocketChannel 的 doWriteBytes，doWriteBytes()方法的代码如下。

```
protected int doWriteBytes(ByteBuf buf) throws Exception {
    final int expectedWrittenBytes = buf.readableBytes();
    return buf.readBytes(javaChannel(), expectedWrittenBytes);
}
```

这里首先获取 Buffer 的可读字节数，然后通过 readBytes 将可读字节写入 JDK 底层的 Channel。回到 doWrite()方法，将内容写入 JDK 底层的 Channel 之后，如果一个字节都没写，说明现在 Channel 可能不可写，将 setOpWrite 设置为 true，用于标识写操作位，并退出循环。如果已经写出字节，则通过 flushedAmount += localFlushedAmount 累加写出的字节数，然后根据 Buffer 是否没有可读字节数判断 Buffer 的数据是否已经写完，如果写完，将 done 设置为 true，说明写操作完成，并退出循环。因为有时候不一定一次就能将 ByteBuf 所有的字节写完，所以会继续通过循环进行写出，直到循环 16 次。如果 ByteBuf 的内容全部写完，会通过 in.remove()将当前 Entry 移除。remove()方法的代码如下。

```
public boolean remove() {
    //获取当前第一个 flush 的 Entry
    Entry e = flushedEntry;
    if (e == null) {
        clearNioBuffers();
        return false;
    }
    Object msg = e.msg;
    ChannelPromise promise = e.promise;
    int size = e.pendingSize;
    removeEntry(e);
    if (!e.cancelled) {
        ReferenceCountUtil.safeRelease(msg);
        safeSuccess(promise);
        decrementPendingOutboundBytes(size, false, true);
    }
    e.recycle();
    return true;
}
```

首先获取当前的 flushedEntry，重点关注 removeEntry 这步。

```
private void removeEntry(Entry e) {
    if (-- flushed == 0) {
        //位置为空
        flushedEntry = null;
        //如果是最后一个节点
        if (e == tailEntry) {
            //全部设置为空
            tailEntry = null;
            unflushedEntry = null;
        }
    } else {
        //移动到下一个节点
        flushedEntry = e.next;
    }
}
```

if (-- flushed == 0) 表示当前节点是否为需要刷新的最后一个节点，如果是，则 flushedEntry 指针设置为空。如果当前节点是 tailEntry 节点，说明当前节点是最后一个节点，将 tailEntry 和 unflushedEntry 两个指针全部设置为空。如果当前节点不是需要刷新的最后一个节点，则通过 flushedEntry = e.next 这步将 flushedEntry 指针移动到下一个节点。以上就是 flush 操作的相关逻辑。

12.4.5 数据输出回调

我们看一段写在 Handler 中的业务代码。

```
public void channelRead(ChannelHandlerContext ctx, Object msg) throws Exception {
    ChannelFuture future = ctx.writeAndFlush("test data");
    future.addListener(new ChannelFutureListener() {
        @Override
        public void operationComplete(ChannelFuture future) throws Exception {
            if (future.isSuccess()){
                System.out.println("写出成功");
            }else{
                System.out.println("写出失败");
            }
        }
    });
}
```

这种写法小伙伴们应该已经不陌生了，首先调用 writeAndFlush()方法将数据写出，然后返回的 Future 添加 Listener，并且重写回调函数。这只是一个最简单的示例，在回调函数中判断 Future 的状态成功与否，成功的话就打印“写出成功”，否则就打印“写出失败”。如果写在 Handler 中通常是 NioEventLoop 线程执行的，在 Future 返回之后才会执行添加 Listener 的操作，如果在用户线程中 writeAndFlush()是异步执行的，在添加监听的时候有可能写出操作没有执行完毕，等写出操作执行完毕之后才会执行回调。以上逻辑在代码中如何体现的呢？首先跟到 writeAndFlush()方法中去，会执行到 AbstractChannelHandlerContext 中的 writeAndFlush()方法中。

```
public ChannelFuture writeAndFlush(Object msg) {
    return writeAndFlush(msg, newPromise());
}
```

这里的逻辑在之前的章节中剖析过，重点关注 newPromise()方法。

```
public ChannelPromise newPromise() {
    return new DefaultChannelPromise(channel(), executor());
}
```

这里直接创建了 DefaultChannelPromise 这个对象并传入了当前 Channel 和当前 Channel 绑定的 NioEventLoop 对象。在 DefaultChannelPromise 构造方法中，也会将 Channel 和 NioEventLoop 对象绑定在自身成员变量中。回到 writeAndFlush()方法。

```
public ChannelFuture writeAndFlush(Object msg, ChannelPromise promise) {
    if (msg == null) {
        throw new NullPointerException("msg");
    }
    if (!validatePromise(promise, true)) {
        ReferenceCountUtil.release(msg);
        return promise;
    }
    write(msg, true, promise);
    return promise;
}
```

这里的逻辑也不陌生，注意最后返回了 Promise，其实就是上一步创建的 DefaultChannelPromise 对象，DefaultChannelPromise 实现了 ChannelFuture 接口，所以返回的 Promise 对象可以被 ChannelFuture 类型接收。继续看 write()方法。

```
private void write(Object msg, boolean flush, ChannelPromise promise) {
    AbstractChannelHandlerContext next = findContextOutbound();
    final Object m = pipeline.touch(msg, next);
    EventExecutor executor = next.executor();
    if (executor.inEventLoop()) {
        if (flush) {
            next.invokeWriteAndFlush(m, promise);
        } else {
            next.invokeWrite(m, promise);
        }
    } else {
        AbstractWriteTask task;
```

```
        if (flush) {
            task = WriteAndFlushTask.newInstance(next, m, promise);
        } else {
            task = WriteTask.newInstance(next, m, promise);
        }
        safeExecute(executor, task, promise, m);
    }
}
```

这里的逻辑我们同样也不陌生，如果是 NioEventLoop 线程，则继续调用 invokeWriteAndFlush()方法；如果不是 NioEventLoop 线程，则将 WriteAndFlush 事件封装成 Task，交给 EventLoop 线程异步。如果是异步执行，则到这一步之后，在业务代码中，writeAndFlush()就会返回并添加监听，有关添加监听的逻辑稍后分析。执行到这里，无论同步还是异步，都会执行到 invokeWriteAndFlush()方法。

```
public void write(ChannelHandlerContext ctx, Object msg, ChannelPromise promise) throws
Exception {
    unsafe.write(msg, promise);
}
```

最终调用 Unsafe 的 write()方法，并传入 Promise 对象。AbstractUnsafe 的 write()方法的代码如下。

```
public final void write(Object msg, ChannelPromise promise) {
    assertEventLoop();

    //负责缓冲写进来的 ByteBuf
    ChannelOutboundBuffer outboundBuffer = this.outboundBuffer;
    if (outboundBuffer == null) {
        safeSetFailure(promise, WRITE_CLOSED_CHANNEL_EXCEPTION);
        ReferenceCountUtil.release(msg);
        return;
    }

    int size;
    try {
        msg = filterOutboundMessage(msg);
        size = pipeline.estimatorHandle().size(msg);
        if (size < 0) {
            size = 0;
        }
    } catch (Throwable t) {
        safeSetFailure(promise, t);
        ReferenceCountUtil.release(msg);
        return;
    }
```

```
    //插入写队列
    outboundBuffer.addMessage(msg, size, promise);
}
```

上面的逻辑之前剖析过，我们关注两个部分，先看 catch 中 safeSetFailure()这步。因为是 catch 块，说明发生了异常，写入缓冲区不成功，safeSetFailure()就是用来设置写出失败的状态的。safeSetFailure()方法的代码如下。

```
protected final void safeSetFailure(ChannelPromise promise, Throwable cause) {
    if (!(promise instanceof VoidChannelPromise) && !promise.tryFailure(cause)) {
        logger.warn("Failed to mark a promise as failure because it's done already: {}", promise,
cause);
    }
}
```

这里看 if 判断，因为 Promise 是 DefaultChannelPromise，所以!(promise instanceof VoidChannelPromise)为 true。重点分析 promise.tryFailure(cause)，这里是设置失败状态，会调用 DefaultPromise 的 tryFailure()方法，代码如下。

```
public boolean tryFailure(Throwable cause) {
    if (setFailure0(cause)) {
        notifyListeners();
        return true;
    }
    return false;
}
```

再看 setFailure0(cause)。

```
private boolean setValue0(Object objResult) {
    if (RESULT_UPDATER.compareAndSet(this, null, objResult) ||
        RESULT_UPDATER.compareAndSet(this, UNCANCELLABLE, objResult)) {
        checkNotifyWaiters();
        return true;
    }
    return false;
}
```

if 块中的 CAS 操作，会将参数 objResult 的值设置到 DefaultPromise 的成员变量 Result 中，表示当前操作为异常状态。

回到 tryFailure()方法，我们关注 notifyListeners()方法，这个方法是执行添加监听的回调函数，当 writeAndFlush()和 addListener()是异步执行的时候，这里有可能已经添加，所以通过这个方法可以调用添加监听后的回调。如果 writeAndFlush()和 addListener()是同步执行的时候，也就是都在

NioEventLoop 线程中执行的时候，那么走到这里 addListener()还没执行，所以不能回调添加监听的回调函数，那么回调是什么时候执行的呢？我们在剖析 addListener()步骤的时候会给大家分析。具体执行回调我们在讲解添加监听的时候进行剖析，以上就是记录异常状态的大概逻辑。

回到 AbstractUnsafe 的 write()方法，我们再关注 outboundBuffer.addMessage(msg, size, promise)这一步，addMessage()方法的代码如下。

```
public void addMessage(Object msg, int size, ChannelPromise promise) {
    Entry entry = Entry.newInstance(msg, size, total(msg), promise);
    //代码省略
}
```

我们只需要关注包装 Entry 的 newInstance()方法，该方法传入 Promise 对象，代码如下。

```
static Entry newInstance(Object msg, int size, long total, ChannelPromise promise) {
    Entry entry = RECYCLER.get();
    entry.msg = msg;
    entry.pendingSize = size;
    entry.total = total;
    entry.promise = promise;
    return entry;
}
```

这里将 Promise 设置到 Entry 的成员变量中了，也就是说，每个 Entry 都关联了唯一的一个 Promise，回到 AbstractChannelHandlerContext 的 invokeWriteAndFlush()方法。

```
private void invokeWriteAndFlush(Object msg, ChannelPromise promise) {
    if (invokeHandler()) {
        invokeWrite0(msg, promise);
        invokeFlush0();
    } else {
        writeAndFlush(msg, promise);
    }
}
```

上面分析了 write()操作中 Promise 的传递以及状态设置的大概过程，继续看 flush()方法中 Promise 的操作过程。这里 invokeFlush0()并没有传入 Promise 对象，是因为 Promise 对象会绑定在缓冲区中 Entry 的成员变量中，可以通过其成员变量获取 Promise 对象。我们之前也分析过 invokeFlush0()，通过事件传递，最终会调用 HeadContext 的 flush()方法。

```
public void flush(ChannelHandlerContext ctx) throws Exception {
    unsafe.flush();
}
```

最后看 AbstractUnsafe 的 flush()方法。

```
public final void flush() {
    assertEventLoop();
    ChannelOutboundBuffer outboundBuffer = this.outboundBuffer;
    if (outboundBuffer == null) {
        return;
    }
    outboundBuffer.addFlush();
    flush0();
}
```

这块逻辑之前已分析过，继续看 flush0()方法。

```
protected void flush0() {
    //代码省略
    try {
        doWrite(outboundBuffer);
    } catch (Throwable t) {
        //代码省略
    } finally {
        inFlush0 = false;
    }
}
```

由于篇幅原因我们省略了大段代码，继续跟进 doWrite()方法。

```
protected void doWrite(ChannelOutboundBuffer in) throws Exception {
    int writeSpinCount = -1;
    boolean setOpWrite = false;
    for (;;) {
        Object msg = in.current();
        if (msg == null) {
            clearOpWrite();
            return;
        }
        if (msg instanceof ByteBuf) {
            //代码省略
            boolean done = false;
            //代码省略
            if (done) {
                //移除当前对象
                in.remove();
            } else {
                break;
            }
        } else if (msg instanceof FileRegion) {
            //代码省略
        } else {
            throw new Error();
```

```
        }
    }
    incompleteWrite(setOpWrite);
}
```

这里也省略了大段代码，我们重点关注 in.remove()。之前介绍过，如果 done 为 true，说明刷新事件已完成，则移除当前 Entry 节点，我们看 remove()方法的代码。

```
public boolean remove() {
    Entry e = flushedEntry;
    if (e == null) {
        clearNioBuffers();
        return false;
    }
    Object msg = e.msg;
    ChannelPromise promise = e.promise;
    int size = e.pendingSize;
    removeEntry(e);
    if (!e.cancelled) {
        ReferenceCountUtil.safeRelease(msg);
        safeSuccess(promise);
        decrementPendingOutboundBytes(size, false, true);
    }
    e.recycle();
    return true;
}
```

我们看上面代码的这一步。

```
ChannelPromise promise = e.promise;
```

之前已经剖析过 Promise 对象会绑定在 Entry 中，而这步就是从 Entry 中获取 Promise 对象，等 remove()操作完成，会执行到下面这一步。

```
safeSuccess(promise);
```

这一步正好和前面分析的 safeSetFailure()相反，这里是设置成功状态，safeSuccess()方法的代码如下。

```
private static void safeSuccess(ChannelPromise promise) {
    if (!(promise instanceof VoidChannelPromise)) {
        PromiseNotificationUtil.trySuccess(promise, null, logger);
    }
}
```

再跟进 trySuccess()方法。

```
public static <V> void trySuccess(Promise<? super V> p, V result, InternalLogger logger) {
```

```
    if (!p.trySuccess(result) && logger != null) {
        //代码省略
    }
}
```

继续跟进 if 中的 trySuccess()方法，最后会执行到 DefaultPromise 的 trySuccess()方法。

```
public boolean trySuccess(V result) {
    if (setSuccess0(result)) {
        notifyListeners();
        return true;
    }
    return false;
}
```

这里看 setSuccess0()方法的代码。

```
private boolean setSuccess0(V result) {
    return setValue0(result == null ? SUCCESS : result);
}
```

上面的逻辑已经剖析过了，这里参数传入一个信号 SUCCESS，表示设置成功状态。继续跟进 setValueo()方法。

```
private boolean setValue0(Object objResult) {
    if (RESULT_UPDATER.compareAndSet(this, null, objResult) ||
        RESULT_UPDATER.compareAndSet(this, UNCANCELLABLE, objResult)) {
        checkNotifyWaiters();
        return true;
    }
    return false;
}
```

同样，在 if 判断中，通过 CAS 操作将参数传入的 SUCCESS 对象赋值到 DefaultPromise 的属性 Result 中，我们看 private volatile Object result 这个属性的类型是 Object 类型，也就是说可以赋值为任意对象。SUCCESS 是一个 Signal 类型的对象，可以简单理解成是一种状态，SUCCESS 表示一种成功的状态。通过上述 CAS 操作，Result 的值将赋值成 SUCCESS，我们回到 trySuccess()方法。

```
public boolean trySuccess(V result) {
    if (setSuccess0(result)) {
        notifyListeners();
        return true;
    }
    return false;
}
```

设置完成功状态之后，会通过 notifyListeners()执行监听中的回调。

```
@Override
public void channelRead(ChannelHandlerContext ctx, Object msg) throws Exception {
    ChannelFuture future = ctx.writeAndFlush("test data");
    future.addListener(new ChannelFutureListener() {
        @Override
        public void operationComplete(ChannelFuture future) throws Exception {
            if (future.isSuccess()){
                System.out.println("写出成功");
            }else{
                System.out.println("写出失败");
            }
        }
    });
}
```

在回调中判断 future.isSuccess()，如果 Promise 设置为成功状态，会返回 true，从而打印“写出成功”。跟到 isSuccess()方法中，会调用 DefaultPromise 的 isSuccess()方法。

```
public boolean isSuccess() {
    Object result = this.result;
    return result != null && result != UNCANCELLABLE && !(result instanceof CauseHolder);
}
```

我们看到，首先会获取 Result 对象，然后判断 Result 不为空，并且不是 UNCANCELLABLE，且不属于 CauseHolder 对象。

前面分析过，如果 Promise 设置为成功状态，则 Result 为 SUCCESS，条件成立，可以执行 if (future.isSuccess()) 中 if 块的逻辑。和设置错误状态的逻辑一样，这里也有同样的问题，如果 writeAndFlush()和 addListener()是异步操作，那么执行到回调的时候，可能 addListener()已经添加完成，可以正常地执行回调。那么，如果 writeAndFlush()和 addListener()是同步操作，writeAndFlush()在执行回调的时候，addListener()并没有执行，无法执行回调方法，那么回调方法是如何执行的呢？我们看 addListener()方法，addListener()传入 ChannelFutureListener 对象，并重写了 operationComplete()方法，也就是执行回调的方法，会执行到 DefaultChannelPromise 的 addListener()方法，代码如下。

```
public ChannelPromise addListener(GenericFutureListener<? extends Future<? super Void>>
listener) {
    super.addListener(listener);
    return this;
}
```

跟进父类的 addListener()。

```
public Promise<V> addListener(GenericFutureListener<? extends Future<? super V>> listener) {
    checkNotNull(listener, "listener");
    synchronized (this) {
        addListener0(listener);
    }
    if (isDone()) {
        notifyListeners();
    }
    return this;
}
```

上面代码通过 addListener0()方法添加 Listener，因为添加 Listener 的操作有可能会在不同的线程中进行，比如用户线程和 NioEventLoop 线程，所以为了防止并发问题，这里增加了 synchronized 关键字。addListener0()方法的代码如下。

```
private void addListener0(GenericFutureListener<? extends Future<? super V>> listener) {
    if (listeners == null) {
        listeners = listener;
    } else if (listeners instanceof DefaultFutureListeners) {
        ((DefaultFutureListeners) listeners).add(listener);
    } else {
        listeners = new DefaultFutureListeners((GenericFutureListener<? extends Future<V>>)
listeners, listener);
    }
}
```

如果是第一次添加 Listener，则成员变量 Listeners 为 null，就把参数传入的 GenericFutureListener 赋值到成员变量 Listeners。如果是第二次添加 Listener，Listeners 不为空，会执行 else if 判断，因为第一次添加的 Listener 是 GenericFutureListener 类型，并不是 DefaultFutureListeners 类型，所以 else if 判断返回 false，进入 else 块。在 else 块中，通过 new 方式创建一个 DefaultFutureListeners 对象并赋值到成员变量 Listeners 中。在 DefaultFutureListeners 的构造方法中，第一个参数传入 DefaultPromise 中的成员变量 Listeners，也就是第一次添加的 GenericFutureListener 对象，第二个参数传入第二次添加的 GenericFutureListener 对象。通过两个 GenericFutureListener 对象包装成一个 DefaultFutureListeners 对象。下面来看 Listeners 的定义。

```
private Object listeners;
```

这里 Listeners 是 Object 类型，可以保存任何类型的对象。再看 DefaultFutureListeners 的构造方法。

```
DefaultFutureListeners(
        GenericFutureListener<? extends Future<?>> first, GenericFutureListener<? extends
Future<?>> second) {
    listeners = new GenericFutureListener[2];
```

```
    //第 0 个
    listeners[0] = first;
    //第 1 个
    listeners[1] = second;
    size = 2;
    //代码省略
}
```

在 DefaultFutureListeners 类中也定义了一个成员变量 Listeners，类型为 GenericFutureListener 数组。在构造方法中初始化 Listeners 这个数组，并且数组中第一个值赋值为第一次添加的 GenericFutureListener，第二个值赋值为第二次添加的 GenericFutureListener。回到 addListener0() 方法。

```
private void addListener0(GenericFutureListener<? extends Future<? super V>> listener) {
    if (listeners == null) {
        listeners = listener;
    } else if (listeners instanceof DefaultFutureListeners) {
        ((DefaultFutureListeners) listeners).add(listener);
    } else {
        listeners = new DefaultFutureListeners((GenericFutureListener<? extends Future<V>>)
listeners, listener);
    }
}
```

经过两次添加 Listener，属性 Listeners 的值就变成了 DefaultFutureListeners 类型的对象，如果第三次添加 Listener，则会执行到 else if 块中，DefaultFutureListeners 对象通过调用 add()方法继续添加 Listener。add()方法的代码如下。

```
public void add(GenericFutureListener<? extends Future<?>> l) {
    GenericFutureListener<? extends Future<?>>[] listeners = this.listeners;
    final int size = this.size;
    if (size == listeners.length) {
        this.listeners = listeners = Arrays.copyOf(listeners, size << 1);
    }
    listeners[size] = l;
    this.size = size + 1;
    //代码省略
}
```

这里的逻辑也比较简单，就是为当前的数组对象 Listeners 追加新的 GenericFutureListener 对象，如果 Listeners 容量不足则进行扩容操作。根据以上逻辑，就完成了 Listener 的添加逻辑。那么再看之前遗留的问题，如果 writeAndFlush()和 addListener()是同步进行的，writeAndFlush()执行回调时，addListener()还没有执行回调，那么回调是如何执行的呢?回到 DefaultPromise 的 addListener()。

```
public Promise<V> addListener(GenericFutureListener<? extends Future<? super V>> listener) {
```

```
    checkNotNull(listener, "listener");
    synchronized (this) {
        addListener0(listener);
    }
    if (isDone()) {
        notifyListeners();
    }
    return this;
}
```

分析完 addListener0()方法，再往下看。这里会有 if 判断 isDone()，就是程序执行到这一步的时候，判断刷新事件是否执行完成。isDone()方法的代码如下。

```
public boolean isDone() {
    return isDone0(result);
}
```

继续跟进 isDone0()，传入了成员变量 Result。

```
private static boolean isDone0(Object result) {
    return result != null && result != UNCANCELLABLE;
}
```

这里判断 Result 不为 null 并且不为 UNCANCELLABLE，就表示完成。因为成功的状态是 SUCCESS，所以 flush 成功会返回 true。

回到 addListener()方法中，如果执行完成，就通过 notifyListeners()方法执行回调，这也解释了前面的问题，在同步操作中，writeAndFlush()在执行回调时并没有添加 Listener，所以添加 Listener 的时候会判断 writeAndFlush()的执行状态，如果状态是完成，则这里会执行回调。同样，在异步操作中，执行到这里 writeAndFlush()可能还没完成，所以不会执行回调，由 writeAndFlush()执行回调。无论 writeAndFlush()和 addListener()谁先完成，都可以执行到回调方法。notifyListeners()方法的代码如下。

```
private void notifyListeners() {
    EventExecutor executor = executor();
    if (executor.inEventLoop()) {
        final InternalThreadLocalMap threadLocals = InternalThreadLocalMap.get();
        final int stackDepth = threadLocals.futureListenerStackDepth();
        if (stackDepth < MAX_LISTENER_STACK_DEPTH) {
            threadLocals.setFutureListenerStackDepth(stackDepth + 1);
            try {
                notifyListenersNow();
            } finally {
                threadLocals.setFutureListenerStackDepth(stackDepth);
            }
```

```
            return;
        }
    }
    safeExecute(executor, new Runnable() {
        @Override
        public void run() {
            notifyListenersNow();
        }
    });
}
```

首先判断是否是 EventLoop 线程，如果是则执行 if 块中的逻辑，如果不是则把执行回调的逻辑封装成 Task 放到 EventLoop 的任务队列中异步执行。重点关注 notifyListenersNow()方法。

```
private void notifyListenersNow() {
    Object listeners;
    synchronized (this) {
        if (notifyingListeners || this.listeners == null) {
            return;
        }
        notifyingListeners = true;
        listeners = this.listeners;
        this.listeners = null;
    }
    for (;;) {
        if (listeners instanceof DefaultFutureListeners) {
            notifyListeners0((DefaultFutureListeners) listeners);
        } else {
            notifyListener0(this, (GenericFutureListener<? extends Future<V>>) listeners);
        }
        //代码省略
    }
}
```

在无限 for 循环中，首先判断 Listeners 是不是 DefaultFutureListeners 类型，根据之前的逻辑，如果只添加了一个 Listener，则 Listener 是 GenericFutureListener 类型。通常在添加的时候只会添加一个 Listener，跟进 else 块中的 notifyListener0()方法。

```
private static void notifyListener0(Future future, GenericFutureListener l) {
    try {
        l.operationComplete(future);
    } catch (Throwable t) {
        logger.warn("An exception was thrown by " + l.getClass().getName() +
".operationComplete()", t);
    }
}
```

我们看到，上面代码中执行了 GenericFutureListener 中重写的回调函数 operationComplete()。以上就是执行回调的相关逻辑。

12.5　自定义编解码

尽管 Netty 预置了丰富的编解码类库功能，但是在实际的业务开发过程中，总是需要对编解码功能做一些定制。使用 Netty 的编解码框架，可以非常方便地进行协议定制。本节将对常用的支持定制的编解码类库进行讲解，以期让读者能够尽快熟悉和掌握编解码框架。

12.5.1　MessageToMessageDecoder 抽象解码器

MessageToMessageDecoder 实际上是 Netty 的二次解码器，它的职责是将一个对象二次解码为其他对象。

为什么称它为二次解码器呢？我们知道，从 SocketChannel 读取的 TCP 数据报是 ByteBuf，实际就是字节数组。首先需要将 ByteBuf 缓冲区中的数据报读取出来，并将其解码为 Java 对象；然后根据某些规则对 Java 对象做二次解码，将其解码为另一个 POJO 对象。因为 MessageToMessageDecoder 在 ByteToMessageDecoder 之后，所以称之为二次解码器。

二次解码器在实际的商业项目中非常有用，以 HTTP+XML 协议栈为例，第一次解码往往是将字节数组解码成 HttpRequest 对象，然后对 HttpRequest 消息中的消息体字符串进行二次解码，将 XML 格式的字符串解码为 POJO 对象，这就用到了二次解码器。类似这样的场景还有很多，不再一一列举。

事实上，做一个超级复杂的解码器，将多个解码器组合成一个大而全的 MessageToMessageDecoder 解码器似乎也能解决多次解码的问题，但是采用这种方式的代码可维护性会非常差。例如，如果我们打算在 HTTP+XML 协议栈中增加一个打印码流的功能，即首次解码获取 HttpRequest 对象之后打印 XML 格式的码流。如果采用多个解码器组合，在中间插入一个打印消息体的 Handler 即可，不需要修改原有的代码；如果做一个大而全的解码器，就需要在解码的方法中增加打印码流的代码，可扩展性和可维护性都会变差。

用户的解码器只需要实现 void decode(ChannelHandlerContext ctx, I msg, List<Object> out)抽象方法即可，由于它是将一个 POJO 对象解码为另一个 POJO 对象，所以一般不会涉及半包的处理，相对于 ByteToMessageDecoder 更加简单些。它的继承关系如下图所示。

```
▼ MessageToMessageDecoder (io.netty.handler.codec)
    Anonymous in decoder in MessageToMessageCodec (io.netty.handler.codec)
    Base64Decoder (io.netty.handler.codec.base64)
    ByteArrayDecoder (io.netty.handler.codec.bytes)
    DatagramDnsQueryDecoder (io.netty.handler.codec.dns)
    DatagramDnsResponseDecoder (io.netty.handler.codec.dns)
    DatagramPacketDecoder (io.netty.handler.codec)
  ▼ HttpContentDecoder (io.netty.handler.codec.http)
        HttpContentDecompressor (io.netty.handler.codec.http)
  ▶ MessageAggregator (io.netty.handler.codec)
    ProtobufDecoder (io.netty.handler.codec.protobuf)
    ProtobufDecoderNano (io.netty.handler.codec.protobuf)
    RedisArrayAggregator (io.netty.handler.codec.redis)
    SctpInboundByteStreamHandler (io.netty.handler.codec.sctp)
    SctpMessageCompletionHandler (io.netty.handler.codec.sctp)
    SctpMessageToMessageDecoder (io.netty.handler.codec.sctp)
    SpdyHttpDecoder (io.netty.handler.codec.spdy)
    StringDecoder (io.netty.handler.codec.string)
  ▶ WebSocketExtensionDecoder (io.netty.handler.codec.http.websocketx.extensions)
  ▶ WebSocketProtocolHandler (io.netty.handler.codec.http.websocketx)
```

12.5.2 MessageToMessageEncoder 抽象编码器

将一个 POJO 对象编码成另一个对象，以 HTTP+XML 协议为例，它的一种实现方式是：先将 POJO 对象编码成 XML 字符串，再将字符串编码为 HTTP 请求或者应答消息。对于复杂协议，往往需要经历多次编码，为了便于功能扩展，可以通过多个编码器组合来实现相关功能。

用户的解码器继承自 MessageToMessageEncoder 解码器，实现 void encode(ChannelHandlerContext ctx, I msg, List<Object> out)方法即可。注意，它与 MessageToByteEncoder 的区别在于输出的是对象列表而不是 ByteBuf，示例代码如下。

```java
public class IntegerToStringEncoder extends MessageToMessageEncoder <Integer> {
        @Override
        public void encode(ChannelHandlerContext ctx, Integer message,
           List<Object> out)
                throws Exception
        {
           out.add(message.toString());
        }
}
```

MessageToMessageEncoder 编码器的实现原理与之前分析的 MessageToByteEncoder 相似，唯一的差别是它编码后的输出是中间对象，并非最终可传输的 ByteBuf。

简单看下它的源码实现：创建 RecyclableArrayList 对象，判断当前需要编码的对象是否是编码器可处理的类型，如果不是，则忽略，执行下一个 ChannelHandler 的 write()方法。

具体的编码方法实现由用户子类编码器负责完成，如果编码后的 RecyclableArrayList 为空，说明编码没有成功，释放 RecyclableArrayList 引用。

如果编码成功，则通过遍历 RecyclableArrayList，循环发送编码后的 POJO 对象，代码如下。

```
public void write(ChannelHandlerContext ctx, Object msg, ChannelPromise promise) throws
Exception {
       CodecOutputList out = null;
       try {
           if (acceptOutboundMessage(msg)) {
               out = CodecOutputList.newInstance();
               @SuppressWarnings("unchecked")
               I cast = (I) msg;
               try {
                   encode(ctx, cast, out);
               } finally {
                   ReferenceCountUtil.release(cast);
               }

               if (out.isEmpty()) {
                   out.recycle();
                   out = null;

                   throw new EncoderException(
                           StringUtil.simpleClassName(this) + " must produce at least one
message.");
               }
           } else {
               ctx.write(msg, promise);
           }
       }
       //省略异常处理代码
   }
```

12.5.3　ObjectEncoder 序列化编码器

ObjectEncoder 是 Java 序列化编码器，它负责将实现 Serializable 接口的对象序列化为 byte[]，然后写入 ByteBuf 用于消息的跨网络传输。下面分析它的实现。它继承自 MessageToByteEncoder，作用是将对象编码成 ByteBuf。

```
public class ObjectEncoder extends MessageToByteEncoder<Serializable>
```

如果要使用 Java 序列化，对象必须实现 Serializable 接口，因此，它的泛型类型为 Serializable。

MessageToByteEncoder 的子类只需要实现 encode(ChannelHandlerContext ctx, I msg, ByteBuf out)方法即可，下面重点关注 encode()方法的实现。

```
protected void encode(ChannelHandlerContext ctx, Serializable msg, ByteBuf out) throws
Exception {
        int startIdx = out.writerIndex();

        ByteBufOutputStream bout = new ByteBufOutputStream(out);
        bout.write(LENGTH_PLACEHOLDER);
        ObjectOutputStream oout = new CompactObjectOutputStream(bout);
        oout.writeObject(msg);
        oout.flush();
        oout.close();

        int endIdx = out.writerIndex();

        out.setInt(startIdx, endIdx - startIdx - 4);
    }
```

首先创建 ByteBufOutputStream 和 ObjectOutputStream，用于将 Object 对象序列化到 ByteBuf 中。值得注意的是，在 writeObject()之前需要先将长度属性（4 字节）预留，用于后续长度属性的更新。

依次写入长度占位符（4 字节）、序列化之后的 Object 对象，再根据 ByteBuf 的 writeIndex()计算序列化之后的码流长度，最后调用 ByteBuf 的 setInt(int index, int value)更新长度占位符为实际的码流长度。

有个细节需要注意，更新码流长度属性使用了 setInt()方法而不是 writeInt()方法，原因就是 setInt()方法只更新内容，而不修改 readerIndex 和 writerIndex。

12.5.4 LengthFieldPrepender 通用编码器

如果协议中的第一个属性为长度属性，Netty 提供了 LengthFieldPrepender 编码器，它可以计算当前待发送消息的二进制字节长度，将该长度添加到 ByteBuf 的缓冲区头中，如下所示。

```
编码前(12 字节)                编码后(14 字节)
+----------------+             +--------+----------------+
|  "HELLO,WORLD" |    ----->   | 0x000C |  "HELLO,WORLD"  |
+----------------+             +--------+----------------+
```

通过 LengthFieldPrepender 可以将待发送消息的长度写入 ByteBuf 的前 2 字节，编码后的消息组成为长度属性+原消息的方式。

通过设置 LengthFieldPrepender 为 true，消息长度将包含长度本身占用的字节数，打开 LengthFieldPrepender 后，上面示例中的编码结果如下所示。

```
编码前(12 字节)                   编码后(14 字节)
+----------------+               +--------+----------------+
|  "HELLO,WORLD" |    ----->     | 0x000E |  "HELLO,WORLD"  |
+----------------+               +--------+----------------+
```

LengthFieldPrepender 工作原理分析如下：首先对长度属性进行设置，如果需要包含消息长度自身，则在原来长度的基础上再加上 lengthFieldLength 的长度。

如果调整后的消息长度小于 0，则抛出参数非法异常。对消息长度自身所占的字节数进行判断，以便采用正确的方法将长度属性写入 ByteBuf，共有 6 种可能。

（1）长度属性所占字节为 1：如果使用 1 字节代表消息长度，则最大长度需要小于 256 字节。对长度进行校验，如果校验失败，则抛出参数非法异常；若校验通过，则创建新的 ByteBuf 并通过 writeByte 将长度值写入 ByteBuf。

（2）长度属性所占字节为 2：如果使用 2 字节代表消息长度，则最大长度需要小于 65 536 字节，对长度进行校验，如果校验失败，则抛出参数非法异常；若校验通过，则创建新的 ByteBuf 并通过 writeShort 将长度值写入 ByteBuf。

（3）长度属性所占字节为 3：如果使用 3 字节代表消息长度，则最大长度需要小于 16 777 216 字节，对长度进行校验，如果校验失败，则抛出参数非法异常；若校验通过，则创建新的 ByteBuf 并通过 writeMedium 将长度值写入 ByteBuf。

（4）长度属性所占字节为 4：创建新的 ByteBuf，并通过 writeInt 将长度值写入 ByteBuf。

（5）长度属性所占字节为 8：创建新的 ByteBuf，并通过 writeLong 将长度值写入 ByteBuf。

（6）其他长度值：直接抛出 Error。

相关代码如下。

```
protected void encode(ChannelHandlerContext ctx, ByteBuf msg, List<Object> out) throws
Exception {
        int length = msg.readableBytes() + lengthAdjustment;
        if (lengthIncludesLengthFieldLength) {
            length += lengthFieldLength;
        }

        if (length < 0) {
            throw new IllegalArgumentException(
```

```
                "Adjusted frame length (" + length + ") is less than zero");
    }

    switch (lengthFieldLength) {
    case 1:
        if (length >= 256) {
            throw new IllegalArgumentException(
                    "length does not fit into a byte: " + length);
        }
        out.add(ctx.alloc().buffer(1).order(byteOrder).writeByte((byte) length));
        break;
    case 2:
        if (length >= 65536) {
            throw new IllegalArgumentException(
                    "length does not fit into a short integer: " + length);
        }
        out.add(ctx.alloc().buffer(2).order(byteOrder).writeShort((short) length));
        break;
    case 3:
        if (length >= 16777216) {
            throw new IllegalArgumentException(
                    "length does not fit into a medium integer: " + length);
        }
        out.add(ctx.alloc().buffer(3).order(byteOrder).writeMedium(length));
        break;
    case 4:
        out.add(ctx.alloc().buffer(4).order(byteOrder).writeInt(length));
        break;
    case 8:
        out.add(ctx.alloc().buffer(8).order(byteOrder).writeLong(length));
        break;
    default:
        throw new Error("should not reach here");
    }
    out.add(msg.retain());
}
```

4

第 4 篇

Netty 实战篇

13

第 13 章 基于 Netty 手写消息推送系统

13.1 环境搭建

配置 pom.xml 文件，主要是配置好 Netty 的依赖、自定义序列化依赖和日志打印的依赖。

```
<dependencies>
        <dependency>
          <groupId>io.netty</groupId>
          <artifactId>netty-all</artifactId>
          <version>4.1.6.Final</version>
        </dependency>
<dependency>
          <groupId>org.msgpack</groupId>
          <artifactId>msgpack</artifactId>
          <version>0.6.12</version>
        </dependency>
         <dependency>
          <groupId>com.alibaba</groupId>
          <artifactId>fastjson</artifactId>
          <version>1.2.4</version>
        </dependency>
        <dependency>
```

```
            <groupId>org.projectlombok</groupId>
            <artifactId>lombok</artifactId>
            <version>1.16.10</version>
        </dependency>
        <dependency>
            <groupId>org.slf4j</groupId>
            <artifactId>slf4j-api</artifactId>
            <version>1.7.25</version>
        </dependency>
        <dependency>
            <groupId>ch.qos.logback</groupId>
            <artifactId>logback-classic</artifactId>
            <version>1.2.3</version>
        </dependency>
</dependencies>
```

13.2　多协议通信设计

13.2.1　自定义协议规则

我们先来规定一下自定义协议的内容。一般来说，协议都分为请求头和请求上下文，为了区别于其他协议，规定请求头元素用[]，每个[]中的内容表示一个头元素。例如：[LOGIN]表示登录动作。

我们约定一下协议规则，如下表所示。

下行命令：指服务端向客户端发送的消息内容	
SYSTEM	系统命令，例如[命令][命令发送时间][接收人] －系统提示内容 例如：[SYSTEM][124343423123][Tom老师] － Student加入聊天室
上行命令：指客户端向服务端发送的命令	
LOGIN	登录动作：[命令][命令发送时间][命令发送人] 例如：[LOGIN][124343423123][Tom老师]
LOGOUT	退出登录动作：[命令][命令发送时间][命令发送人] 例如：[LOGOUT][124343423123][Tom老师]
CHAT	聊天：[命令][命令发送时间][命令发送人][命令接收人] － 聊天内容 例如：[CHAT][124343423123][Tom老师][ALL] －大家好，我是Tom老师！
FLOWER	发送鲜花特效：[命令][命令发送时间][命令发送人][命令接收人] 例如：[FLOWER][124343423123][you][ALL]

IMP（Instant Messaging Protocol，即时通信协议）枚举类定义消息指令如下。

```
package com.gupaoedu.vip.netty.chat.protocol;


/**
 * 自定义 IMP
 *
 */
public enum IMP {
    /** 系统消息 */
    SYSTEM("SYSTEM"),
    /** 登录指令 */
    LOGIN("LOGIN"),
    /** 退出指令 */
    LOGOUT("LOGOUT"),
    /** 聊天消息 */
    CHAT("CHAT"),
    /** 送鲜花 */
    FLOWER("FLOWER");

    private String name;

    public static boolean isIMP(String content){
        return content.matches("^\\[(SYSTEM|LOGIN|LOGIN|CHAT)\\]");
    }

    IMP(String name){
        this.name = name;
    }

    public String getName(){
        return this.name;
    }

    public String toString(){
        return this.name;
    }

}
```

IMMessage 类，用于封装自定义协议的消息内容。

```
package com.gupaoedu.vip.netty.chat.protocol;

import lombok.Data;
import org.msgpack.annotation.Message;
```

```java
/**
 * 自定义消息实体类
 *
 */
@Message
@Data
public class IMMessage{

    private String addr;        //IP 地址及端口
    private String cmd;         //命令类型[LOGIN]或者[SYSTEM]或者[LOGOUT]
    private long time;          //命令发送时间
    private int online;         //当前在线人数
    private String sender;      //发送人
    private String receiver;    //接收人
    private String content;     //消息内容
    private String terminal;    //终端

    public IMMessage(){}

    public IMMessage(String cmd,long time,int online,String content){
        this.cmd = cmd;
        this.time = time;
        this.online = online;
        this.content = content;
        this.terminal = terminal;
    }

    public IMMessage(String cmd,String terminal,long time,String sender){
        this.cmd = cmd;
        this.time = time;
        this.sender = sender;
        this.terminal = terminal;
    }


    public IMMessage(String cmd,long time,String sender,String content){
        this.cmd = cmd;
        this.time = time;
        this.sender = sender;
        this.content = content;
        this.terminal = terminal;
    }

    @Override
    public String toString() {
        return "IMMessage{" +
```

```
                "addr='" + addr + '\'' +
                ", cmd='" + cmd + '\'' +
                ", time=" + time +
                ", online=" + online +
                ", sender='" + sender + '\'' +
                ", receiver='" + receiver + '\'' +
                ", content='" + content + '\'' +
                '}';
    }
}
```

13.2.2 自定义编解码器

IMDecoder 类，主要功能是接收消息后对自定义协议内容进行解码处理。

```
package com.gupaoedu.vip.netty.chat.protocol;

import io.netty.buffer.ByteBuf;
import io.netty.channel.ChannelHandlerContext;
import io.netty.handler.codec.ByteToMessageDecoder;

import java.util.List;
import java.util.regex.Matcher;
import java.util.regex.Pattern;

import org.msgpack.MessagePack;
import org.msgpack.MessageTypeException;

/**
 * 自定义 IMP 的编码器
 */
public class IMDecoder extends ByteToMessageDecoder {

    //解析 IM 写一下请求内容的正则
    private Pattern pattern = Pattern.compile("^\\[(.*)\\](\\s\\-\\s(.*))?");

    @Override
    protected void decode(ChannelHandlerContext ctx, ByteBuf in,List<Object> out) throws
Exception {
        try{
            //先获取可读字节数
            final int length = in.readableBytes();
            final byte[] array = new byte[length];
            String content = new String(array,in.readerIndex(),length);

            //空消息不解析
            if(!(null == content || "".equals(content.trim()))){
```

```
            if(!IMP.isIMP(content)){
                ctx.channel().pipeline().remove(this);
                return;
            }
           }

           in.getBytes(in.readerIndex(), array, 0, length);
           out.add(new MessagePack().read(array,IMMessage.class));
           in.clear();
       }catch(MessageTypeException e){
            ctx.channel().pipeline().remove(this);
       }
   }

   /**
    * 字符串解析成自定义即时通信协议
    * @param msg
    * @return
    */
   public IMMessage decode(String msg){
       if(null == msg || "".equals(msg.trim())){ return null; }
       try{
            Matcher m = pattern.matcher(msg);
            String header = "";
            String content = "";
            if(m.matches()){
                header = m.group(1);
                content = m.group(3);
            }

            String [] heards = header.split("\\]\\[");
            long time = 0;
            try{ time = Long.parseLong(heards[1]); } catch(Exception e){}
            String nickName = heards[2];
            //昵称最多十个字符长度
            nickName = nickName.length() < 10 ? nickName : nickName.substring(0, 9);

            if(msg.startsWith("[" + IMP.LOGIN.getName() + "]")){
                return new IMMessage(heards[0],heards[3],time,nickName);
            }else if(msg.startsWith("[" + IMP.CHAT.getName() + "]")){
                return new IMMessage(heards[0],time,nickName,content);
            }else if(msg.startsWith("[" + IMP.FLOWER.getName() + "]")){
                return new IMMessage(heards[0],heards[3],time,nickName);
            }else{
                return null;
            }
       }catch(Exception e){
```

```
            e.printStackTrace();
            return null;
        }
    }
}
```

IMEncoder 类，主要功能是发送消息前对自定义协议内容进行编码处理。

```
package com.gupaoedu.vip.netty.chat.protocol;

import org.msgpack.MessagePack;

import io.netty.buffer.ByteBuf;
import io.netty.channel.ChannelHandlerContext;
import io.netty.handler.codec.MessageToByteEncoder;

/**
 * 自定义 IMP 的编码器
 */
public class IMEncoder extends MessageToByteEncoder<IMMessage> {

    @Override
    protected void encode(ChannelHandlerContext ctx, IMMessage msg, ByteBuf out)
            throws Exception {
        out.writeBytes(new MessagePack().write(msg));
    }

    public String encode(IMMessage msg){
        if(null == msg){ return ""; }
        String prex = "[" + msg.getCmd() + "]" + "[" + msg.getTime() + "]";
        if(IMP.LOGIN.getName().equals(msg.getCmd()) ||
            IMP.FLOWER.getName().equals(msg.getCmd())){
            prex += ("[" + msg.getSender() + "][" + msg.getTerminal() + "]");
        }else if(IMP.CHAT.getName().equals(msg.getCmd())){
            prex += ("[" + msg.getSender() + "]");
        }else if(IMP.SYSTEM.getName().equals(msg.getCmd())){
            prex += ("[" + msg.getOnline() + "]");
        }
        if(!(null == msg.getContent() || "".equals(msg.getContent()))){
            prex += (" - " + msg.getContent());
        }
        return prex;
    }

}
```

13.2.3　对 HTTP 的支持

HttpServerHandler 类，主要处理服务端分发请求的逻辑。

```
package com.gupaoedu.vip.netty.chat.server.handler;

import java.io.File;
import java.io.RandomAccessFile;
import java.net.URL;

import lombok.extern.slf4j.Slf4j;

import io.netty.*;

@Slf4j
public class HttpServerHandler extends SimpleChannelInboundHandler<FullHttpRequest> {

    //获取 Class 路径
    private URL baseURL = HttpServerHandler.class.getResource("");
    private final String webroot = "webroot";

    private File getResource(String fileName) throws Exception{
        String basePath = baseURL.toURI().toString();
        int start = basePath.indexOf("classes/");
        basePath = (basePath.substring(0,start) + "/" + "classes/").replaceAll("/+","/");

        String path = basePath + webroot + "/" + fileName;
//        log.info("BaseURL:" + basePath);
        path = !path.contains("file:") ? path : path.substring(5);
        path = path.replaceAll("//", "/");
        return new File(path);
    }

    @Override
    public void channelRead0(ChannelHandlerContext ctx, FullHttpRequest request) throws
Exception {
         String uri = request.getUri();

        RandomAccessFile file = null;
        try{
            String page = uri.equals("/") ? "chat.html" : uri;
            file =  new RandomAccessFile(getResource(page), "r");
        }catch(Exception e){
            ctx.fireChannelRead(request.retain());
            return;
        }
```

```
        HttpResponse response = new DefaultHttpResponse(request.getProtocolVersion(),
HttpResponseStatus.OK);
        String contextType = "text/html;";
        if(uri.endsWith(".css")){
             contextType = "text/css;";
        }else if(uri.endsWith(".js")){
             contextType = "text/javascript;";
        }else if(uri.toLowerCase().matches(".*\\.(jpg|png|gif)$")){
             String ext = uri.substring(uri.lastIndexOf("."));
             contextType = "image/" + ext;
        }
        response.headers().set(HttpHeaders.Names.CONTENT_TYPE, contextType +
"charset=utf-8;");

        boolean keepAlive = HttpHeaders.isKeepAlive(request);

        if (keepAlive) {
            response.headers().set(HttpHeaders.Names.CONTENT_LENGTH, file.length());
            response.headers().set(HttpHeaders.Names.CONNECTION,
HttpHeaders.Values.KEEP_ALIVE);
        }
        ctx.write(response);

        ctx.write(new DefaultFileRegion(file.getChannel(), 0, file.length()));

        ChannelFuture future = ctx.writeAndFlush(LastHttpContent.EMPTY_LAST_CONTENT);
        if (!keepAlive) {
            future.addListener(ChannelFutureListener.CLOSE);
        }

        file.close();
    }

    @Override
    public void exceptionCaught(ChannelHandlerContext ctx, Throwable cause)
            throws Exception {
        Channel client = ctx.channel();
        log.info("Client:"+client.remoteAddress()+"异常");
        //当出现异常就关闭连接
        cause.printStackTrace();
        ctx.close();
    }
}
```

13.2.4　对自定义协议的支持

TerminalServerHandler 类，用于处理 Java 控制台发过来的 Java Object 消息体。

```
package com.gupaoedu.vip.netty.chat.server.handler;

import lombok.extern.slf4j.Slf4j;

import com.gupaoedu.vip.netty.chat.processor.MsgProcessor;
import com.gupaoedu.vip.netty.chat.protocol.IMMessage;

import io.netty.channel.ChannelHandlerContext;
import io.netty.channel.SimpleChannelInboundHandler;
import org.json.simple.JSONObject;

@Slf4j
public class TerminalServerHandler extends SimpleChannelInboundHandler<IMMessage>{

    private MsgProcessor processor = new MsgProcessor();

    @Override
    protected void channelRead0(ChannelHandlerContext ctx, IMMessage msg) throws Exception {
        processor.sendMsg(ctx.channel(), msg);
    }

    /**
     * 异常处理
     */
    @Override
    public void exceptionCaught(ChannelHandlerContext ctx, Throwable cause) {
        log.info("Socket Client: 与客户端断开连接:" + cause.getMessage());
        cause.printStackTrace();
        ctx.close();
    }

}
```

13.2.5　对 WebSocket 协议的支持

WebSocketServerHandler 类，处理浏览器发送过来的 WebSocket 请求。

```
package com.gupaoedu.vip.netty.chat.server.handler;

import lombok.extern.slf4j.Slf4j;

import com.gupaoedu.vip.netty.chat.processor.MsgProcessor;
```

```
import io.netty.channel.Channel;
import io.netty.channel.ChannelHandlerContext;
import io.netty.channel.SimpleChannelInboundHandler;
import io.netty.handler.codec.http.websocketx.TextWebSocketFrame;

@Slf4j
public class WebSocketServerHandler extends SimpleChannelInboundHandler<TextWebSocketFrame>
{

    private MsgProcessor processor = new MsgProcessor();

    @Override
    protected void channelRead0(ChannelHandlerContext ctx,TextWebSocketFrame msg) throws
Exception {
        processor.sendMsg(ctx.channel(), msg.text());
    }

    @Override
    public void exceptionCaught(ChannelHandlerContext ctx, Throwable cause)
            throws Exception {
        Channel client = ctx.channel();
        String addr = processor.getAddress(client);
        log.info("WebSocket Client:" + addr + "异常");
        // 当出现异常就关闭连接
        cause.printStackTrace();
        ctx.close();
    }

}
```

13.3 服务端逻辑处理

服务端基本思路是，所有客户端的消息全部发送到服务端的消息容器，每一条消息都携带了客户端的标识信息，然后由服务端转发给所有在线的客户端。先来看支持多协议的顶层设计。

13.3.1 多协议串行处理

编写 ChatServer 类，将各类协议请求的 Handler 处理器串联起来加入 Pipeline 中，并开启监听端口。

```
package com.gupaoedu.vip.netty.chat.server;

import io.netty.*;
```

```
import java.io.IOException;

import com.gupaoedu.vip.netty.chat.protocol.IMDecoder;
import com.gupaoedu.vip.netty.chat.protocol.IMEncoder;
import com.gupaoedu.vip.netty.chat.server.handler.HttpServerHandler;
import com.gupaoedu.vip.netty.chat.server.handler.TerminalServerHandler;
import com.gupaoedu.vip.netty.chat.server.handler.WebSocketServerHandler;
import lombok.extern.slf4j.Slf4j;

@Slf4j
public class ChatServer{

    private int port = 8080;
    public void start(int port){
        EventLoopGroup bossGroup = new NioEventLoopGroup();
        EventLoopGroup workerGroup = new NioEventLoopGroup();
        try {
            ServerBootstrap b = new ServerBootstrap();
            b.group(bossGroup, workerGroup)
                    .channel(NioServerSocketChannel.class)
                    .option(ChannelOption.SO_BACKLOG, 1024)
                    .childHandler(new ChannelInitializer<SocketChannel>() {
                        @Override
                        public void initChannel(SocketChannel ch) throws Exception {

                            ChannelPipeline pipeline = ch.pipeline();

                            /** 解析自定义协议 */
                            pipeline.addLast(new IMDecoder());  //Inbound
                            pipeline.addLast(new IMEncoder());  //Outbound
                            pipeline.addLast(new TerminalServerHandler());  //Inbound

                            /** 解析 HTTP 请求 */
                            pipeline.addLast(new HttpServerCodec());  //Outbound
                            //主要是将同一个 HTTP 请求或响应的多个消息对象变成一个 fullHttpRequest
完整的消息对象
                            pipeline.addLast(new HttpObjectAggregator(64 * 1024));//Inbound
                            //主要用于处理大数据流，比如 1GB 的文件如果直接传输肯定会占满 JVM 内存
                                                  //加上这个 Handler 就不用考虑这个问题了
                            pipeline.addLast(new ChunkedWriteHandler());//Inbound、Outbound
                            pipeline.addLast(new HttpServerHandler());//Inbound

                            /** 解析 WebSocket 请求 */
                            pipeline.addLast(new WebSocketServerProtocolHandler("/im"));
//Inbound
                            pipeline.addLast(new WebSocketServerHandler()); //Inbound
```

```
                    }
                });
            ChannelFuture f = b.bind(this.port).sync();
            log.info("服务已启动,监听端口" + this.port);
            f.channel().closeFuture().sync();
        } catch (InterruptedException e) {
            e.printStackTrace();
        } finally {
            workerGroup.shutdownGracefully();
            bossGroup.shutdownGracefully();
        }
    }
    public void start() {
        start(this.port);
    }


    public static void main(String[] args) throws IOException{
        if(args.length > 0) {
            new ChatServer().start(Integer.valueOf(args[0]));
        }else{
            new ChatServer().start();
        }
    }

}
```

13.3.2 服务端用户中心

MsgProcessor 类，主要处理用户登录、退出、上线、下线、发送消息等行为动作的逻辑。

```
package com.gupaoedu.vip.netty.chat.processor;

import com.alibaba.fastjson.JSONObject;
import com.gupaoedu.vip.netty.chat.protocol.IMDecoder;
import com.gupaoedu.vip.netty.chat.protocol.IMEncoder;
import com.gupaoedu.vip.netty.chat.protocol.IMMessage;
import com.gupaoedu.vip.netty.chat.protocol.IMP;

import io.netty.channel.Channel;
import io.netty.channel.group.ChannelGroup;
import io.netty.channel.group.DefaultChannelGroup;
import io.netty.handler.codec.http.websocketx.TextWebSocketFrame;
import io.netty.util.AttributeKey;
import io.netty.util.concurrent.GlobalEventExecutor;
```

```
/**
 * 主要用于自定义协议内容的逻辑处理
 *
 */
public class MsgProcessor {

    //记录在线用户
    private static ChannelGroup onlineUsers = new
DefaultChannelGroup(GlobalEventExecutor.INSTANCE);

    //定义一些扩展属性
    public static final AttributeKey<String> NICK_NAME = AttributeKey.valueOf("nickName");
    public static final AttributeKey<String> IP_ADDR = AttributeKey.valueOf("ipAddr");
    public static final AttributeKey<JSONObject> ATTRS = AttributeKey.valueOf("attrs");
    public static final AttributeKey<String> FROM = AttributeKey.valueOf("from");

    //自定义解码器
    private IMDecoder decoder = new IMDecoder();
    //自定义编码器
    private IMEncoder encoder = new IMEncoder();

    /**
     * 获取用户昵称
     * @param client
     * @return
     */
    public String getNickName(Channel client){
        return client.attr(NICK_NAME).get();
    }
    /**
     * 获取用户远程 IP 地址
     * @param client
     * @return
     */
    public String getAddress(Channel client){
        return client.remoteAddress().toString().replaceFirst("/","");
    }

    /**
     * 获取扩展属性
     * @param client
     * @return
     */
    public JSONObject getAttrs(Channel client){
        try{
            return client.attr(ATTRS).get();
        }catch(Exception e){
```

```
            return null;
        }
    }

    /**
     * 获取扩展属性
     * @param client
     * @return
     */
    private void setAttrs(Channel client,String key,Object value){
        try{
            JSONObject json = client.attr(ATTRS).get();
            json.put(key, value);
            client.attr(ATTRS).set(json);
        }catch(Exception e){
            JSONObject json = new JSONObject();
            json.put(key, value);
            client.attr(ATTRS).set(json);
        }
    }

    /**
     * 退出通知
     * @param client
     */
    public void logout(Channel client){
        //如果 nickName 为 null，没有遵从聊天协议的连接，表示为非法登录
        if(getNickName(client) == null){ return; }
        for (Channel channel : onlineUsers) {
            IMMessage request = new IMMessage(IMP.SYSTEM.getName(), sysTime(),
onlineUsers.size(), getNickName(client) + "离开");
            String content = encoder.encode(request);
            channel.writeAndFlush(new TextWebSocketFrame(content));
        }
        onlineUsers.remove(client);
    }

    /**
     * 发送消息
     * @param client
     * @param msg
     */
    public void sendMsg(Channel client,IMMessage msg){

        sendMsg(client,encoder.encode(msg));
    }
```

```
    /**
     * 发送消息
     * @param client
     * @param msg
     */
    public void sendMsg(Channel client,String msg){

        IMMessage request = decoder.decode(msg);
        if(null == request){ return; }

        String addr = getAddress(client);

        if(request.getCmd().equals(IMP.LOGIN.getName())){
            client.attr(NICK_NAME).getAndSet(request.getSender());
            client.attr(IP_ADDR).getAndSet(addr);
            client.attr(FROM).getAndSet(request.getTerminal());
//          System.out.println(client.attr(FROM).get());
            onlineUsers.add(client);

            for (Channel channel : onlineUsers) {
                boolean isself = (channel == client);
                if(!isself){
                    request = new IMMessage(IMP.SYSTEM.getName(), sysTime(),
onlineUsers.size(), getNickName(client) + "加入");
                }else{
                    request = new IMMessage(IMP.SYSTEM.getName(), sysTime(),
onlineUsers.size(), "已与服务器建立连接！");
                }

                if("Console".equals(channel.attr(FROM).get())){
                    channel.writeAndFlush(request);
                    continue;
                }
                String content = encoder.encode(request);
                channel.writeAndFlush(new TextWebSocketFrame(content));
            }
        }else if(request.getCmd().equals(IMP.CHAT.getName())){
            for (Channel channel : onlineUsers) {
                boolean isself = (channel == client);
                if (isself) {
                    request.setSender("you");
                }else{
                    request.setSender(getNickName(client));
                }
                request.setTime(sysTime());

                if("Console".equals(channel.attr(FROM).get()) & !isself){
```

```
                    channel.writeAndFlush(request);
                    continue;
                }
                String content = encoder.encode(request);
                channel.writeAndFlush(new TextWebSocketFrame(content));
            }
        }else if(request.getCmd().equals(IMP.FLOWER.getName())){
            JSONObject attrs = getAttrs(client);
            long currTime = sysTime();
            if(null != attrs){
                long lastTime = attrs.getLongValue("lastFlowerTime");
                //60秒之内不允许重复送鲜花
                int secends = 10;
                long sub = currTime - lastTime;
                if(sub < 1000 * secends){
                    request.setSender("you");
                    request.setCmd(IMP.SYSTEM.getName());
                    request.setContent("您送鲜花太频繁," + (secends - Math.round(sub / 1000))
+ "秒后再试");

                    String content = encoder.encode(request);
                    client.writeAndFlush(new TextWebSocketFrame(content));
                    return;
                }
            }

            //正常送花
            for (Channel channel : onlineUsers) {
                if (channel == client) {
                    request.setSender("you");
                    request.setContent("你给大家送了一波鲜花雨");
                    setAttrs(client, "lastFlowerTime", currTime);
                }else{
                    request.setSender(getNickName(client));
                    request.setContent(getNickName(client) + "送来一波鲜花雨");
                }
                request.setTime(sysTime());

                String content = encoder.encode(request);
                channel.writeAndFlush(new TextWebSocketFrame(content));
            }
        }
    }

    /**
     * 获取系统时间
     * @return
```

```
     */
    private Long sysTime(){
        return System.currentTimeMillis();
    }

}
```

13.4　客户端控制台处理

13.4.1　控制台接入代码

编写 ChatClient 类，完成 Java 控制台输入与服务端的交互逻辑。

```
package com.gupaoedu.vip.netty.chat.client;

import io.netty.bootstrap.Bootstrap;
import io.netty.channel.ChannelFuture;
import io.netty.channel.ChannelInitializer;
import io.netty.channel.ChannelOption;
import io.netty.channel.EventLoopGroup;
import io.netty.channel.nio.NioEventLoopGroup;
import io.netty.channel.socket.SocketChannel;
import io.netty.channel.socket.nio.NioSocketChannel;

import java.io.IOException;

import com.gupaoedu.vip.netty.chat.client.handler.ChatClientHandler;
import com.gupaoedu.vip.netty.chat.protocol.IMDecoder;
import com.gupaoedu.vip.netty.chat.protocol.IMEncoder;

/**
 * 客户端
 * @author Tom
 *
 */
public class ChatClient  {

    private ChatClientHandler clientHandler;
    private String host;
    private int port;

    public ChatClient(String nickName){
        this.clientHandler = new ChatClientHandler(nickName);
    }
```

```
    public void connect(String host,int port){
         this.host = host;
         this.port = port;

        EventLoopGroup workerGroup = new NioEventLoopGroup();
        try {
            Bootstrap b = new Bootstrap();
            b.group(workerGroup);
            b.channel(NioSocketChannel.class);
            b.option(ChannelOption.SO_KEEPALIVE, true);
            b.handler(new ChannelInitializer<SocketChannel>() {
                @Override
                public void initChannel(SocketChannel ch) throws Exception {
                    ch.pipeline().addLast(new IMDecoder());
                    ch.pipeline().addLast(new IMEncoder());
                    ch.pipeline().addLast(clientHandler);
                }
            });
            ChannelFuture f = b.connect(this.host, this.port).sync();
            f.channel().closeFuture().sync();
        } catch (InterruptedException e) {
            e.printStackTrace();
        } finally {
            workerGroup.shutdownGracefully();
        }
    }


    public static void main(String[] args) throws IOException{
         new ChatClient("Cover").connect("127.0.0.1",8080);

     String url = "http://localhost:8080/images/a.png";
     System.out.println(url.toLowerCase().matches(".*\\.(gif|png|jpg)$"));

    }

}
```

13.4.2 控制台消息处理

ChatClientHandler 类的主要功能是完成客户端和服务端的会话，将控制台输入的文本信息转换为 IMMessage 对象发送到服务端，将服务端传过来的 Java Object 转换为文本输出到控制台。服务端消息包括从 HTML 页面发送过来的消息，也包括其他客户端从 Java 控制台发送过来的消息。

```
package com.gupaoedu.vip.netty.chat.client.handler;

import io.netty.channel.ChannelHandlerContext;


import java.io.IOException;
import java.util.Scanner;
import java.util.regex.Matcher;

import io.netty.channel.SimpleChannelInboundHandler;
import lombok.extern.slf4j.Slf4j;

import com.gupaoedu.vip.netty.chat.protocol.IMMessage;
import com.gupaoedu.vip.netty.chat.protocol.IMP;

import java.util.regex.Pattern;

/**
 * 聊天客户端逻辑实现
 * @author Tom
 *
 */
@Slf4j
public class ChatClientHandler extends SimpleChannelInboundHandler<IMMessage> {

    private ChannelHandlerContext ctx;
    private String nickName;
    public ChatClientHandler(String nickName){
        this.nickName = nickName;
    }

    /**启动客户端控制台*/
    private void session() throws IOException {
        new Thread(){
            public void run(){
                    System.out.println(nickName + ",你好，请在控制台输入对话内容");
                IMMessage message = null;
                Scanner scanner = new Scanner(System.in);
                do{
                    if(scanner.hasNext()){
                      String input = scanner.nextLine();
                      if("exit".equals(input)){
                          message = new IMMessage(IMP.LOGOUT.getName(),"Console",System.
currentTimeMillis(),nickName);
                      }else{
                          message = new IMMessage(IMP.CHAT.getName(),System.
currentTimeMillis(),nickName,input);
```

```
                    }
                }
            }
            while (sendMsg(message));
            scanner.close();
        }
    }.start();
}

/**
 * TCP 链路建立成功后调用
 */
@Override
public void channelActive(ChannelHandlerContext ctx) throws Exception {
     this.ctx = ctx;
    IMMessage message = new
IMMessage(IMP.LOGIN.getName(),"Console",System.currentTimeMillis(),this.nickName);
    sendMsg(message);
    log.info("成功连接服务器,已执行登录动作");
    session();
}
/**
 * 发送消息
 * @param msg
 * @return
 * @throws IOException
 */
private boolean sendMsg(IMMessage msg){
    ctx.channel().writeAndFlush(msg);
     System.out.println("继续输入开始对话...");
    return msg.getCmd().equals(IMP.LOGOUT) ? false : true;
}
/**
 * 收到消息后调用
 * @throws IOException
 */
@Override
public void channelRead0(ChannelHandlerContext ctx, IMMessage msg) throws IOException {
 IMMessage m = (IMMessage)msg;
     System.out.println((null == m.getSender() ? "" : (m.getSender() + ":")) +
removeHtmlTag(m.getContent()));
}


 public static String removeHtmlTag(String htmlStr){
     String regEx_script="<script[^>]*?>[\\s\\S]*?<\\/script>"; //定义 Script 的正则表达式
     String regEx_style="<style[^>]*?>[\\s\\S]*?<\\/style>"; //定义 Style 的正则表达式
```

```
        String regEx_html="<[^>]+>"; //定义 HTML 标签的正则表达式

        Pattern p_script=Pattern.compile(regEx_script,Pattern.CASE_INSENSITIVE);
        Matcher m_script=p_script.matcher(htmlStr);
        htmlStr=m_script.replaceAll(""); //过滤 Script 标签

        Pattern p_style=Pattern.compile(regEx_style,Pattern.CASE_INSENSITIVE);
        Matcher m_style=p_style.matcher(htmlStr);
        htmlStr=m_style.replaceAll(""); //过滤 Style 标签

        Pattern p_html=Pattern.compile(regEx_html,Pattern.CASE_INSENSITIVE);
        Matcher m_html=p_html.matcher(htmlStr);
        htmlStr=m_html.replaceAll(""); //过滤 HTML 标签

        return htmlStr.trim(); //返回文本字符串
    }

    /**
     * 发生异常时调用
     */
    @Override
    public void exceptionCaught(ChannelHandlerContext ctx, Throwable cause) {
     log.info("与服务器断开连接:"+cause.getMessage());
        ctx.close();
    }
}
```

13.5　客户端 Web 页面交互实现

13.5.1　Web 页面设计

编写 chat.html 文件，完成页面交互组件，包括登录界面、聊天界面。

```
<!DOCTYPE html>
<html>
<head>
    <meta charset="utf-8">
    <meta content="width=device-width, initial-scale=1.0, maximum-scale=1.0,
minimum-scale=1.0, user-scalable=0" name="viewport">
    <title>咕泡学院在线聊天室</title>
    <link rel="stylesheet" type="text/css" href="css/style.css" />
    <script type="text/javascript" src="/js/lib/jquery.min.js"></script>
     <script type="text/javascript" src="/js/lib/jquery.snowfall.js"></script>
     <script type="text/javascript" src="/js/chat.util.js"></script>
</head>
```

```
<body>
    <div id="loginbox">
        <div style="width:300px;margin:200px auto;">
              欢迎进入咕泡学院 WebSocket 聊天室
            <br/>
            <br/>
            <input type="text" style="width:180px;" placeholder="进入前，请先输入昵称"
id="nickname" name="nickname" />
            <input type="button" style="width:50px;" value="进入" onclick="CHAT.login();" />
            <div id="error-msg" style="color:red;"></div>
        </div>
    </div>
    <div id="chatbox" style="display: none;">
        <div style="background:#3d3d3d;height: 28px; width: 100%;font-size:12px;position:
fixed;top: 0px;z-index: 999;">
            <div style="line-height: 28px;color:#fff;">
                <span style="text-align:left;margin-left:10px;">咕泡学院聊天室</span>
                <span style="float:right; margin-right:10px;">
                  <span>当前在线<span id="onlinecount">0</span>人</span> |
                    <span id="shownikcname">匿名</span> |
                    <a href="javascript:;" onclick="CHAT.logout()" style="color:#fff;">退出
</a>
                </span>
            </div>
        </div>
        <div id="doc">
            <div id="chat">
                <div id="message" class="message">
                    <div id="onlinemsg" style="background:#EFEFF4; font-size:12px;
margin-top:40px; margin-left:10px; color:#666;">
                    </div>
                </div>
                <form onsubmit="return false;">
                  <div class="tool-box">
                      <div class="face-box" id="face-box"></div>
                      <span class="face" onclick="CHAT.openFace()" title="选择表情"></span>
                      <!--
                        <span class="img" id="tool-img-btn" title="发送图片"></span>
                        <span class="file" id="tool-file-btn" title="上传文件"></span>
                       -->
                      <span class="flower" onclick="CHAT.sendFlower()" title="送鲜花"></span>
                  </div>
                  <div class="input-box">
                        <div class="input" contenteditable="true" id="send-message"></div>
                        <div class="action">
                            <input class="button" type="button" id="mjr_send"
```

```
onclick="CHAT.sendText()" value="发送"/>
                                </div>
                         </div>
                            <div class="copyright">咕泡学院&copy;版权所有</div>
                      </form>
                  </div>
              </div>
           </div>
</body>

</html>
```

HTML 页面最终效果如下图所示。

登录界面

聊天界面

13.5.2　WebSocket 接入

接下来，实现 JavaScript 的逻辑，来完成 HTML 页面的交互功能。先实现 init()方法，完成浏览器 WebSocket 组件的初始化工作。

```
//初始化聊天组件
        init:function(nickname){

            ...


            if (!window.WebSocket) {
                window.WebSocket = window.MozWebSocket;
            }
            if (window.WebSocket) {
                CHAT.socket = new WebSocket(CHAT.serverAddr);
                CHAT.socket.onmessage = function(e) {
                 appendToPanel(e.data);
                };
                CHAT.socket.onopen = function(e) {
                 CHAT.socket.send("[LOGIN][" + new Date().getTime() +"][" + nickname +
"][WebSocket]");
                };
                CHAT.socket.onclose = function(e) {
                    appendToPanel("[SYSTEM][" + new Date().getTime() + "][0] - 服务器关闭,暂
时不能聊天!");
                };
            } else {
                alert("你的浏览器不支持 WebSocket! ");
            }
        }
```

13.5.3 登录和退出

实现 login()方法，可以实现用户登录逻辑。实现 logout()方法，可以实现退出逻辑，退出逻辑处理比较简单，就是重新刷新页面，从而断开 WebSocket 与服务端的连接。

```
//登录到聊天室
        login:function(){
            $("#error-msg").empty();
            var _reg = /^\S{1,10}$/;
            var nickname = $("#nickname").val();
          if (nickname != "") {
                if (!(_reg.test($.trim(nickname)))) {
                   $('#error-msg').html("昵称长度必须在 10 个字符以内");
                   return false;
                }
                $("#nickname").val('');
              $("#loginbox").hide();
              $("#chatbox").show();
              this.init(nickname);
```

```
            }else{
                 $('#error-msg').html("先输入昵称才能进入聊天室");
                return false;
            }
            return false;
        },
        //退出登录
        logout:function(){
             location.reload();
        }
```

13.5.4　发送文字信息

实现 sendText()方法，发送文本聊天信息到服务器。

```
//发送聊天消息
        sendText:function() {
            var message = $("#send-message");
            //去掉空格
            if(message.html().replace(/\s/ig,"") == ""){ return; }
            if (!window.WebSocket) { return; }
            if (CHAT.socket.readyState == WebSocket.OPEN) {
            var msg = ("[CHAT][" + new Date().getTime() + "]" + "[" + CHAT.nickname + "] - "
+ message.html().replace(/\n/ig,"<br/>"));
            CHAT.socket.send(msg);
            message.empty();
            message.focus();
            } else {
                alert("与服务器连接失败.");
            }
        }
```

scrollToBottom()方法实现浏览器滚动条自动滚动功能，使最新的消息始终保持在用户窗口的底部。

```
//将滚动条设置到顶部，以便能看到最新的消息
        scrollToBottom:function(){
            window.scrollTo(0, $("#onlinemsg")[0].scrollHeight);
        },
```

clear()方法实现清空聊天记录功能。

```
        //清空聊天记录
        clear:function(){
            CHAT.box.innerHTML = "";
        },
```

13.5.5 发送图片表情

selectFace()方法实现发送表情的功能。我们提前把表情窗口需要预设的表情图片预置到 Netty 服务端的静态文件夹中，然后在网页中引入静态图片的 URL 即可。我们发送表情实际上发送的是 img 标签的全部代码，这样，HTML 客户端接收到代码后就能正常显示为表情图片。如果在控制台显示，则需要将 HTML 代码用正则表达式剔除。

```
//选择表情
        selectFace:function(img){
            var faceBox = $("#face-box");
            faceBox.hide();
            faceBox.removeClass("open");
        var i = '<img src="' + img + '" />';
        $("#send-message").html($("#send-message").html() + i);
        $("#send-message").focus();
        },
        //打开表情弹窗
        openFace:function(e){
            var faceBox = $("#face-box");
            if(faceBox.hasClass("open")){
                faceBox.hide();
                faceBox.removeClass("open");
                return;
            }
            faceBox.addClass("open");
            faceBox.show();
            var box = '';
            for(var i = 1;i <= 130; i ++){
                var img = '/images/face/' + i + '.gif';
                box += '<span class="face-item" onclick="CHAT.selectFace(\'' + img + '\');">';
                box += '<img src="' + img + '"/>';
                box += '</span>';
            }
            faceBox.html(box);
        }
```

发送图片表情的页面效果如下图所示。

13.5.6　发送鲜花雨特效

鲜花雨特效是在 sendFlower()方法中实现的，主要是调用一个 JavaScript 插件的功能，当接收到鲜花指令时，就在网页中触发特效。

```
//发送鲜花
        sendFlower:function(){
            if (!window.WebSocket) { return; }
            if (CHAT.socket.readyState == WebSocket.OPEN) {
            var message = ("[FLOWER][" + new Date().getTime() + "]" + "[" + CHAT.nickname +
"][WebSocket]");
            CHAT.socket.send(message);
            $("#send-message").focus();
            } else {
                alert("与服务器连接失败.");
            }
        }
```

鲜花雨特效的页面效果如下图所示。

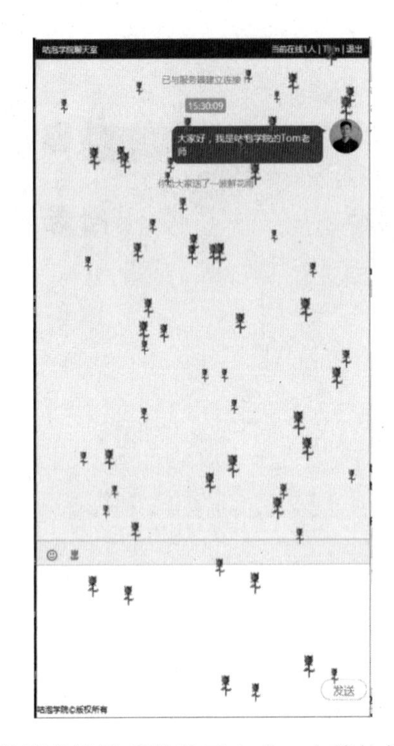

至此，完整的支持多协议的消息推送系统就已完成，完整的代码可以在随书源码中下载。

14

第 14 章 Netty 高性能调优工具类解析

14.1 多线程共享 FastThreadLocal

我们在剖析堆外内存分配的时候简单介绍过 FastThreadLocal，它类似于 JDK 的 ThreadLocal，也是用于多线程条件下，保证统一线程的对象共享，只是 Netty 中定义的 FastThreadLocal 性能要高于 JDK 的 ThreadLocal，本章开始我们来分析其具体原因。

14.1.1 FastThreadLocal 的使用和创建

首先看一个最简单的 Demo。

```
public class FastThreadLocalDemo {

   final class FastThreadLocalTest extends FastThreadLocal<Object>{
      @Override
      protected Object initialValue() throws Exception {
         return new Object();
      }
   }
```

```
    private final FastThreadLocalTest fastThreadLocalTest;

    public FastThreadLocalDemo(){
        fastThreadLocalTest = new FastThreadLocalTest();
    }

    public static void main(String[] args){

        FastThreadLocalDemo fastThreadLocalDemo = new FastThreadLocalDemo();

        new Thread(new Runnable() {
            @Override
            public void run() {
                Object obj  = fastThreadLocalDemo.fastThreadLocalTest.get();
                try {
                    for (int i=0;i<10;i++){
                        fastThreadLocalDemo.fastThreadLocalTest.set(new Object());
                        Thread.sleep(1000);
                    }
                }catch (Exception e){
                    e.printStackTrace();
                }
            }
        }).start();

        new Thread(new Runnable() {
            @Override
            public void run() {
                try {
                    Object obj  = fastThreadLocalDemo.fastThreadLocalTest.get();
                    for (int i=0;i<10;i++){
                        System.out.println(obj ==
fastThreadLocalDemo.fastThreadLocalTest.get());
                        Thread.sleep(1000);
                    }
                }catch (Exception e){

                }
            }
        }).start();
    }
}
```

从上面示例中看出，首先声明一个内部类 FastThreadLocalTest 继承 FastThreadLocal，并重写 initialValue()方法。initialValue()方法就是用来初始化线程共享对象的。然后声明一个成员变量 fastThreadLocalTest，类型是内部类 FastThreadLocalTest，在构造方法中初始化 fastThreadLocalTest。

main()方法中创建当前类 FastThreadLocalDemo 的对象 fastThreadLocalDemo，然后启动两个线程，每个线程通过 fastThreadLocalDemo.fastThreadLocalTest.get()方式获取线程共享对象，因为 fastThreadLocalDemo 是相同的，所以 fastThreadLocalTest 对象也是同一个，同一个对象在不同线程中进行 get()。第一个线程循环通过 set()方法修改共享对象的值，第二个线程则循环判断 fastThreadLocalTest.get()获得的对象和第一次 get()获得的对象是否相等。这里输出结果都是 true，说明其他线程虽然不断修改共享对象的值，但都不影响当前线程共享对象的值，这样就实现了线程共享对象的功能。

根据上述示例，我们剖析 FastThreadLocal 的创建，FastThreadLocal 的构造方法的代码如下。

```
public FastThreadLocal() {
    index = InternalThreadLocalMap.nextVariableIndex();
}
```

这里的 index 代表 FastThreadLocal 对象的一个下标，每创建一个 FastThreadLocal 都会有一个唯一的自增的下标，跟进 nextVariableIndex()方法。

```
public static int nextVariableIndex() {
    int index = nextIndex.getAndIncrement();
    if (index < 0) {
        nextIndex.decrementAndGet();
        throw new IllegalStateException("too many thread-local indexed variables");
    }
    return index;
}
```

上述代码中，获取 nextIndex 通过 getAndIncrement()进行原子自增，创建第一个 FastThreadLocal 对象时，nextIndex 为 0；创建第二个 FastThreadLocal 对象时，nextIndex 为 1；依此类推，第 n 个 FastThreadLocal 对象时的 nextIndex 为 n-1，如下图所示。

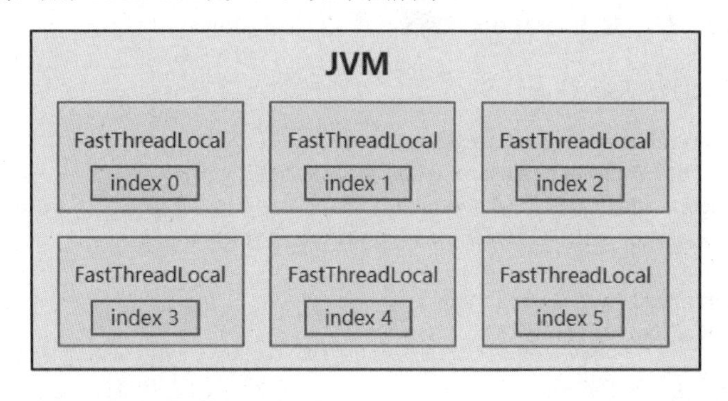

回到 Demo 中，看线程中的这一句。

```
Object obj = fastThreadLocalDemo.fastThreadLocalTest.get();
```

这是调用了 FastThreadLocal 对象的 get()方法，作用是创建一个线程共享对象。get()方法的代码如下。

```
public final V get() {
    return get(InternalThreadLocalMap.get());
}
```

这里调用了一个重载的 get()方法，参数中通过 InternalThreadLocalMap 的 get()方法获取了一个 InternalThreadLocalMap 对象。我们跟到 InternalThreadLocalMap 的 get()方法中，分析其是如何获取 InternalThreadLocalMap 对象的。

```
public static InternalThreadLocalMap get() {
    Thread thread = Thread.currentThread();
    if (thread instanceof FastThreadLocalThread) {
        return fastGet((FastThreadLocalThread) thread);
    } else {
        return slowGet();
    }
}
```

这里首先获取当前线程，然后判断当前线程是否为 FastThreadLocalThread 线程，通常 NioEventLoop 线程都是 FastThreadLocalThread，用于线程则不是 FastThreadLocalThread。

如果是 FastThreadLocalThread 线程，则调用 fastGet()方法获取 InternalThreadLocalMap，从名字上我们能知道，这是一种效率极高的获取方式。

如果不是 FastThreadLocalThread 线程，则调用 slowGet()方法获取 InternalThreadLocalMap，同样根据名字，我们知道这是一种效率不太高的获取方式。

因为我们的 Demo 并不是 EventLoop 线程，所以调用 slowGet()方法。

首先剖析 slowGet()方法。

```
private static InternalThreadLocalMap slowGet() {
    ThreadLocal<InternalThreadLocalMap> slowThreadLocalMap =
UnpaddedInternalThreadLocalMap.slowThreadLocalMap;
    InternalThreadLocalMap ret = slowThreadLocalMap.get();
    if (ret == null) {
        ret = new InternalThreadLocalMap();
        slowThreadLocalMap.set(ret);
    }
    return ret;
}
```

通过 UnpaddedInternalThreadLocalMap.slowThreadLocalMap 获取一个 ThreadLocal 对象 slowThreadLocalMap，slowThreadLocalMap 是 UnpaddedInternalThreadLocalMap 类的一个静态属性，类型是 ThreadLocal 类型。这里的 ThreadLocal 是 JDK 的 ThreadLocal。

然后通过 slowThreadLocalMap 对象的 get()方法，获取一个 InternalThreadLocalMap。如果是第一次获取，InternalThreadLocalMap 有可能是 null，则在 if 块中新建一个 InternalThreadLocalMap 对象，并设置在 ThreadLocal 对象中。

因为 Netty 实现的 FastThreadLocal 要比 JDK 的 ThreadLocal 快，所以这里的方法叫作 slowGet()方法。

回到 InternalThreadLocalMap 的 get()方法。

```
public static InternalThreadLocalMap get() {
    Thread thread = Thread.currentThread();
    if (thread instanceof FastThreadLocalThread) {
        return fastGet((FastThreadLocalThread) thread);
    } else {
        return slowGet();
    }
}
```

继续剖析 fastGet()方法，通常 EventLoop 线程创建 FastThreadLocalThread 线程，所以 EventLoop 线程执行到这一步的时候会调用 fastGet()方法。

```
private static InternalThreadLocalMap fastGet(FastThreadLocalThread thread) {
    InternalThreadLocalMap threadLocalMap = thread.threadLocalMap();
    if (threadLocalMap == null) {
        thread.setThreadLocalMap(threadLocalMap = new InternalThreadLocalMap());
    }
    return threadLocalMap;
}
```

首先 FastThreadLocalThread 对象直接通过 ThreadLocalMap 获取 ThreadLocalMap 对象。如果 ThreadLocalMap 为 null，则创建一个 InternalThreadLocalMap 对象设置到 FastThreadLocalThread 的成员变量中。

我们知道，FastThreadLocalThread 对象中维护了一个 InternalThreadLocalMap 类型的成员变量，可以直接通过 threadLocalMap()方法获取该变量的值，也就是 InternalThreadLocalMap。

跟进 InternalThreadLocalMap 的构造方法。

```
private InternalThreadLocalMap() {
    super(newIndexedVariableTable());
}
```

这里调用了父类的构造方法，传入一个 newIndexedVariableTable()，代码如下。

```
private static Object[] newIndexedVariableTable() {
    Object[] array = new Object[32];
    Arrays.fill(array, UNSET);
    return array;
}
```

这里创建了一个长度为 32 的数组，并将数组中的每一个对象都设置为 UNSET，UNSET 是一个 Object 的对象，表示该下标的值没有被设置。

回到 InternalThreadLocalMap 的构造方法，看其父类的构造方法。

```
UnpaddedInternalThreadLocalMap(Object[] indexedVariables) {
    this.indexedVariables = indexedVariables;
}
```

这里初始化了一个数组类型的成员变量 IndexedVariables，就是 newIndexedVariableTable 返回 Object 的数组。可以知道，每个 InternalThreadLocalMap 对象中都维护了一个 Object 类型的数组，那么这个数组有什么作用呢？继续往下剖析。

回到 FastThreadLocal 的 get()方法。

```
public final V get() {
    return get(InternalThreadLocalMap.get());
}
```

剖析完 InternalThreadLocalMap.get()的相关逻辑，继续看重载的 get()方法。

```
public final V get(InternalThreadLocalMap threadLocalMap) {
    Object v = threadLocalMap.indexedVariable(index);
    if (v != InternalThreadLocalMap.UNSET) {
        return (V) v;
    }
    return initialize(threadLocalMap);
}
```

首先看这一步。

```
Object v = threadLocalMap.indexedVariable(index);
```

这一步是获取当前 Index 下标的 Object，其实就是获取每个 FastThreadLocal 对象绑定的线程共享对象。Index 已经分析过，是每一个 FastThreadLocal 的唯一下标。

跟进 indexedVariable()方法。

```
public Object indexedVariable(int index) {
    Object[] lookup = indexedVariables;
    return index < lookup.length? lookup[index] : UNSET;
}
```

首先获取 indexedVariables。indexedVariables 是 InternalThreadLocalMap 对象中维护的数组，初始大小是 32。然后在 return 中判断当前 Index 是不是小于当前数组的长度，如果小于则获取当前下标 Index 的数组元素，否则返回 UNSET，代表没有设置的对象。

其实每一个 FastThreadLocal 对象中所绑定的线程共享对象，都存放在 ThreadLocalMap 对象中的一个对象数组中，数组中元素的下标对应着 FastThreadLocal 中的 Index 属性，对应关系如下图所示。

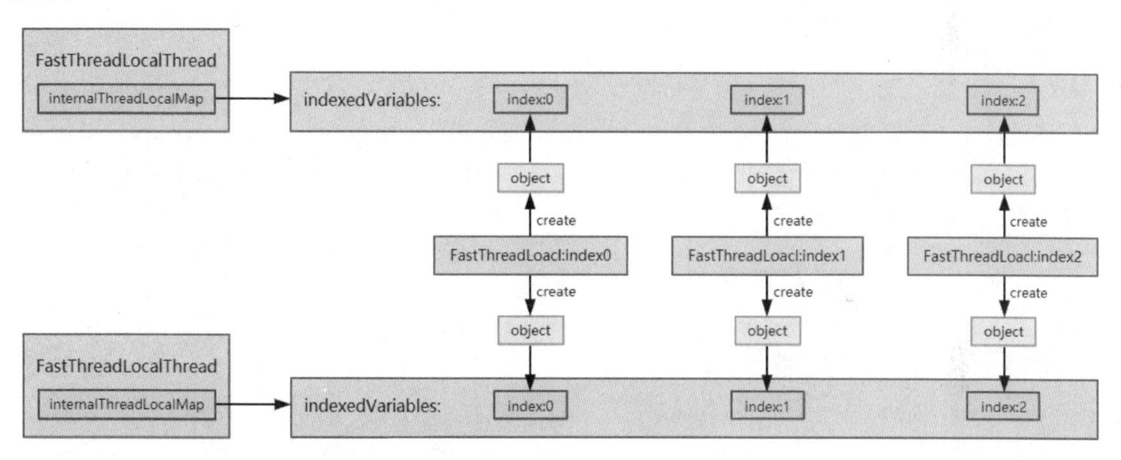

回到 FastThreadLocal 重载的 get()方法。

```
public final V get(InternalThreadLocalMap threadLocalMap) {
    Object v = threadLocalMap.indexedVariable(index);
    if (v != InternalThreadLocalMap.UNSET) {
        return (V) v;
    }
    return initialize(threadLocalMap);
}
```

根据以上逻辑，第一次获取对象 v 时只能获取到 UNSET 对象，因为该对象并没有保存在 ThreadLocalMap 中的数组 IndexedVariables 中，所以第一次获取在 if 判断中为 false 时，会执行到 initialize()方法中。跟到 initialize()方法中。

```
private V initialize(InternalThreadLocalMap threadLocalMap) {
    V v = null;
    try {
        v = initialValue();
    } catch (Exception e) {
        PlatformDependent.throwException(e);
    }
    threadLocalMap.setIndexedVariable(index, v);
    addToVariablesToRemove(threadLocalMap, this);
    return v;
}
```

首先调用 initialValue()方法，这里的 initialValue 实际上调用的是 FastThreadLocal 子类的重写 initialValue()方法。在 Demo 中对应这个方法的代码如下。

```
@Override
protected Object initialValue() throws Exception {
    return new Object();
}
```

通过这个方法会创建一个线程共享对象。通过 ThreadLocalMap 对象的 setIndexedVariable()方法将创建的线程共享对象设置到 ThreadLocalMap 维护的数组中，参数为 FastThreadLocal 和创建的对象本身。

setIndexedVariable()方法的代码如下。

```
public boolean setIndexedVariable(int index, Object value) {
    Object[] lookup = indexedVariables;
    if (index < lookup.length) {
        Object oldValue = lookup[index];
        lookup[index] = value;
        return oldValue == UNSET;
    } else {
        expandIndexedVariableTableAndSet(index, value);
        return true;
    }
}
```

先判断 FastThreadLocal 对象的 Index 是否超过数组 IndexedVariables 的长度，如果没有超过，则直接通过下标设置新创建的线程共享对象。通过这个操作，下次获取该对象的时候就可以直接通过数组下标进行取出。

如果 Index 超过了数组 IndexedVariables 的长度，则通过 expandIndexedVariableTableAndSet()方法将数组扩容，并且根据 Index 通过数组下标的方式将线程共享对象设置到数组 IndexedVariables 中。

以上就是线程共享对象的创建和获取的过程。

14.1.2　FastThreadLocal 的设值

FastThreadLocal 的设值是由 set()方法完成的，其实就是通过调用 set()方法修改线程共享对象，作用域是当前线程，我们回顾上一节 Demo 中的一个线程 set 对象的过程。

```
new Thread(new Runnable() {
    @Override
    public void run() {
        Object obj  = fastThreadLocalDemo.fastThreadLocalTest.get();
        try {
            for (int i=0;i<10;i++){
                fastThreadLocalDemo.fastThreadLocalTest.set(new Object());
                Thread.sleep(1000);
            }
        }catch (Exception e){
            e.printStackTrace();
        }
    }
}).start();
```

set()方法的代码如下。

```
public final void set(V value) {
    if (value != InternalThreadLocalMap.UNSET) {
        set(InternalThreadLocalMap.get(), value);
    } else {
        remove();
    }
}
```

首先判断当前设置的对象是不是 UNSET。因为不是 UNSET，所以进到 if 块中。

if 块调用了重载的 set()方法，参数仍然为 InternalThreadLocalMap，同时，参数也传入了 set 的 Value 值。

跟进重载的 set()方法。

```
public final void set(InternalThreadLocalMap threadLocalMap, V value) {
    if (value != InternalThreadLocalMap.UNSET) {
        if (threadLocalMap.setIndexedVariable(index, value)) {
            addToVariablesToRemove(threadLocalMap, this);
        }
    } else {
        remove(threadLocalMap);
    }
}
```

这里重点关注 if(threadLocalMap.setIndexedVariable(index, value))这部分。通过 ThreadLocalMap 调用 setIndexedVariable()方法进行对象的设置，传入了当前 FastThreadLocal 的下标和 Value。setIndexedVariable()方法的代码如下。

```
public boolean setIndexedVariable(int index, Object value) {
    Object[] lookup = indexedVariables;
    if (index < lookup.length) {
        Object oldValue = lookup[index];
        lookup[index] = value;
        return oldValue == UNSET;
    } else {
        expandIndexedVariableTableAndSet(index, value);
        return true;
    }
}
```

这里的逻辑其实和 get 非常类似，都是直接通过索引操作的，根据索引值，直接通过数组下标的方式对元素进行设置。

回到 FastThreadLocal 的 set()方法。

```
public final void set(V value) {
    if (value != InternalThreadLocalMap.UNSET) {
        set(InternalThreadLocalMap.get(), value);
    } else {
        remove();
    }
}
```

如果修改的对象是 UNSET 对象，则会调用 remove()方法，代码如下。

```
public final void remove(InternalThreadLocalMap threadLocalMap) {
    if (threadLocalMap == null) {
        return;
    }
    Object v = threadLocalMap.removeIndexedVariable(index);
    removeFromVariablesToRemove(threadLocalMap, this);
    if (v != InternalThreadLocalMap.UNSET) {
        try {
            onRemoval((V) v);
        } catch (Exception e) {
            PlatformDependent.throwException(e);
        }
    }
}
```

Object v = threadLocalMap.removeIndexedVariable(index) 这一步是根据索引 Index 将值设置成 UNSET 的。

跟进 removeIndexedVariable()方法。

```
public Object removeIndexedVariable(int index) {
    Object[] lookup = indexedVariables;
    if (index < lookup.length) {
        Object v = lookup[index];
        lookup[index] = UNSET;
        return v;
    } else {
        return UNSET;
    }
}
```

这里的逻辑也比较简单，根据 Index 通过数组下标的方式将元素设置成 UNSET 对象。

回到 remove()方法中，if(v != InternalThreadLocalMap.UNSET)判断如果设置的值不是 UNSET 对象，则调用 onRemoval()方法。

跟进 onRemoval()方法。

```
protected void onRemoval(@SuppressWarnings("UnusedParameters") V value) throws Exception { }
```

它是个空实现，用于交给子类去完成。

14.2　Recycler 对象回收站

Recycler 我们应该不陌生，因为在前面章节中，有很多地方使用了 Recycler。Recycler 是 Netty 实现的一个轻量级对象回收站，很多对象在使用完毕之后，并没有直接交给 GC 去处理，而是通过对象回收站将对象回收，目的是为了对象重用和减少 GC 压力。比如 ByteBuf 对象的回收，因为 ByteBuf 对象在 Netty 中会被频繁创建，并且占用比较大的内存空间，所以使用完毕后会通过对象回收站的方式进行回收，以达到资源重用的目的。

14.2.1　Recycler 的使用和创建

在 Netty 中，Recycler 的使用是相当频繁的。Recycler 的作用是保证对象的循环利用，对象使用完可以通过 Recycler 回收，需要再次使用则从对象池中取出，不用每次都创建新对象，从而减少对系统资源的占用，同时也减轻了 GC 的压力。先看一个示例。

```
public class RecyclerDemo {
    private static final Recycler<User> RECYCLER = new Recycler<User>() {
        @Override
        protected User newObject(Handle<User> handle) {
            return new User(handle);
        }
    };
    static class User{
        private final Recycler.Handle<User> handle;
        public User(Recycler.Handle<User> handle){
            this.handle=handle;
        }
        public void recycle(){
            handle.recycle(this);
        }
    }
    public static void main(String[] args){
        User user1 = RECYCLER.get();
        user1.recycle();
        User user2 = RECYCLER.get();
        user2.recycle();
        System.out.println(user1==user2);
    }
}
```

首先定义了一个 Recycler 的成员变量 RECYCLER，在匿名内部类中重写了 newObject()方法，也就是创建对象的方法，该方法是用户自定义的。newObject()方法返回的 new User(handle)代表当回收站没有此类对象的时候，可以通过这种方式创建对象。成员变量 RECYCLER 可以用来对此类对象回收和再利用。然后定义了一个静态内部类 User，User 中有个成员变量 Handle，在构造方法中为其赋值，Handle 的作用是用于对象回收。并且定义了一个方法 recycle()，方法中通过 handle.recycle(this)这种方式将自身对象进行回收，通过这步操作，可以将对象回收到 Recycler 中。以上逻辑先做了解，之后会进行详细分析。

在 main()方法中，通过 RECYCLER 的 get()方法获取一个 User，然后进行回收，再通过 get()方法将回收站的对象取出，再次进行回收，最后判断两次取出的对象是否为同一个对象，最后结果输出为 true。以上 Demo 就可以说明 Recycler 的回收再利用的功能。

简单介绍完 Demo，我们就来详细地分析 Recycler 的机制。在 Recycler 的类的源码中，我们会看到这样一段逻辑。

```
private final FastThreadLocal<Stack<T>> threadLocal = new FastThreadLocal<Stack<T>>() {
    @Override
    protected Stack<T> initialValue() {
```

```
        return new Stack<T>(Recycler.this, Thread.currentThread(), maxCapacityPerThread,
maxSharedCapacityFactor,
                ratioMask, maxDelayedQueuesPerThread);
    }
};
```

这一段逻辑用于保存线程共享对象，而这里的共享对象，就是一个 Stack 类型的对象。每个 Stack 中都维护着一个 DefaultHandle 类型的数组，用于盛放回收的对象，有关 Stack 和线程的关系如下图所示。

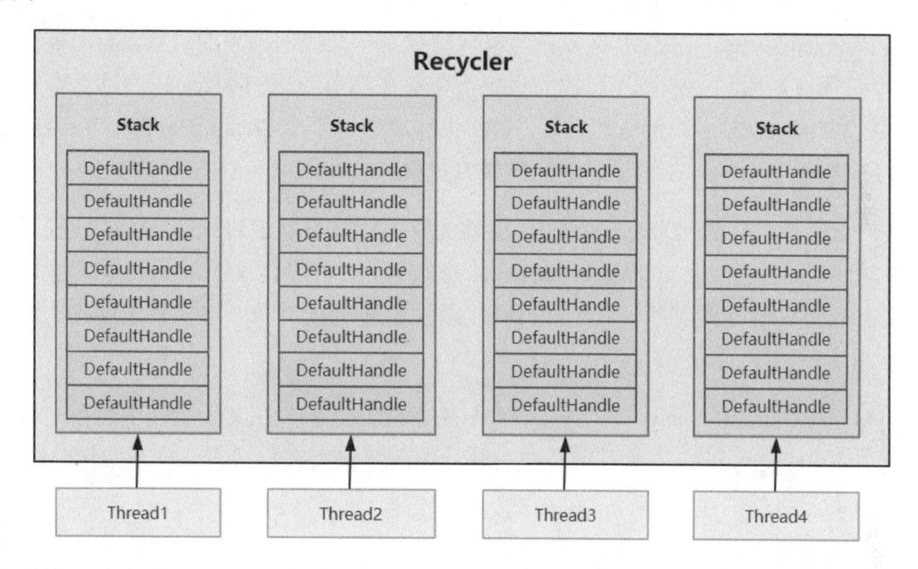

也就是说，在每个 Recycler 中，都维护着一个线程共享的栈，用于对一类对象的回收。Stack 的构造方法的代码如下。

```
Stack(Recycler<T> parent, Thread thread, int maxCapacity, int maxSharedCapacityFactor,
      int ratioMask, int maxDelayedQueues) {
    this.parent = parent;
    this.thread = thread;
    this.maxCapacity = maxCapacity;
    availableSharedCapacity = new AtomicInteger(max(maxCapacity / maxSharedCapacityFactor,
LINK_CAPACITY));
    elements = new DefaultHandle[min(INITIAL_CAPACITY, maxCapacity)];
    this.ratioMask = ratioMask;
    this.maxDelayedQueues = maxDelayedQueues;
}
```

首先介绍几个构造方法中初始化的关键属性。

- Parent：表示 Recycler 对象自身。
- Thread：表示当前 Stack 绑定的哪个线程。
- maxCapacity：表示当前 Stack 的最大容量，表示 Stack 最多能盛放多少个元素。
- elements：表示 Stack 中存储的对象，类型为 DefaultHandle，可以被外部对象引用，从而实现回收。
- ratioMask：用来控制对象回收的频率，也就是说每次通过 Recycler 回收对象的时候，不是每次都会进行回收，而是通过该参数控制回收频率。
- maxDelayedQueues：稍微有些复杂，很多时候，一个线程创建的对象，有可能会被另一个线程所释放，而另一个线程释放的对象不会放在当前线程的 Stack 中，而是存放在一个叫作 WeakOrderQueue 的数据结构中，里面也存放着一个个 DefaultHandle，WeakOrderQueue 会存放线程 1 创建且在线程 2 进行释放的对象。

现在我们已经知道，maxDelayedQueues 属性的意思就是设置该线程能回收的线程对象的最大值。假设当前线程是线程 A，maxDelayedQueues 值设置为 2，那么线程 A 回收了线程 B 创建的对象，又回收了线程 C 创建的对象，就不能再回收线程 D 创建的对象，最多只能回收 2 个线程创建的对象。

属性 availableSharedCapacity 表示在线程 A 中创建的对象，在其他线程中缓存的最大个数，同样，相关逻辑会在之后的内容进行剖析，另外介绍两个没有在构造方法中初始化的属性。

```
private WeakOrderQueue cursor, prev;
private volatile WeakOrderQueue head;
```

这里 cursor、prev 和 head 是存放了其他线程的链表的指针，用于指向 WeakOrderQueue，也是稍作了解，之后会进行详细剖析。有关 Stack 异线程之间对象的关系如下图所示。

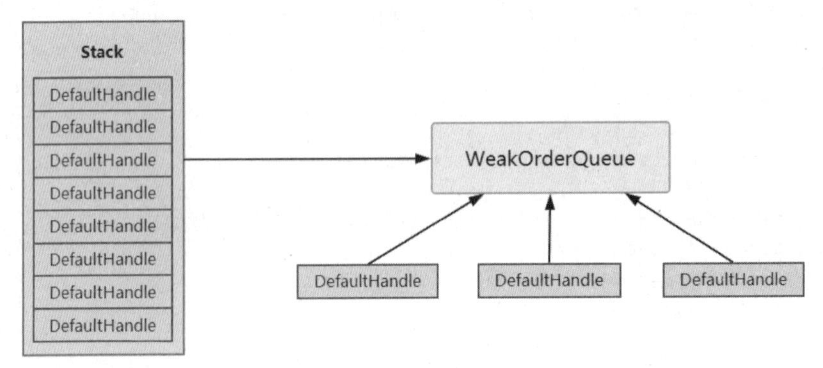

继续介绍 Recycler 的构造方法，同时熟悉有关 Stack 各个参数的默认值。

```
protected Recycler() {
    this(DEFAULT_MAX_CAPACITY_PER_THREAD);
}
```

这里调用了重载的构造方法，并传入了参数 DEFAULT_MAX_CAPACITY_PER_THREAD。DEFAULT_MAX_CAPACITY_PER_THREAD 的默认值是 32 768，是在 static 块中被初始化的，可以跟进去自行分析。这个值就代表每个线程中 Stack 中最多回收的元素的个数。继续跟进重载的构造方法。

```
protected Recycler(int maxCapacityPerThread) {
    this(maxCapacityPerThread, MAX_SHARED_CAPACITY_FACTOR);
}
```

这里又调用了重载的构造方法，并且传入 32 768 和 MAX_SHARED_CAPACITY_ FACTOR。

MAX_SHARED_CAPACITY_FACTOR 的默认值是 2，同样在 static 块中进行了初始化，有关该属性的用处稍后讲解。继续跟进构造方法。

```
protected Recycler(int maxCapacityPerThread, int maxSharedCapacityFactor) {
    this(maxCapacityPerThread, maxSharedCapacityFactor, RATIO, MAX_DELAYED_QUEUES_
PER_THREAD);
}
```

这里同样调用了重载的构造方法，传入了 32 768 和 2，还有两个属性 RATIO 和 MAX_DELAYED_QUEUES_PER_THREAD。RATIO 也在 static 中被初始化，默认值是 8。同上，MAX_DELAYED_QUEUES_PER_THREAD 的默认值是 2 倍 CPU 核数。继续跟进构造方法。

```
protected Recycler(int maxCapacityPerThread, int maxSharedCapacityFactor,
                   int ratio, int maxDelayedQueuesPerThread) {
    ratioMask = safeFindNextPositivePowerOfTwo(ratio) - 1;
    if (maxCapacityPerThread <= 0) {
        this.maxCapacityPerThread = 0;
        this.maxSharedCapacityFactor = 1;
        this.maxDelayedQueuesPerThread = 0;
    } else {
        this.maxCapacityPerThread = maxCapacityPerThread;
        this.maxSharedCapacityFactor = max(1, maxSharedCapacityFactor);
        this.maxDelayedQueuesPerThread = max(0, maxDelayedQueuesPerThread);
    }
}
```

分析 Recycler 构造方法，主要就是将几个属性进行了初始化。

- ratioMask：它是获取 safeFindNextPositivePowerOfTwo() 方法的返回值。在此处 safeFindNextPositivePowerOfTwo()方法的返回值是 8，因此 ratioMask 的最终赋值是 7。

- maxCapacityPerThread：它是一个大于 0 的数，如果不满足判断条件就进入 else 代码块，最终会被赋值为 32 768。
- 被赋值为 32 768。
- maxSharedCapacityFactor：最终会被赋值为 2。
- maxDelayedQueuesPerThread：被赋值为 CPU 核数×2。

我们再回到 Stack 的构造方法。

```
Stack(Recycler<T> parent, Thread thread, int maxCapacity, int maxSharedCapacityFactor,
      int ratioMask, int maxDelayedQueues) {
    this.parent = parent;
    this.thread = thread;
    this.maxCapacity = maxCapacity;
    availableSharedCapacity = new AtomicInteger(max(maxCapacity / maxSharedCapacityFactor,
LINK_CAPACITY));
    elements = new DefaultHandle[min(INITIAL_CAPACITY, maxCapacity)];
    this.ratioMask = ratioMask;
    this.maxDelayedQueues = maxDelayedQueues;
}
```

根据 Recycler 初始化属性的逻辑，可以知道 Stack 中几个属性的值。

- maxCapacity：默认值为 32 768。
- ratioMask：默认值为 7。
- maxDelayedQueues：默认值为 CPU 核数×2。
- availableSharedCapacity：默认值是 32 768/2，也就是 16 384。

14.2.2 从 Recycler 中获取对象

回顾上节 Demo 中的 main()方法，从回收站获取对象。

```
public static void main(String[] args){
    User user1 = RECYCLER.get();
    user1.recycle();
    User user2 = RECYCLER.get();
    user2.recycle();
    System.out.println(user1==user2);
}
```

通过 Recycler 的 get()方法获取对象，代码如下。

```
public final T get() {
    if (maxCapacityPerThread == 0) {
        return newObject((Handle<T>) NOOP_HANDLE);
```

```
    }
    Stack<T> stack = threadLocal.get();
    DefaultHandle<T> handle = stack.pop();
    if (handle == null) {
        handle = stack.newHandle();
        handle.value = newObject(handle);
    }
    return (T) handle.value;
}
```

首先判断 maxCapacityPerThread 是否为 0，maxCapacityPerThread 代表 Stack 最多能缓存多少个对象，如果缓存 0 个，说明对象将一个都不会回收。通过调用 newObject 创建一个对象，并传入一个 NOOP_HANDLE，NOOP_HANDLE 是一个 Handle，我们看其定义。

```
private static final Handle NOOP_HANDLE = new Handle() {
    @Override
    public void recycle(Object object) {

    }
};
```

这里的 recycle()方法是一个空实现，代表不进行任何对象回收。回到 get()方法中，我们看第二步 Stack<T> stack = threadLocal.get()，这里通过 FastThreadLocal 对象获取当前线程的 Stack。获取 Stack 之后，从 Stack 中 pop 出一个 Handle，其作用稍后分析。如果取出的对象为 null，说明当前回收站内没有任何对象，通常第一次执行到这里对象还没回收，就会是 null，这样则会通过 stack.newHandle()创建一个 Handle。创建出来的 Handle 的 Value 属性，通过重写的 newObject()方法进行赋值，也就是 Demo 中的 User。跟进 newHandle()方法。

```
DefaultHandle<T> newHandle() {
    return new DefaultHandle<T>(this);
}
```

这里创建一个 DefaultHandle 对象，并传入 this，这里的 this 是当前 Stack。DefaultHandle 的构造方法的代码如下。

```
DefaultHandle(Stack<?> stack) {
    this.stack = stack;
}
```

这里初始化了 stack 属性。DefaultHandle 中还有一个 Value 的成员变量。

```
    private Object value;
```

这里的 Value 用来绑定回收的对象本身。回到 get()方法中，分析 Handle，我们回到上一步。

```
DefaultHandle<T> handle = stack.pop();
```

我们分析从 Stack 中弹出一个 Handle 的逻辑，跟进 pop()方法。

```
DefaultHandle<T> pop() {
    int size = this.size;
    if (size == 0) {
        if (!scavenge()) {
            return null;
        }
        size = this.size;
    }
    size --;
    DefaultHandle ret = elements[size];
    elements[size] = null;
    if (ret.lastRecycledId != ret.recycleId) {
        throw new IllegalStateException("recycled multiple times");
    }
    ret.recycleId = 0;
    ret.lastRecycledId = 0;
    this.size = size;
    return ret;
}
```

首先获取 size，size 表示当前 Stack 的对象数。如果 size 为 0，则调用 scavenge()方法，这个方法是异线程回收对象的方法，我们放在之后的小节进行分析。如果 size 大于 0，则 size 进行自减，代表取出一个元素，然后通过 size 的数组下标的方式将 Handle 取出，之后将当前下标设置为 null，最后将属性 recycleId、lastRecycledId、size 进行赋值。recycleId 和 lastRecycledId 会在之后的小节进行分析，回到 get()方法。

```
public final T get() {
    if (maxCapacityPerThread == 0) {
        return newObject((Handle<T>) NOOP_HANDLE);
    }
    Stack<T> stack = threadLocal.get();
    DefaultHandle<T> handle = stack.pop();
    if (handle == null) {
        handle = stack.newHandle();
        handle.value = newObject(handle);
    }
    return (T) handle.value;
}
```

无论是从 Stack 中弹出的 Handle，还是创建的 Handle，最后都要通过 handle.value 获取实际使用的对象。

14.2.3　相同线程内的对象回收

上节中剖析了从 Recycler 中获取一个对象，本节分析在创建和回收是同线程的前提下，Recycler 是如何进行回收的。回顾前面章节 Demo 中的 main()方法。

```
public static void main(String[] args){
    User user1 = RECYCLER.get();
    user1.recycle();
    User user2 = RECYCLER.get();
    user2.recycle();
    System.out.println(user1==user2);
}
```

这是一个同线程回收对象的典型场景，在一个线程中将对象创建并且回收，我们的 User 对象定义了 recycle 方法。

```
static class User{
    private final Recycler.Handle<User> handle;
    public User(Recycler.Handle<User> handle){
        this.handle=handle;
    }
    public void recycle(){
        handle.recycle(this);
    }
}
```

这里的 recycle 是通过 Handle 对象的 recycle()方法实现对象回收的，实际调用的是 DefaultHandle 的 recycle()方法。跟进 recycle()方法。

```
public void recycle(Object object) {
    if (object != value) {
        throw new IllegalArgumentException("object does not belong to handle");
    }
    stack.push(this);
}
```

如果回收的对象为 null，则抛出异常。如果不为 null，则通过自身绑定 Stack 的 push()方法将自身 push 到 Stack 中。push()方法的代码如下。

```
void push(DefaultHandle<?> item) {
    Thread currentThread = Thread.currentThread();
    if (thread == currentThread) {
        pushNow(item);
    } else {
        pushLater(item, currentThread);
```

```
    }
}
```

首先判断当前线程和创建 Stack 的时候保存的线程是否是同一线程。如果是，说明是同线程回收对象，则执行 pushNow()方法将对象放入 Stack 中。pushNow()方法的代码如下。

```
private void pushNow(DefaultHandle<?> item) {
    if ((item.recycleId | item.lastRecycledId) != 0) {
        throw new IllegalStateException("recycled already");
    }
    item.recycleId = item.lastRecycledId = OWN_THREAD_ID;
    int size = this.size;
    if (size >= maxCapacity || dropHandle(item)) {
        return;
    }
    if (size == elements.length) {
        elements = Arrays.copyOf(elements, min(size << 1, maxCapacity));
    }
    elements[size] = item;
    this.size = size + 1;
}
```

如果第一次回收，item.recycleId 和 item.lastRecycledId 都为 0，则不会进入 if 块。继续往下看，item.recycleId = item.lastRecycledId = OWN_THREAD_ID 这一步将 Handle 的 recycleId 和 lastRecycledId 赋值为 OWN_THREAD_ID，OWN_THREAD_ID 在每一个 recycle 中都是唯一固定的，这里我们只需要记住这个概念就行。然后获取当前 size，如果 size 超过上限大小，则直接返回。这里还有个判断 dropHandle。

```
boolean dropHandle(DefaultHandle<?> handle) {
    if (!handle.hasBeenRecycled) {
        if ((++handleRecycleCount & ratioMask) != 0) {
            return true;
        }
        handle.hasBeenRecycled = true;
    }
    return false;
}
```

if (!handle.hasBeenRecycled) 表示当前对象之前是否没有被回收过，如果是第一次回收，会返回 true，然后进入 if。再看 if 中的判断 if ((++handleRecycleCount & ratioMask) != 0)，handleRecycleCount 表示当前位置 Stack 回收了多少次对象（回收了多少次，不代表回收了多少个对象，因为不是每次回收都会被成功地保存在 Stack 中），我们之前分析过 ratioMask 是 7，这里 (++handleRecycleCount & ratioMask) != 0 表示回收的对象数如果不是 8 的倍数，则返回 true，表示只回收 1/8 的对象，然后将 hasBeenRecycled 设置为 true，表示已经被回收。回到 pushNow()方法中，如果 size 的大小等于 Stack 中的数组 Elements 的大小，则将数组 Elements 进行扩容，最后 size 通过数组下标的方式将当前 Handle 设置到 Elements 的元素中，并将 size 进行自增。

14.2.4　不同线程间的对象回收

异线程回收对象，就是创建对象和回收对象不在同一条线程的情况下对象回收的逻辑。在 14.2.1 节简单介绍过，异线程回收对象，是不会放在当前线程的 Stack 中的，而是放在一个 WeakOrderQueue 的数据结构中，回顾我们之前的示意图如下图所示。

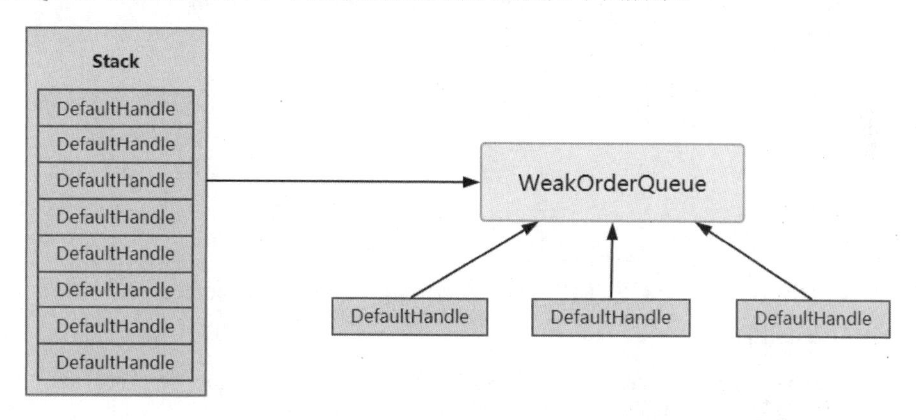

相关的逻辑，我们跟到源码中，首先从回收对象的入口方法开始。DefaultHandle 的 recycle()方法代码如下。

```
public void recycle(Object object) {
    if (object != value) {
        throw new IllegalArgumentException("object does not belong to handle");
    }
    stack.push(this);
}
```

继续看 push()方法的代码。

```
void push(DefaultHandle<?> item) {
    Thread currentThread = Thread.currentThread();
    if (thread == currentThread) {
        pushNow(item);
    } else {
        pushLater(item, currentThread);
    }
}
```

上节分析过，同线程会执行到 pushNow()，有关具体逻辑也进行了分析。如果不是同线程，则会执行到 pushLater()方法，传入 Handle 对象和当前线程对象，跟进 pushLater()方法。

```
private void pushLater(DefaultHandle<?> item, Thread thread) {
    Map<Stack<?>, WeakOrderQueue> delayedRecycled = DELAYED_RECYCLED.get();
    WeakOrderQueue queue = delayedRecycled.get(this);
    if (queue == null) {
        if (delayedRecycled.size() >= maxDelayedQueues) {
            delayedRecycled.put(this, WeakOrderQueue.DUMMY);
            return;
        }
        if ((queue = WeakOrderQueue.allocate(this, thread)) == null) {
            return;
        }
        delayedRecycled.put(this, queue);
    } else if (queue == WeakOrderQueue.DUMMY) {
        return;
    }
    queue.add(item);
}
```

首先通过 DELAYED_RECYCLED.get()方法获取一个 delayedRecycled 对象，我们看 DELAYED_RECYCLED 的代码。

```
private static final FastThreadLocal<Map<Stack<?>, WeakOrderQueue>> DELAYED_RECYCLED =
        new FastThreadLocal<Map<Stack<?>, WeakOrderQueue>>() {
    @Override
    protected Map<Stack<?>, WeakOrderQueue> initialValue() {
        return new WeakHashMap<Stack<?>, WeakOrderQueue>();
    }
};
```

我们看到 DELAYED_RECYCLED 是一个 FastThreadLocal 对象，initialValue()方法创建一个 WeakHashMap 对象，WeakHashMap 是一个 Map，Key 为 Stack，Value 为前面提到过的 WeakOrderQueue。从中可以分析到，每个线程都维护了一个 WeakHashMap 对象。WeakHashMap 中的元素，是一个 Stack 和 WeakOrderQueue 的映射，说明不同的 Stack 对应不同的 WeakOrderQueue。这里的映射关系可以举例说明。

比如线程 1 创建了一个对象，在线程 3 进行了回收；线程 2 创建了一个对象，同样也在线程 3 进行了回收，那么线程 3 对应的 WeakHashMap 中保存了两组关系：线程 1 对应的 Stack 和 WeakOrderQueue，以及线程 2 对应的 Stack 和 WeakOrderQueue。我们回到 pushLater()方法中，继续往下看。

```
WeakOrderQueue queue = delayedRecycled.get(this)
```

获取了当前线程的 WeakHashMap 对象 delayedRecycled 之后，通过 delayedRecycled 创建对象的线程的 Stack，获取 WeakOrderQueue。这里的 this，就是创建对象的那个线程所属的 Stack，这个 Stack 是绑定在 Handle 中的，在创建 Handle 对象的时候进行的绑定。假设当前线程是线程 2，创建 Handle 的线程是线程 1，通过 Handle 的 Stack 获取线程 1 的 WeakOrderQueue。if (queue == null) 说明线程 2 没有回收过线程 1 的对象，则进入 if 块的逻辑，首先判断 if (delayedRecycled.size() >= maxDelayedQueues)。

- delayedRecycled.size()表示当前线程回收其他创建对象的线程的个数，也就是有几个其他的线程在当前线程回收对象。
- maxDelayedQueues 表示最多能回收的线程个数，如果超过这个值，就表示当前线程不能再回收其他线程的对象了。

通过 delayedRecycled.put(this, WeakOrderQueue.DUMMY) 标识创建对象的线程的 Stack 所对应的 WeakOrderQueue 不可用，DUMMY 可以理解为不可用。如果没有超过 maxDelayedQueues，则通过 if 判断中的 WeakOrderQueue.allocate(this, thread)方式创建一个 WeakOrderQueue。allocate 传入 this，也就是创建对象的线程对应的 Stack，跟进 allocate()方法。

```
static WeakOrderQueue allocate(Stack<?> stack, Thread thread) {
    return reserveSpace(stack.availableSharedCapacity, LINK_CAPACITY)
            ? new WeakOrderQueue(stack, thread) : null;
}
```

reserveSpace(stack.availableSharedCapacity, LINK_CAPACITY)表示线程 1 的 Stack 还能不能分配 LINK_CAPACITY 个元素，如果可以，则直接通过 new 的方式创建一个 WeakOrderQueue 对象。回到 reserveSpace()方法。

```
private static boolean reserveSpace(AtomicInteger availableSharedCapacity, int space) {
    assert space >= 0;
    for (;;) {
        int available = availableSharedCapacity.get();
        if (available < space) {
            return false;
        }
        if (availableSharedCapacity.compareAndSet(available, available - space)) {
            return true;
        }
    }
}
```

参数 availableSharedCapacity 表示线程 1 的 Stack 允许外部线程给其缓存多少个对象，之前我们分析过是 16 384，space 默认是 16。方法中通过一个 CAS 操作，将 16 384 减去 16，表示 Stack

可以给其他线程缓存的对象数为 16 384-16，而这 16 个元素，将由线程 2 缓存。

回到 pushLater() 方法，创建之后通过 delayedRecycled.put(this, queue) 将 Stack 和 WeakOrderQueue 进行关联，通过 queue.add(item)将创建的 WeakOrderQueue 添加一个 Handle。讲解 WeakOrderQueue 之前，先了解下 WeakOrderQueue 的数据结构。WeakOrderQueue 维护了多个 Link，Link 之间通过链表进行连接，每个 Link 可以盛放 16 个 Handle。我们分析过，在 reserveSpace() 方法中将 stack.availableSharedCapacity-16，其实就表示先分配 16 个空间放在 Link 里，下次回收的时候，如果这 16 个空间没有填满，则可以继续往里盛放。如果 16 个空间都已填满，则通过继续添加 Link 的方式继续分配 16 个空间用于盛放 Handle。WeakOrderQueue 和 WeakOrderQueue 之间也通过链表进行关联，可以根据下图理解上述逻辑。

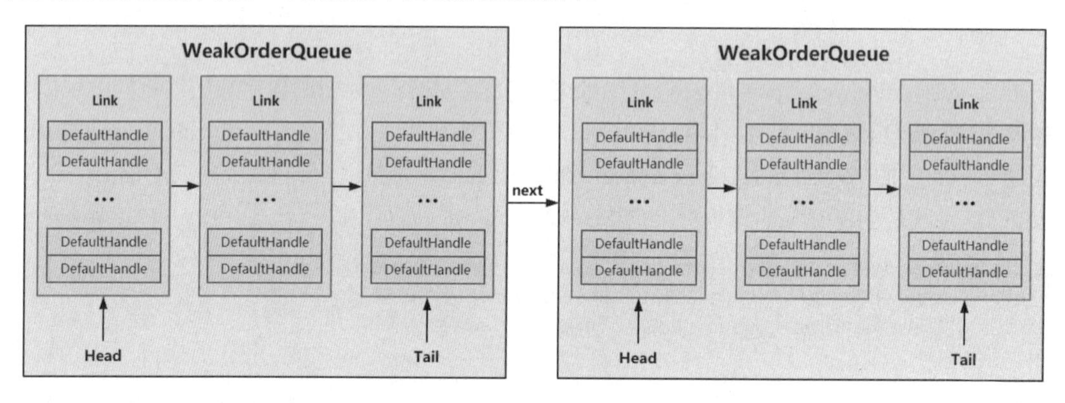

根据以上思路，我们看 WeakOrderQueue 的构造方法。

```
private WeakOrderQueue(Stack<?> stack, Thread thread) {
    head = tail = new Link();
    owner = new WeakReference<Thread>(thread);
    synchronized (stack) {
        next = stack.head;
        stack.head = this;
    }
    availableSharedCapacity = stack.availableSharedCapacity;
}
```

这里有个 Head 和 Tail，都指向一个 Link 对象，其实在 WeakOrderQueue 中维护了一个链表，Head 和 Tail 分别代表头节点和尾节点，初始状态下，头节点和尾节点都指向同一个节点。简单看下 Link 的类的定义。

```
private static final class Link extends AtomicInteger {
    private final DefaultHandle<?>[] elements = new DefaultHandle[LINK_CAPACITY];
    private int readIndex;
```

```
    private Link next;
}
```

每次创建一个 Link，都会创建一个 DefaultHandle 类型的数组用于盛放 DefaultHandle 对象，默认大小是 16 个。

readIndex 是一个读指针，之后小节会进行分析。next 节点则指向下一个 Link。回到 WeakOrderQueue 的构造方法中，owner 是对当前线程进行一个包装，代表了当前线程。

接下来在一个同步块中，将当前创建的 WeakOrderQueue 插入 Stack 指向的第一个 WeakOrderQueue，也就是 Stack 的 Head 属性，指向我们创建的 WeakOrderQueue，如下图所示。

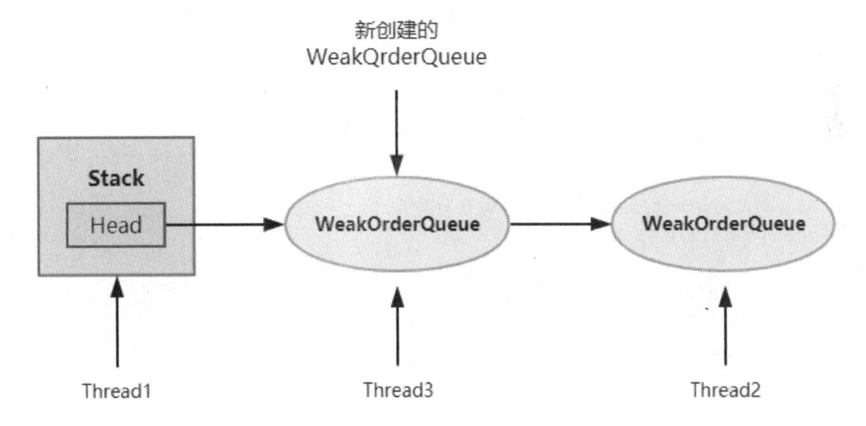

如果线程 2 创建一个和 Stack 关联的 WeakOrderQueue，Stack 的头节点就会指向线程 2 创建的 WeakOrderQueue。如果之后线程 3 也创建了一个和 Stack 关联的 WeakOrderQueue，Stack 的头节点就会指向新创建的线程 3 的 WeakOrderQueue，然后线程 3 的 WeakOrderQueue 再指向线程 2 的 WeakOrderQueue。也就是无论哪个线程创建一个和同一个 Stack 关联的 WeakOrderQueue 的时候，都插入 Stack 指向的 WeakOrderQueue 列表的头部，这样就可以将 Stack 和其他线程释放对象的容器 WeakOrderQueue 进行绑定。回到 pushLater()方法。

```
private void pushLater(DefaultHandle<?> item, Thread thread) {
    Map<Stack<?>, WeakOrderQueue> delayedRecycled = DELAYED_RECYCLED.get();
    WeakOrderQueue queue = delayedRecycled.get(this);
    if (queue == null) {
        if (delayedRecycled.size() >= maxDelayedQueues) {
            delayedRecycled.put(this, WeakOrderQueue.DUMMY);
            return;
        }
        if ((queue = WeakOrderQueue.allocate(this, thread)) == null) {
            return;
```

```
        }
        delayedRecycled.put(this, queue);
    } else if (queue == WeakOrderQueue.DUMMY) {
        return;
    }
    queue.add(item);
}
```

根据之前分析的 WeakOrderQueue 的数据结构，分析最后一步，也就是 WeakOrderQueue 的 add()方法。

```
void add(DefaultHandle<?> handle) {
    handle.lastRecycledId = id;
    Link tail = this.tail;
    int writeIndex;
    if ((writeIndex = tail.get()) == LINK_CAPACITY) {
        if (!reserveSpace(availableSharedCapacity, LINK_CAPACITY)) {
            return;
        }
        this.tail = tail = tail.next = new Link();
        writeIndex = tail.get();
    }
    tail.elements[writeIndex] = handle;
    handle.stack = null;
    tail.lazySet(writeIndex + 1);
}
```

首先看 handle.lastRecycledId = id ，lastRecycledId 表示 Handle 上次回收的 id，而 id 表示 WeakOrderQueue 的 id，weakOrderQueue 每次创建的时候，会自增一个唯一的 id。Link tail = this.tail 表示获取当前 WeakOrderQueue 中指向最后一个 Link 的指针，也就是尾指针。再看 if ((writeIndex = tail.get()) == LINK_CAPACITY) ，tail.get()表示获取当前 Link 中已经填充元素的个数，如果等于 16，说明元素已经填充满。然后通过 reserveSpace()方法判断当前 WeakOrderQueue 是否还能缓存 Stack 的对象，reserveSpace()方法会根据 Stack 的属性 availableSharedCapacity-16 的方式判断还能否缓存 Stack 的对象，如果不能再缓存 Stack 的对象，则返回。如果还能继续缓存，则再创建一个 Link，并将尾节点指向新创建的 Link，并且原来尾节点的 next 节点指向新创建的 Link，然后获取当前 Link 的 writeIndex，也就是写指针。如果新创建的 Link 中没有元素，writeIndex 为 0，之后将尾部的 Link 的 Elements 属性，也就是一个 DefaultHandle 类型的数组，通过数组下标的方式将第 writeIndex 个节点赋值为要回收的 Handle，然后将 Handle 的 Stack 属性设置为 null，表示当前 Handle 不是通过 Stack 进行回收的，最后将 Tail 节点的元素个数进行+1，表示下一次将从 writeIndex+1 的位置往里写。

14.2.5　获取不同线程间释放的对象

上节分析了异线程回收对象，原理是通过与 Stack 关联的 WeakOrderQueue 进行回收。如果对象经过异线程回收之后，当前线程需要取出对象进行二次利用，当前 Stack 为空，则会通过当前 Stack 关联的 WeakOrderQueue 进行取出，这也是本节要分析的获取异线程释放对象的内容。

在介绍之前，先看 Stack 类中的两个属性。

```
private WeakOrderQueue cursor, prev;
private volatile WeakOrderQueue head;
```

这里的 cursor、prev 和 head 都是指向链表中 WeakOrderQueue 的指针，其中 head 指向最近创建的与 Stack 关联的 WeakOrderQueue，也就是头节点。cursor 代表的是寻找当前 WeakOrderQueue，prev 则是 cursor 的上一个节点，如下图所示。

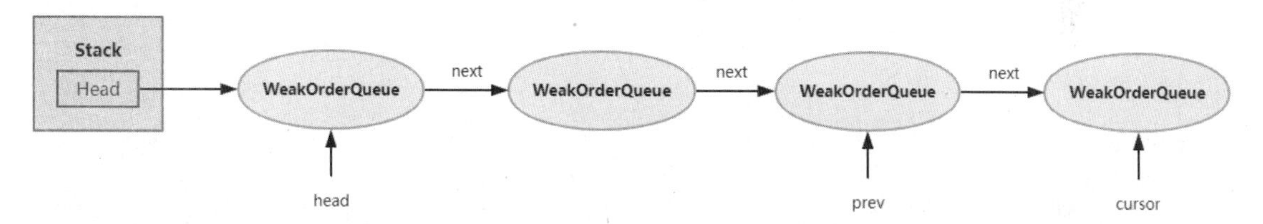

我们从获取对象的入口方法 Handle 的 get()方法开始分析。

```
public final T get() {
    if (maxCapacityPerThread == 0) {
        return newObject((Handle<T>) NOOP_HANDLE);
    }
    Stack<T> stack = threadLocal.get();
    DefaultHandle<T> handle = stack.pop();
    if (handle == null) {
        handle = stack.newHandle();
        handle.value = newObject(handle);
    }
    return (T) handle.value;
}
```

这块逻辑我们并不陌生，Stack 对象通过 pop()方法弹出一个 Handle，跟进 pop()方法。

```
DefaultHandle<T> pop() {
    int size = this.size;
    if (size == 0) {
        if (!scavenge()) {
            return null;
        }
        size = this.size;
```

```
    }
    size --;
    DefaultHandle ret = elements[size];
    elements[size] = null;
    if (ret.lastRecycledId != ret.recycleId) {
        throw new IllegalStateException("recycled multiple times");
    }
    ret.recycleId = 0;
    ret.lastRecycledId = 0;
    this.size = size;
    return ret;
}
```

这里重点关注的是，如果 size 为空，也就是当前 Stack 为空的情况下，会执行到 scavenge()方法。这个方法就是从 WeakOrderQueue 获取对象的方法。跟进 scavenge()方法。

```
boolean scavenge() {
    if (scavengeSome()) {
        return true;
    }
    prev = null;
    cursor = head;
    return false;
}
```

scavengeSome()方法表示如果已经回收到了对象，则直接返回；如果没有回收到对象，则将 prev 和 cursor 两个指针进行重置。继续跟进 scavengeSome()方法。

```
boolean scavengeSome() {
    WeakOrderQueue cursor = this.cursor;
    if (cursor == null) {
        cursor = head;
        if (cursor == null) {
            return false;
        }
    }
    boolean success = false;
    WeakOrderQueue prev = this.prev;
    do {
        if (cursor.transfer(this)) {
            success = true;
            break;
        }
        WeakOrderQueue next = cursor.next;
        if (cursor.owner.get() == null) {
            if (cursor.hasFinalData()) {
                for (;;) {
```

```
                if (cursor.transfer(this)) {
                    success = true;
                } else {
                    break;
                }
            }
        }
        if (prev != null) {
            prev.next = next;
        }
    } else {
        prev = cursor;
    }
    cursor = next;
} while (cursor != null && !success);
this.prev = prev;
this.cursor = cursor;
return success;
}
```

首先获取 cursor 指针，cursor 指针代表要回收的 WeakOrderQueue。如果 cursor 为空，则让其指向头节点，如果头节点也为空，说明当前 Stack 没有与其关联的 WeakOrderQueue，则返回 false。通过一个布尔值 success 标记回收状态，然后获取 prev 指针，也就是 cursor 的上一个节点，之后进入一个 do-while 循环。do-while 循环的终止条件是没有遍历到最后一个节点并且回收的状态为 false。我们仔细来看 do-while 循环中的代码逻辑，首先 cursor 指针会调用 transfer()方法，该方法表示从当前指针指向的 WeakOrderQueue 中将元素放入当前 Stack 中，如果取出成功则将 success 设置为 true 并跳出循环，transfer()方法我们稍后分析。

继续往下看，如果没有获得元素，则会通过 next 属性获取下一个 WeakOrderQueue，然后进入一个判断 if (cursor.owner.get() == null)。owner 属性是与当前 WeakOrderQueue 关联的一个线程，get()方法获得关联的线程对象，如果这个对象为 null 说明该线程不存在，则进入 if 块，也就是一些清理的工作。if 块中又进入一个判断 if (cursor.hasFinalData())，表示当前的 WeakOrderQueue 中是否还有数据，如果有数据则通过 for 循环将数据通过 transfer()方法传输到当前 Stack 中，传输成功的，将 success 标记为 true。Transfer()方法是将 WeakOrderQueue 中一个 Link 中的 Handle 往 Stack 传输，这里通过 for 循环将每个 Link 中的数据都传输到 Stack 中。

继续往下看，如果 prev 节点不为空，则通过 prev.next = next 将 cursor 节点进行释放，也就是 prev 的下一个节点指向 cursor 的下一个节点。继续往下看 else 块中的 prev = cursor，表示如果当前线程还在，则将 prev 赋值为 cursor，代表 prev 后移一个节点，最后通过 cursor = next 将 cursor 后移一位，然后继续进行循环。循环结束之后，将 Stack 的 prev 和 cursor 属性进行保存。我们跟到 transfer()方法中，分析如何将 WeakOrderQueue 中的 Handle 传输到 Stack 中。

```
boolean transfer(Stack<?> dst) {
    Link head = this.head;
    if (head == null) {
        return false;
    }
    if (head.readIndex == LINK_CAPACITY) {
        if (head.next == null) {
            return false;
        }
        this.head = head = head.next;
    }
    final int srcStart = head.readIndex;
    int srcEnd = head.get();
    final int srcSize = srcEnd - srcStart;
    if (srcSize == 0) {
        return false;
    }
    final int dstSize = dst.size;
    final int expectedCapacity = dstSize + srcSize;
    if (expectedCapacity > dst.elements.length) {
        final int actualCapacity = dst.increaseCapacity(expectedCapacity);
        srcEnd = min(srcStart + actualCapacity - dstSize, srcEnd);
    }
    if (srcStart != srcEnd) {
        final DefaultHandle[] srcElems = head.elements;
        final DefaultHandle[] dstElems = dst.elements;
        int newDstSize = dstSize;
        for (int i = srcStart; i < srcEnd; i++) {
            DefaultHandle element = srcElems[i];
            if (element.recycleId == 0) {
                element.recycleId = element.lastRecycledId;
            } else if (element.recycleId != element.lastRecycledId) {
                throw new IllegalStateException("recycled already");
            }
            srcElems[i] = null;
            if (dst.dropHandle(element)) {
                continue;
            }
            element.stack = dst;
            dstElems[newDstSize ++] = element;
        }
        if (srcEnd == LINK_CAPACITY && head.next != null) {
            reclaimSpace(LINK_CAPACITY);
            this.head = head.next;
        }
```

```
        head.readIndex = srcEnd;
        if (dst.size == newDstSize) {
            return false;
        }
        dst.size = newDstSize;
        return true;
    } else {
        return false;
    }
}
```

剖析之前，我们回顾 WeakOrderQueue 的数据结构，如下图所示。

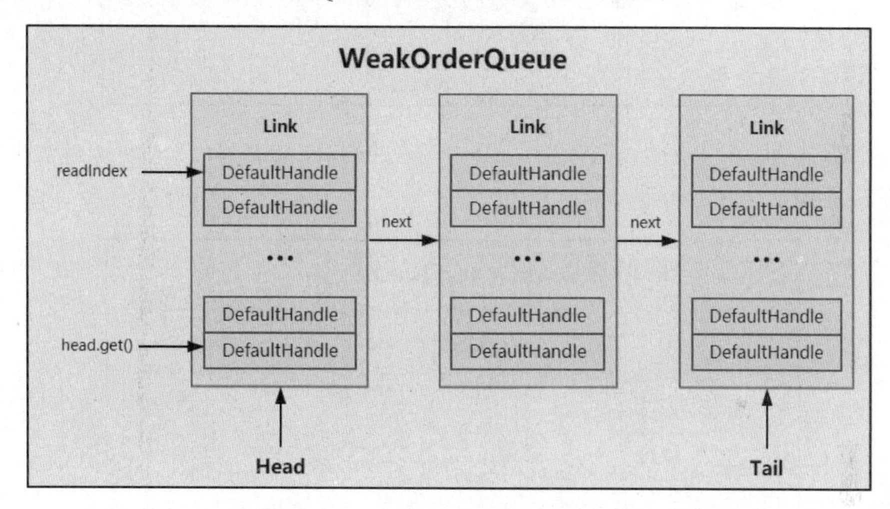

我们上节分析过，WeakOrderQueue 是由多个 Link 组成的，每个 Link 通过链表的方式进行关联，其中 Head 属性指向第一个 Link，Tail 属性指向最后一个 Link。在每个 Link 中有多个 Handle，Link 中维护了一个读指针 readIndex，标识着读取 Link 中 Handle 的位置。继续分析 transfer()方法，首先获取头节点，并判断头节点是否为空，如果头节点为空，说明当前 WeakOrderQueue 并没有 Link，返回 false。if (head.readIndex == LINK_CAPACITY)判断读指针是否为 16，因为 Link 中元素最大数量就是 16，所以如果读指针为 16，说明当前 Link 中的数据都被取走了。接着判断 head.next == null，表示是否还有下一个 Link，如果没有，则说明当前 WeakOrderQueue 没有元素了，返回 false。如果当前 Head 的下一个节点不为 null，则将当前头节点指向下一个节点，将原来的头节点进行释放，其移动关系如下图所示。

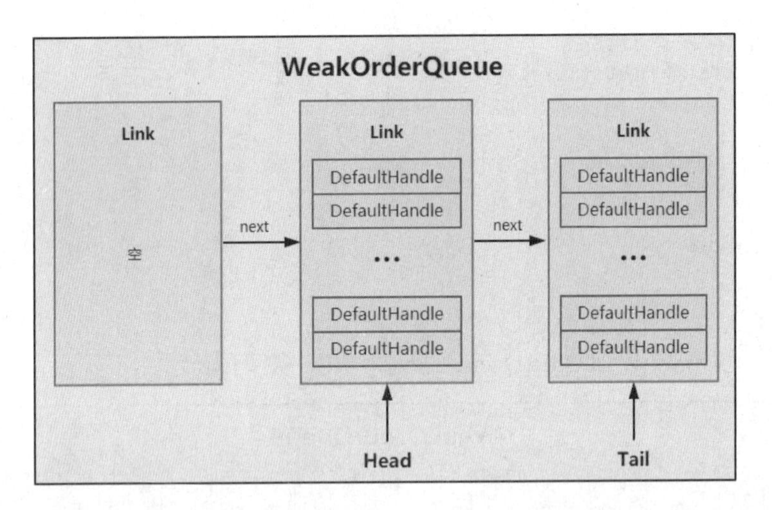

继续往下看，获取头节点的读指针和 Head 中元素的数量，计算可以传输元素的大小，如果大小为 0，则返回 false，如下图所示。

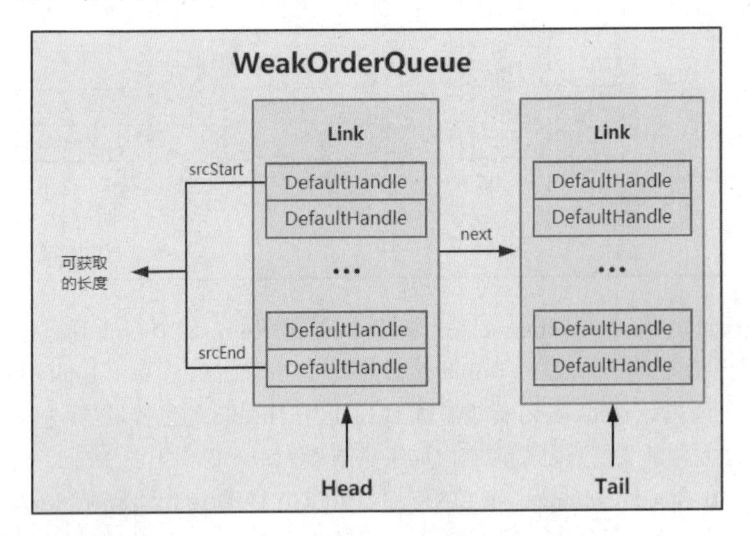

接着，获取当前 Stack 的大小，当前 Stack 大小加上可以传输的大小表示 Stack 中所需要的容量。if (expectedCapacity > dst.elements.length) 表示如果需要的容量大于当前 Stack 中所维护的数组的大小，则将 Stack 中维护的数组进行扩容，进入 if 块中。扩容之后会返回 actualCapacity，表示扩容之后的大小。再看 srcEnd = min(srcStart + actualCapacity - dstSize, srcEnd) 这步，srcEnd 表示可以从 Link 中获取的最后一个元素的下标。

这里对 srcStart + actualCapacity - dstSize 进行拆分，actualCapacity - dstSize 表示扩容后的大小-

原 Stack 的大小，也就是最多能往 Stack 中传输多少元素。读指针+可以往 Stack 传输的数量和，表示往 Stack 中传输的最后一个下标，这里的下标和 srcEnd 中取一个较小的值，也就是既不能超过 Stack 的容量，也不能造成当前 Link 中下标越界。

继续往下看，int newDstSize = dstSize 表示初始化 Stack 的下标，表示 Stack 中从这个下标开始添加数据。然后判断 srcStart != srcEnd，表示能不能从 Link 中获取内容，如果不能，则返回 false，如果可以，则进入 if 块中，接着获取当前 Link 的数组 Elements 和 Stack 中的数组 Elements，通过 for 循环，以数组下标的方式不断将当前 Link 中的数据放入 Stack 中，for 循环中首先获取 Link 的第 *i* 个元素，接下来关注一个细节。

```
if (element.recycleId == 0) {
    element.recycleId = element.lastRecycledId;
} else if (element.recycleId != element.lastRecycledId) {
    throw new IllegalStateException("recycled already");
}
```

这里 element.recycleId == 0 表示对象没有被回收过，则赋值为 lastRecycledId，lastRecycledId 是 WeakOrderQueue 中的唯一下标，通过赋值标记 Element 被回收过，然后继续判断 element.recycleId != element.lastRecycledId，表示该对象被回收过，但是回收的 recycleId 却不是最后一次回收 lastRecycledId，这是一种异常情况，表示一个对象在不同的地方被回收过两次，这种情况则抛出异常，接着将 Link 的第 *i* 个元素设置为 null，继续往下看。

```
if (dst.dropHandle(element)) {
    continue;
}
```

这里表示控制回收站回收的频率，之前的章节中分析过，这里不再赘述。

- element.stack = dst 表示将 Handle 的 Stack 属性设置到当前 Stack。
- dstElems[newDstSize ++] = element 表示通过数组下标的方式将 Link 中的 Handle 赋值到 Stack 的数组中。

继续往下看：

```
if (srcEnd == LINK_CAPACITY && head.next != null) {
    reclaimSpace(LINK_CAPACITY);
    this.head = head.next;
}
```

这里的 if 表示循环结束后，如果 Link 中的数据已经回收完毕，并且还有下一个节点则会进到 reclaimSpace()方法。跟到 reclaimSpace()方法。

```
private void reclaimSpace(int space) {
    assert space >= 0;
    availableSharedCapacity.addAndGet(space);
}
```

将 availableSharedCapacity 加上 16，表示 WeakOrderQueue 还可以继续插入 Link。继续看 transfer()方法。

- this.head = head.next 表示将头节点后移一个元素。
- head.readIndex = srcEnd 表示将读指针指向 srcEnd，下一次读取可以从 srcEnd 开始。
- if (dst.size == newDstSize) 表示没有向 Stack 传输任何对象，则返回 false，否则就通过 dst.size = newDstSize 更新 Stack 的大小为 newDstSize，并返回 true。

以上就是从 Link 中往 Stack 中传输数据的过程。

15

第 15 章 单机百万连接性能调优

15.1 模拟 Netty 单机连接瓶颈

我们知道，通常启动一个服务端会绑定一个端口例如 8000 端口当然客户端连接端口是有限制的，除去最大端口 65535 和默认的 1024 及以下的端口，就只剩下 1 024~65 535 个，再扣除一些常用端口，实际可用端口只有 6 万个左右。那么，我们如何实现单机百万连接呢？

假设在服务端启动[8 000,8 100）这 100 个端口，100×6 万就可以实现 600 万个左右的连接，这是 TCP 的一个基础知识，虽然对于客户端来说是同一个端口号，但是对于服务端来说是不同的端口号，由于 TCP 是一个私源组概念，也就是说它是由源 IP 地址、源端口号、目的 IP 地址和目的端口号确定的，当源 IP 地址和源端口号是一样的，但是目的端口号不一样，那么最终系统底层会把它当作两条 TCP 连接来处理，所以这里取巧给服务端开启了 100 个端口号，这就是单机百万连接的准备工作，如下图所示。

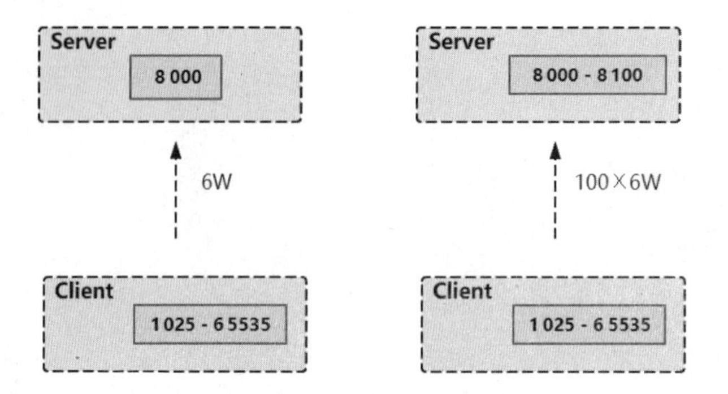

单机 1024 及以下的端口只能给 ROOT 保留使用，客户端端口范围为 1 025~65 535，接下来用代码实现单机百万连接的模拟场景。先看服务端类，循环开启[8 000,8 100）这 100 个监听端口，等待客户端连接，代码如下。

```
package com.gupaoedu.vip.netty.connection;

import io.netty.bootstrap.ServerBootstrap;
import io.netty.channel.ChannelFuture;
import io.netty.channel.ChannelFutureListener;
import io.netty.channel.ChannelOption;
import io.netty.channel.EventLoopGroup;
import io.netty.channel.nio.NioEventLoopGroup;
import io.netty.channel.socket.nio.NioServerSocketChannel;

/**
 * @author Tom
 */
public final class Server {
    public static final int BEGIN_PORT = 8000;
    public static final int N_PORT = 8100;

    public static void main(String[] args) {
        new Server().start(Server.BEGIN_PORT, Server.N_PORT);
    }

    public void start(int beginPort, int nPort) {
        System.out.println("服务端启动中...");

        EventLoopGroup bossGroup = new NioEventLoopGroup();
        EventLoopGroup workerGroup = new NioEventLoopGroup();

        ServerBootstrap bootstrap = new ServerBootstrap();
        bootstrap.group(bossGroup, workerGroup);
        bootstrap.channel(NioServerSocketChannel.class);
```

```
        bootstrap.childOption(ChannelOption.SO_REUSEADDR, true);

        bootstrap.childHandler(new ConnectionCountHandler());


        for (int i = 0; i <= (nPort - beginPort); i++) {
            final int port = beginPort + i;

            bootstrap.bind(port).addListener(new ChannelFutureListener() {
                public void operationComplete(ChannelFuture channelFuture) throws Exception {
                    System.out.println("成功绑定监听端口: " + port);
                }
            });
        }
        System.out.println("服务端已启动!");
    }
}
```

然后看 ConnectionCountHandler 类的实现逻辑，主要用来统计单位时间内的请求数，每接入一个连接则自增一个数字，每 2s 统计一次，代码如下。

```
package com.gupaoedu.vip.netty.connection;


import io.netty.channel.ChannelHandler;
import io.netty.channel.ChannelHandlerContext;
import io.netty.channel.ChannelInboundHandlerAdapter;

import java.util.concurrent.Executors;
import java.util.concurrent.TimeUnit;
import java.util.concurrent.atomic.AtomicInteger;

/**
 * Created by Tom.
 */
@ChannelHandler.Sharable
public class ConnectionCountHandler extends ChannelInboundHandlerAdapter {

    private AtomicInteger nConnection = new AtomicInteger();

    public ConnectionCountHandler() {
        Executors.newSingleThreadScheduledExecutor().scheduleAtFixedRate(new Runnable() {
            public void run() {
                System.out.println("当前客户端连接数: " + nConnection.get());
            }
        },0, 2, TimeUnit.SECONDS);
```

```
    }

    @Override
    public void channelActive(ChannelHandlerContext ctx) {
        nConnection.incrementAndGet();
    }

    @Override
    public void channelInactive(ChannelHandlerContext ctx) {
        nConnection.decrementAndGet();
    }

}
```

再看客户端类代码，主要功能是循环依次往服务端开启的 100 个端口发起请求，直到服务端无响应、线程挂起为止，代码如下。

```
package com.gupaoedu.vip.netty.connection;


import io.netty.bootstrap.Bootstrap;
import io.netty.channel.*;
import io.netty.channel.nio.NioEventLoopGroup;
import io.netty.channel.socket.SocketChannel;
import io.netty.channel.socket.nio.NioSocketChannel;

/**
 * Created by Tom.
 */
public class Client {

    private static final String SERVER_HOST = "127.0.0.1";

    public static void main(String[] args) {
        new Client().start(Server.BEGIN_PORT, Server.N_PORT);
    }

    public void start(final int beginPort, int nPort) {
        System.out.println("客户端已启动...");
        EventLoopGroup eventLoopGroup = new NioEventLoopGroup();
        final Bootstrap bootstrap = new Bootstrap();
        bootstrap.group(eventLoopGroup);
        bootstrap.channel(NioSocketChannel.class);
        bootstrap.option(ChannelOption.SO_REUSEADDR, true);
        bootstrap.handler(new ChannelInitializer<SocketChannel>() {
            @Override
            protected void initChannel(SocketChannel ch) {
```

```
            }
        });

        int index = 0;
        int port;
        while (!Thread.interrupted()) {

            port = beginPort + index;
            try {
                ChannelFuture channelFuture = bootstrap.connect(SERVER_HOST, port);
                channelFuture.addListener(new ChannelFutureListener() {
                    public void operationComplete(ChannelFuture future) throws Exception {
                        if (!future.isSuccess()) {
                            System.out.println("连接失败，程序关闭!");
                            System.exit(0);
                        }
                    }
                });
                channelFuture.get();
            } catch (Exception e) {
            }

            if (port == nPort) { index = 0; }else { index ++; }
        }
    }
}
```

最后，将服务端程序打包发布到 Linux 服务器上，同样将客户端程序打包发布到另一台 Linux 服务器上。接下来分别启动服务端和客户端程序。运行一段时间之后，会发现服务端监听的连接数定格在一个值不再变化，如下所示。

```
当前客户端连接数: 870
当前客户端连接数: 870
当前客户端连接数: 870
当前客户端连接数: 870
当前客户端连接数: 870
当前客户端连接数: 870
当前客户端连接数: 870
当前客户端连接数: 870
当前客户端连接数: 870
...
```

并且抛出如下异常。

```
Exception in thread "nioEventLoopGroup-2-1" java.lang.InternalError:
java.io.FileNotFoundException: /usr/java/jdk1.8.0_121/jre/lib/ext/cldrdata.jar (Too many
```

```
open files)
        at sun.misc.URLClassPath$JarLoader.getResource(URLClassPath.java:1040)
        at sun.misc.URLClassPath.getResource(URLClassPath.java:239)
        at java.net.URLClassLoader$1.run(URLClassLoader.java:365)
        at java.net.URLClassLoader$1.run(URLClassLoader.java:362)
        at java.security.AccessController.doPrivileged(Native Method)
        at java.net.URLClassLoader.findClass(URLClassLoader.java:361)
        at java.lang.ClassLoader.loadClass(ClassLoader.java:424)
        at java.lang.ClassLoader.loadClass(ClassLoader.java:411)
        at sun.misc.Launcher$AppClassLoader.loadClass(Launcher.java:331)
        at java.lang.ClassLoader.loadClass(ClassLoader.java:357)
        at java.util.ResourceBundle$RBClassLoader.loadClass(ResourceBundle.java:503)
        at java.util.ResourceBundle$Control.newBundle(ResourceBundle.java:2640)
        at java.util.ResourceBundle.loadBundle(ResourceBundle.java:1501)
        at java.util.ResourceBundle.findBundle(ResourceBundle.java:1465)
        at java.util.ResourceBundle.findBundle(ResourceBundle.java:1419)
        at java.util.ResourceBundle.getBundleImpl(ResourceBundle.java:1361)
        at java.util.ResourceBundle.getBundle(ResourceBundle.java:845)
        at java.util.logging.Level.computeLocalizedLevelName(Level.java:265)
        at java.util.logging.Level.getLocalizedLevelName(Level.java:324)
        at java.util.logging.SimpleFormatter.format(SimpleFormatter.java:165)
        at java.util.logging.StreamHandler.publish(StreamHandler.java:211)
        at java.util.logging.ConsoleHandler.publish(ConsoleHandler.java:116)
        at java.util.logging.Logger.log(Logger.java:738)
        at io.netty.util.internal.logging.JdkLogger.log(JdkLogger.java:606)
        at io.netty.util.internal.logging.JdkLogger.warn(JdkLogger.java:482)
        at io.netty.util.concurrent.SingleThreadEventExecutor$5.run
(SingleThreadEventExecutor.java:876)
        at io.netty.util.concurrent.DefaultThreadFactory$DefaultRunnableDecorator.run
(DefaultThreadFactory.java:144)
        at java.lang.Thread.run(Thread.java:745)
```

这个时候，我们就应该要知道，这已经是服务器所能接受客户端连接数量的瓶颈值，也就是服务端最大支持 870 个连接。接下来要做的事情是想办法突破这个瓶颈，让单台服务器也能支持 100 万个连接，这是一件多么激动人心的事情。

15.2 单机百万连接调优解决思路

15.2.1 突破局部文件句柄限制

首先在服务端输入命令，看一下单个进程所能支持的最大句柄数。

```
ulimit -n
```

输入命令后，会出现 1 024 的数字，表示 Linux 系统中一个进程能够打开的最大文件数，由于开启一个 TCP 连接就会在 Linux 系统中对应创建一个文件，所以就是受这个文件的最大文件数限制。那为什么前面演示的服务端连接数最终定格在 870，比 1 024 小呢？其实是因为除了连接数，还有 JVM 打开的文件 Class 类也算作进程内打开的文件，所以，1 024 减去 JVM 打开的文件数剩下的就是 TCP 所能支持的连接数。

接下来想办法突破这个限制，首先在服务器命令行输入以下命令，打开/etc/security/limits.conf 文件。

```
sudo vi /etc/security/limits.conf
```

然后在这个文件末尾加上下面两行代码。

```
* hard nofile 1000000
* soft nofile 1000000
```

前面的*表示当前用户，hard 和 soft 分别表示限制和警告限制，nofile 表示最大的文件数标识，后面的数字 1 000 000 表示任何用户都能打开 100 万个文件，这也是操作系统所能支持的最大值，如下图所示。

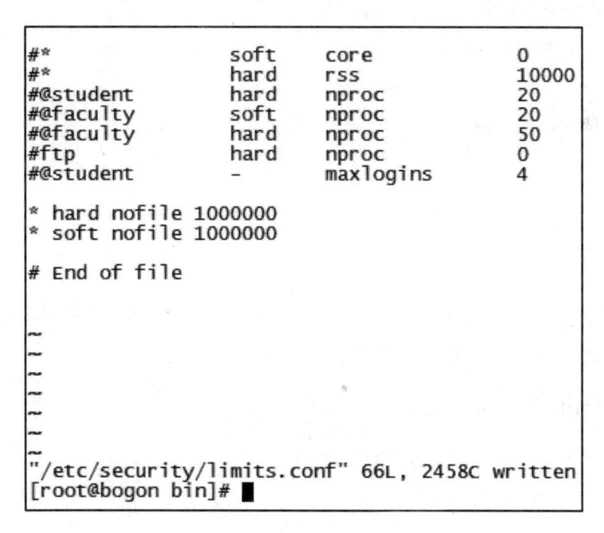

接下来，输入以下命令。

```
ulimit -n
```

这时候，我们发现还是 1 024，没变，重启服务器。将服务端程序和客户端程序分别重新运行，这时候只需静静地观察连接数的变化，最终连接数停留在 137 920，同时抛出了异常，如下所示。

```
当前客户端连接数: 137920
当前客户端连接数: 137920
当前客户端连接数: 137920
当前客户端连接数: 137920
当前客户端连接数: 137920
Exception in thread "nioEventLoopGroup-2-1" java.lang.InternalError:
java.io.FileNotFoundException: /usr/java/jdk1.8.0_121/jre/lib/ext/cldrdata.jar (Too many
open files)
...
```

这又是为什么呢？肯定还有地方限制了连接数，想要突破这个限制，就需要突破全局文件句柄数的限制。

15.2.2 突破全局文件句柄限制

首先在 Linux 命令行输入以下命令，可以查看 Linux 系统所有用户进程所能打开的文件数。

```
cat /proc/sys/fs/file-max
```

通过上面这个命令可以看到全局的限制，发现得到的结果是 10 000。可想而知，局部文件句柄数不能大于全局的文件句柄数。所以，必须将全局的文件句柄数限制调大，突破这个限制。首先切换为 ROOT 用户，不然没有权限。

```
sudo  -s
echo 2000> /proc/sys/fs/file-max
exit
```

我们改成 20 000 来测试一下，继续试验。分别启动服务端程序和客户端程序，发现连接数已经超出了 20 000 的限制。

前面使用 echo 来配置/proc/sys/fs/file-max 的话，重启服务器就会失效，还会变回原来的 10 000，因此，直接用 vi 命令修改，输入以下命令行。

```
sodu vi /etc/sysctl.conf
```

在/etc/sysctl.conf 文件末尾加上下面的内容。

```
fs.file-max=1000000
```

结果如下图所示。

```
[root@bogon ~]# sudo vi  /etc/sysctl.conf

# sysctl settings are defined through files in
# /usr/lib/sysctl.d/, /run/sysctl.d/, and /etc/sysctl.d/.
#
# Vendors settings live in /usr/lib/sysctl.d/.
# To override a whole file, create a new file with the same in
# /etc/sysctl.d/ and put new settings there. To override
# only specific settings, add a file with a lexically later
# name in /etc/sysctl.d/ and put new settings there.
#
# For more information, see sysctl.conf(5) and sysctl.d(5).

fs.file-max=1000000
~
~
~
~
~
~
~
~
~
~
"/etc/sysctl.conf" 12L, 470C written
```

接下来重启 Linux 服务器，再启动服务端程序和客户端程序。

```
当前客户端连接数: 9812451
当前客户端连接数: 9812462
当前客户端连接数: 9812489
当前客户端连接数: 9812501
当前客户端连接数: 9812503
...
```

最终连接数定格在 98 万左右。我们发现主要受限于本机本身的性能。用 htop 命令查看一下，发现 CPU 都接近 100%，如下图所示。

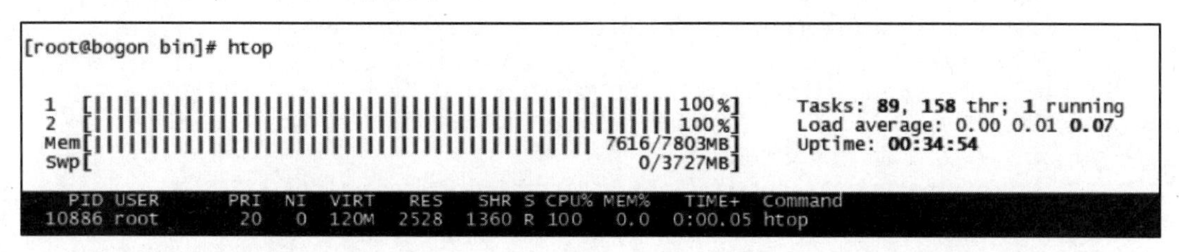

以上是操作系统层面的调优和性能提升，下面主要介绍基于 Netty 应用层面的调优。

15.3　Netty 应用级别的性能调优

15.3.1　Netty 应用级别的性能瓶颈复现

首先来看一下应用场景，下面是一段标准的服务端应用程序代码。

```
package com.gupaoedu.vip.netty.thread;


import io.netty.bootstrap.ServerBootstrap;
import io.netty.channel.*;
import io.netty.channel.nio.NioEventLoopGroup;
import io.netty.channel.socket.SocketChannel;
import io.netty.channel.socket.nio.NioServerSocketChannel;
import io.netty.handler.codec.FixedLengthFrameDecoder;

/**
 * Created by Tom.
 */
public class Server {

    private static final int port = 8000;

    public static void main(String[] args) {

        EventLoopGroup bossGroup = new NioEventLoopGroup();
        EventLoopGroup workerGroup = new NioEventLoopGroup();
        final EventLoopGroup businessGroup = new NioEventLoopGroup(1000);

        ServerBootstrap bootstrap = new ServerBootstrap();
        bootstrap.group(bossGroup, workerGroup)
                .channel(NioServerSocketChannel.class)
                .childOption(ChannelOption.SO_REUSEADDR, true);


        bootstrap.childHandler(new ChannelInitializer<SocketChannel>() {
            @Override
            protected void initChannel(SocketChannel ch) {
                //自定义长度的解码，每次发送一个 long 类型的长度数据
                //每次传递一个系统的时间戳
                ch.pipeline().addLast(new FixedLengthFrameDecoder(Long.BYTES));
                ch.pipeline().addLast(businessGroup, ServerHandler.INSTANCE);
            }
        });


        ChannelFuture channelFuture = bootstrap.bind(port).addListener(new
ChannelFutureListener() {
            public void operationComplete(ChannelFuture channelFuture) throws Exception {
                System.out.println("服务端启动成功，绑定端口为: " + port);
            }
        });
```

```
    }
}
```

我们重点关注服务端的逻辑处理 ServerHandler 类。

```
package com.gupaoedu.vip.netty.thread;


import io.netty.buffer.ByteBuf;
import io.netty.buffer.Unpooled;
import io.netty.channel.ChannelHandler;
import io.netty.channel.ChannelHandlerContext;
import io.netty.channel.SimpleChannelInboundHandler;

import java.util.concurrent.ThreadLocalRandom;

/**
 * Created by Tom.
 */
@ChannelHandler.Sharable
public class ServerHandler extends SimpleChannelInboundHandler<ByteBuf> {
    public static final ChannelHandler INSTANCE = new ServerHandler();


    //channelread0 是主线程
    @Override
    protected void channelRead0(ChannelHandlerContext ctx, ByteBuf msg) {
        ByteBuf data = Unpooled.directBuffer();
        //从客户端读一个时间戳
        data.writeBytes(msg);
        //模拟一次业务处理，有可能是数据库操作，也有可能是逻辑处理
        Object result = getResult(data);
        //重新写回给客户端
        ctx.channel().writeAndFlush(result);
    }

    //模拟去数据库获取一个结果
    protected Object getResult(ByteBuf data) {

        int level = ThreadLocalRandom.current().nextInt(1, 1000);

        //计算出每次响应需要的时间，用来作为 QPS 的参考数据

        //90.0% == 1ms   1000 100 > 1ms
        int time;
        if (level <= 900) {
```

```
            time = 1;
        //95.0% == 10ms    1000 50 > 10ms
        } else if (level <= 950) {
            time = 10;
        //99.0% == 100ms    1000 10 > 100ms
        } else if (level <= 990) {
            time = 100;
        //99.9% == 1000ms    1000 1 > 1000ms
        } else {
            time = 1000;
        }

        try {
            Thread.sleep(time);
        } catch (InterruptedException e) {
        }

        return data;
    }

}
```

上面代码中有一个 getResult()方法。可以把 getResult()方法看作是在数据库中查询数据的一个方法，把每次查询的结果返回给客户端。实际上，为了模拟查询数据性能，getResult()传入的参数是由客户端传过来的时间戳，最终返回的还是客户端传过来的值。只不过返回之前做了一次随机的线程休眠处理，以模拟真实的业务处理性能。如下表所示是模拟场景的性能参数。

数据处理的业务接口占比	处理所耗的时间
90%	1ms
95%	10ms
99%	100ms
99.9%	1 000ms

下面来看客户端，也是一段标准的代码。

```
package com.gupaoedu.vip.netty.thread;


import io.netty.bootstrap.Bootstrap;
import io.netty.channel.ChannelInitializer;
import io.netty.channel.ChannelOption;
import io.netty.channel.EventLoopGroup;
import io.netty.channel.nio.NioEventLoopGroup;
import io.netty.channel.socket.SocketChannel;
import io.netty.channel.socket.nio.NioSocketChannel;
```

```
import io.netty.handler.codec.FixedLengthFrameDecoder;

/**
 * Created by Tom.
 */
public class Client {

    private static final String SERVER_HOST = "127.0.0.1";

    public static void main(String[] args) throws Exception {
        new Client().start(8000);
    }

    public void start(int port) throws Exception {
        EventLoopGroup eventLoopGroup = new NioEventLoopGroup();
        final Bootstrap bootstrap = new Bootstrap();
        bootstrap.group(eventLoopGroup)
                .channel(NioSocketChannel.class)
                .option(ChannelOption.SO_REUSEADDR, true)
                .handler(new ChannelInitializer<SocketChannel>() {
                    @Override
                    protected void initChannel(SocketChannel ch) {
                        ch.pipeline().addLast(new FixedLengthFrameDecoder(Long.BYTES));
                        ch.pipeline().addLast(ClientHandler.INSTANCE);
                    }
        });

        //客户端每秒钟向服务端发起 1 000 次请求
        for (int i = 0; i < 1000; i++) {
            bootstrap.connect(SERVER_HOST, port).get();
        }
    }
}
```

从上面代码中看到，客户端会向服务端发起 1 000 次请求。重点来看客户端逻辑处理 ClientHandler 类。

```
package com.gupaoedu.vip.netty.thread;


import io.netty.buffer.ByteBuf;
import io.netty.channel.ChannelHandler;
import io.netty.channel.ChannelHandlerContext;
import io.netty.channel.SimpleChannelInboundHandler;

import java.util.concurrent.TimeUnit;
import java.util.concurrent.atomic.AtomicInteger;
```

```
import java.util.concurrent.atomic.AtomicLong;

/**
 * Created by Tom.
 */
@ChannelHandler.Sharable
public class ClientHandler extends SimpleChannelInboundHandler<ByteBuf> {
    public static final ChannelHandler INSTANCE = new ClientHandler();

    private static AtomicLong beginTime = new AtomicLong(0);
    //总响应时间
    private static AtomicLong totalResponseTime = new AtomicLong(0);
    //总请求数
    private static AtomicInteger totalRequest = new AtomicInteger(0);

    public static final Thread THREAD = new Thread(){
        @Override
        public void run() {
            try {
                while (true) {
                    long duration = System.currentTimeMillis() - beginTime.get();
                    if (duration != 0) {
                        System.out.println("QPS: " + 1000 * totalRequest.get() / duration + ", 
" + "平均响应时间: " + ((float) totalResponseTime.get()) / totalRequest.get() + "ms.");
                        Thread.sleep(2000);
                    }
                }

            } catch (InterruptedException ignored) {
            }
        }
    };

    @Override
    public void channelActive(final ChannelHandlerContext ctx) {
        ctx.executor().scheduleAtFixedRate(new Runnable() {
            public void run() {
                ByteBuf byteBuf = ctx.alloc().ioBuffer();
                //将当前系统时间发送到服务端
                byteBuf.writeLong(System.currentTimeMillis());
                ctx.channel().writeAndFlush(byteBuf);
            }
        }, 0, 1, TimeUnit.SECONDS);
    }

    @Override
    protected void channelRead0(ChannelHandlerContext ctx, ByteBuf msg) {
```

```
        //获取一个响应时间差，本次请求的响应时间
        totalResponseTime.addAndGet(System.currentTimeMillis() - msg.readLong());
        //每次自增
        totalRequest.incrementAndGet();

        if (beginTime.compareAndSet(0, System.currentTimeMillis())) {
            THREAD.start();
        }
    }

}
```

上面代码主要模拟了 Netty 真实业务环境下的处理耗时情况，QPS 大概在 1 000 次，每 2s 统计一次。接下来，启动服务端和客户端查看控制台日志。首先运行服务端，看到控制台日志如下图所示。

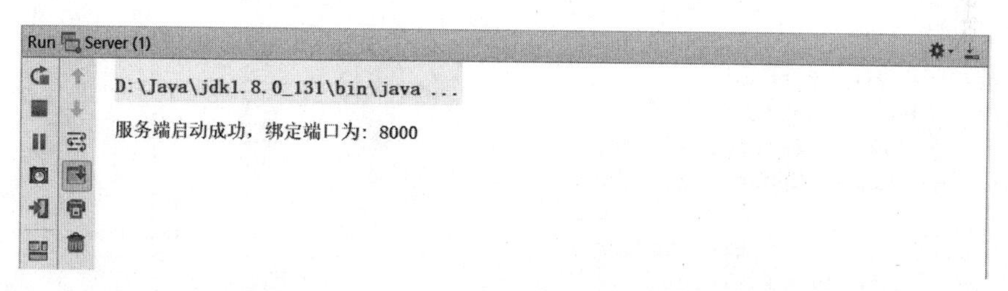

然后运行客户端，看到控制台日志如下图所示，一段时间之后，发现 QPS 保持在 1 000 次以内，平均响应时间越来越长。

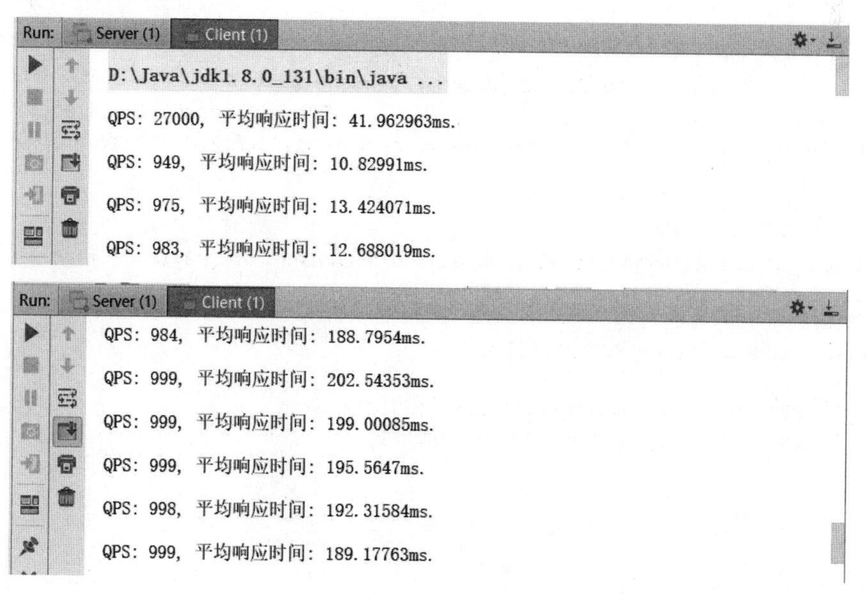

回到服务端 ServerHander 的 getResul()方法，在 getResult()方法中有线程休眠导致阻塞，不难发现，它最终会阻塞主线程，导致所有的请求挤压在一个线程中。如果把下面的代码放入线程池中，效果将完全不同。

```
Object result =getResult(data);
ctx.channel().wrteAndFlush(result);
```

把这两行代码放到业务线程池里，不断在后台运行，运行完成后即时返回结果。

15.3.2 Netty 应用级别的性能调优方案

下面来改造一下代码，在服务端的代码中新建一个 ServerThreadPoolHander 类。

```
package com.gupaoedu.vip.netty.thread;


import io.netty.buffer.ByteBuf;
import io.netty.buffer.Unpooled;
import io.netty.channel.ChannelHandler;
import io.netty.channel.ChannelHandlerContext;

import java.util.concurrent.ExecutorService;
import java.util.concurrent.Executors;

/**
 * Created by Tom.
 */
@ChannelHandler.Sharable
public class ServerThreadPoolHandler extends ServerHandler {
    public static final ChannelHandler INSTANCE = new ServerThreadPoolHandler();
    private static ExecutorService threadPool = Executors.newFixedThreadPool(1000);


    @Override
    protected void channelRead0(final ChannelHandlerContext ctx, ByteBuf msg) {
        final ByteBuf data = Unpooled.directBuffer();
        data.writeBytes(msg);
        threadPool.submit(new Runnable() {
            public void run() {
                Object result = getResult(data);
                ctx.channel().writeAndFlush(result);
            }
        });

    }
}
```

然后在服务端的 Handler 处理注册为 ServerThreadPoolHander，删除原来的 ServerHandler，代码如下。

```
ch.pipeline().addLast(ServerThreadPoolHandler.INSTANCE);
```

随后，启动服务端和客户端程序，查看控制台日志，如下图所示。

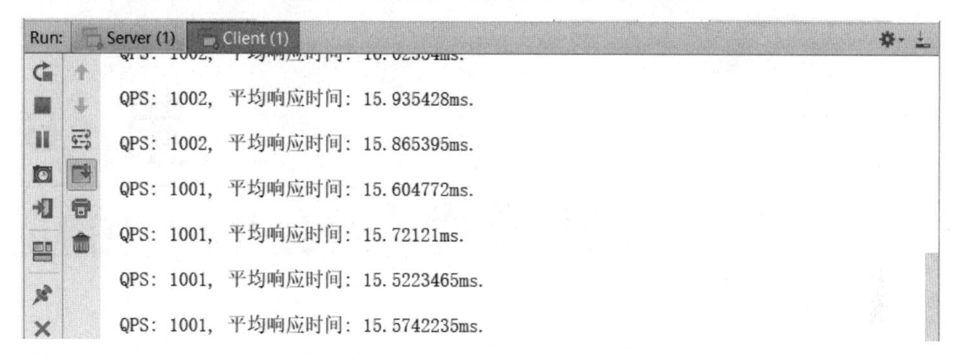

最终耗时稳定在 15ms 左右，QPS 也超过了 1 000 次。实际上这个结果还不是最优的状态，继续调整。将 ServerThreadPoolHander 的线程个数调整到 20，代码如下。

```
public static final ChannelHandler INSTANCE = new ServerThreadPoolHandler();
private static ExecutorService threadPool = Executors.newFixedThreadPool(20);
```

然后启动程序，发现平均响应时间相差也不是太多，如下图所示。

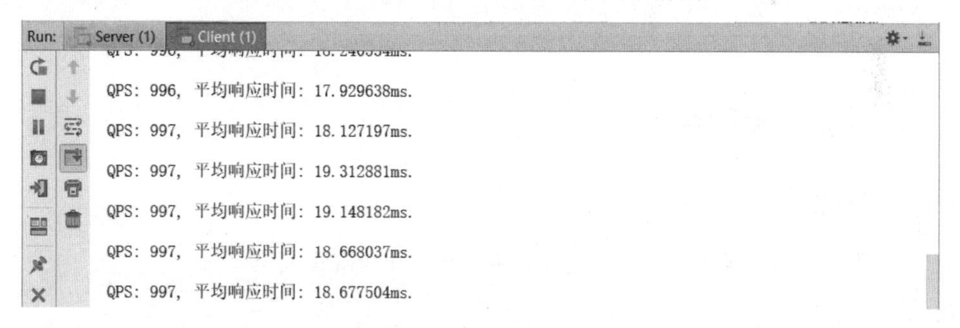

由此得出的结论是：具体的线程数需要在真实的环境下不断地调整、测试，才能确定最合适的数值。本章旨在告诉大家优化的方法，而不是结果。

16

第 16 章 设计模式在 Netty 中的应用

如果你对设计模式不是很熟悉，建议补习一下设计模式再来阅读本章的内容。本章主要分析设计模式的应用案例，不对设计模式本身做详细讲解。

16.1 单例模式源码举例

单例模式要点回顾如下。

（1）一个类在任何情况下只有一个对象，并提供一个全局访问点。

（2）延迟创建。

（3）避免线程安全问题。

Netty 中的单例模式应用非常广泛，单例模式本身也比较简单，下面列举 MqttEncoder 类，采用饿汉式单例的写法。饿汉式单例最大的优点就是简单，绝对线程安全。具体代码如下。

```
@Sharable
public final class MqttEncoder extends MessageToMessageEncoder<MqttMessage> {
    public static final MqttEncoder INSTANCE = new MqttEncoder();
```

```
    private MqttEncoder() {
    }

    protected void encode(ChannelHandlerContext ctx, MqttMessage msg, List<Object> out) throws
Exception {
        out.add(doEncode(ctx.alloc(), msg));
    }
...
}
```

16.2　策略模式源码举例

策略模式要点回顾如下。

（1）封装一系列可相互替换的算法家族。

（2）动态选择某一个策略。

Netty 在根据 CPU 核数分配线程数量的一个优化时，如果是 2 的平方则采用 PowerOfTwoEventExecutorChooser 来创建 EventExecutorChooser，如果不是 2 的平方则采用 GenericEventExecutorChooser 来创建 EventExecutorChooser，这里用的是三元运算选择策略，具体代码如下。

```
public final class DefaultEventExecutorChooserFactory implements EventExecutorChooserFactory
{
    public static final DefaultEventExecutorChooserFactory INSTANCE = new
DefaultEventExecutorChooserFactory();

    private DefaultEventExecutorChooserFactory() {
    }

    public EventExecutorChooser newChooser(EventExecutor[] executors) {
        return (EventExecutorChooser)(isPowerOfTwo(executors.length)?new
DefaultEventExecutorChooserFactory.PowerOfTwoEventExecutorChooser(executors):new
DefaultEventExecutorChooserFactory.GenericEventExecutorChooser(executors));
    }

    private static boolean isPowerOfTwo(int val) {
        return (val & -val) == val;
    }
...
}
```

16.3 装饰者模式源码举例

装饰者模式要点回顾如下。

（1）装饰者和被装饰者实现同一个接口。

（2）装饰者通常继承被装饰者，同宗同源。

（3）动态修改、重载被装饰者的方法。

从 Netty 的 ByteBuf 类结构图可以看到，ByteBuf 的直接实现类有五个，忽略 WrappedByteBuf 这个类，其实直接实现类有四个。为什么要忽略掉 WrappedByteBuf 呢？因为它是 ByteBuf 装饰者的基类，本身没有任何实现功能。来看 WrappedByteBuf 的代码，主要功能就是保存被装饰者的引用。

```
class WrappedByteBuf extends ByteBuf {
    protected final ByteBuf buf;

    protected WrappedByteBuf(ByteBuf buf) {
        if(buf == null) {
            throw new NullPointerException("buf");
        } else {
            this.buf = buf;
        }
    }
...
}
```

具体的装饰者，继承上面的装饰者的顶级类，在自己的构造函数中接收 ByteBuf 的类型的参数，并把它传递给它的父类，用户在调用装饰者时，会把创建的最上面的四种待装饰的组件类以构造方法的形式传递进去，整个体系就运行起来了。 而且装饰者可以按照自己的需求重写父类的方法，或者在现在的基础上添加新的方法调用进行增强。例如 UnreleasableByteBuf 类，重写了 release()方法，并返回 false 表示 UnreleasableByteBuf 不支持被释放。

```
final class UnreleasableByteBuf extends WrappedByteBuf {
    private SwappedByteBuf swappedBuf;

    UnreleasableByteBuf(ByteBuf buf) {
        super(buf);
    }

    ...
```

```
    public boolean release() {
        return false;
    }

    public boolean release(int decrement) {
        return false;
    }
}
```

再比如 SimpLeakAwareByteBuf 类，从字面意思来看，其实是一个内存泄露感知的 ByteBuf，同样继承自 WrappedByteBuf。从构造方法来看，其构造参数多了一个 ResourceLeak，主要用于对内存泄露的跟踪。主要差异在于增加了 release()方法，代码如下。

```
final class SimpleLeakAwareByteBuf extends WrappedByteBuf {
    private final ResourceLeak leak;

    SimpleLeakAwareByteBuf(ByteBuf buf, ResourceLeak leak) {
        super(buf);
        this.leak = leak;
    }

    ...

    public boolean release() {
        boolean deallocated = super.release();
        if(deallocated) {
            this.leak.close();
        }

        return deallocated;
    }

    public boolean release(int decrement) {
        boolean deallocated = super.release(decrement);
        if(deallocated) {
            this.leak.close();
        }

        return deallocated;
    }
}
```

看到上面的代码，还是来关注 release()方法。它调用 release()方法，根据返回结果额外增加了一个资源泄露的监控行为。

16.4 观察者模式源码举例

观察者模式要点回顾如下。

（1）两个角色：观察者和被观察者。

（2）观察者订阅消息，被观察者发布消息。

（3）订阅则能收到消息，取消订阅则收不到消息。

Netty 里面的观察者和被观察者模式一般用 Promise 和 Future 来实现。项目中用得比较多的一个方法就是 channel.writeAndFlush()方法。当调用 channel.writeAndFlush()方法的时候，实际上就是创建了一个被观察者 ChannelFuture，来看源码。

```
public abstract class AbstractChannel extends DefaultAttributeMap implements Channel {

...

    public ChannelFuture writeAndFlush(Object msg) {
        return this.pipeline.writeAndFlush(msg);
    }

    public ChannelFuture writeAndFlush(Object msg, ChannelPromise promise) {
        return this.pipeline.writeAndFlush(msg, promise);
    }
    ...
}
```

writeAndFlush()方法的返回值是 ChannelFuture。当调用 ChannelFuture 的 addListener()方法的时候，其实就是往 ChannelFuture 中添加被一个 ChannelPromise，继续往下跟踪源码。

```
    public ChannelFuture writeAndFlush(Object msg, ChannelPromise promise) {
        if (msg == null) {
            throw new NullPointerException("msg");
        }

        if (!validatePromise(promise, true)) {
            ReferenceCountUtil.release(msg);
            // cancelled
            return promise;
        }

        write(msg, true, promise);
```

```
        return promise;
    }
```

上面的 writeAndFlush()方法还有一个重载方法，其中一个参数就是 ChannelPromise，通常情况下 Future 和 Promise 是成对出现的。我们发现 ChannelPromise 就是 ChannelFuture 的子类，在 Promise 中定义了非常多的回调方法，提供给用户去重载，用户用自己的逻辑通常实现各种 Listener 接口来重载达到回调通知的目的。

16.5　迭代器模式源码举例

迭代器模式要点回顾如下。

（1）实现迭代器接口。

（2）实现对容器中的各个对象逐个访问的方法。

Netty 里面的 CompositeByteBuf 这个零拷贝的实现，就使用了迭代器模式。首先看一段代码。

```
    public static void main(String[] args) {
        ByteBuf header = Unpooled.wrappedBuffer(new byte[]{1, 2, 3});
        ByteBuf body = Unpooled.wrappedBuffer(new byte[]{4, 5, 6});

        ByteBuf merge = merge(header, body);
        merge.forEachByte(value -> {
            System.out.println(value);
            return true;
        });
    }

    public static ByteBuf merge(ByteBuf header, ByteBuf body) {
        CompositeByteBuf byteBuf = ByteBufAllocator.DEFAULT.compositeBuffer(2);
        byteBuf.addComponent(true, header);
        byteBuf.addComponent(true, body);

        return byteBuf;
}
```

这段代码把两个 ByteBuf 添加到一起，forEachByte 就是实现了迭代器模式。那么为什么说它是零拷贝呢？

我们找到 forEachByte()的实现，在 AbstractByteBuf 里面，有下面这样一段代码。

```
    @Override
    public int forEachByte(ByteProcessor processor) {
```

```
        ensureAccessible();
        try {
            return forEachByteAsc0(readerIndex, writerIndex, processor);
        } catch (Exception e) {
            PlatformDependent.throwException(e);
            return -1;
        }
}
```

从 readerIndex 开始读，读到 writeIndex。继续跟进 forEachByteAsc0()方法，查看源码。

```
private int forEachByteAsc0(int start, int end, ByteProcessor processor) throws Exception {
        for (; start < end; ++start) {
            if (!processor.process(_getByte(start))) {
                return start;
            }
        }

        return -1;
}
```

继续看 getByte()方法的实现，找到 CompositeByteBuf 类的实现。

```
public class CompositeByteBuf extends AbstractReferenceCountedByteBuf implements
Iterable<ByteBuf> {
    protected byte_getByte(int index) {
        CompositeByteBuf.Component c = this.findComponent(index);
        return c.buf.getByte(index - c.offset);
    }
    ...
}
```

先找到 Index 对应的 Component，然后迭代的时候直接返回 Component 的 Byte 内容，就实现了零拷贝。其他的 ByteBuf 如果迭代的话，可能会把所有的数据都拷贝一遍。

16.6 责任链模式源码举例

责任链是指多个对象都有机会处理同一个请求，从而避免请求的发送者和接收者之间的耦合关系。然后，将这些对象连成一条链，并且沿着这条链往下传递请求，直到有一个对象可以处理它为止。在处理过程中，每个对象只处理它自己关心的那一部分，不相关的部分可以继续往下传递，直到链中的某个对象不想处理，可以将请求终止或丢弃。

责任链模式要点回顾如下。

（1）需要有一个顶层责任处理接口。

（2）需要有动态创建链、添加和删除责任处理器的接口。

（3）需要有上下文机制。

（4）需要有责任终止机制。

Netty 的 Pipeline 就是采用了责任链设计模式，底层采用双向链表的数据结构，将链上的各个处理器串联起来。客户端每一个请求到来，Netty 都认为是 Pipeline 中所有的处理器都有机会处理它。因此，对于入栈的请求，全部从头节点开始往后传播，一直传播到尾节点（来到尾节点的 msg 才会被释放）。

Netty 的责任链模式中的组件如下。

（1）责任处理器接口 ChannelHandler。

（2）添加删除责任处理器的接口 ChannelPipeline。

（3）上下文组件 ChannelHandlerContext，可获得用户记需要的数据。

（4）终止责任链的 ctx.fireXXX()方法，可终止传播事件。

Pipeline 中所有的 Handler 为顶级抽象接口，它规定了所有的 Handler 统一要有添加、移除、异常捕获的行为。

```
public interface ChannelPipeline
        extends ChannelInboundInvoker, ChannelOutboundInvoker, Iterable<Entry<String,
ChannelHandler>> {

    ChannelPipeline addFirst(String name, ChannelHandler handler);

    ChannelPipeline addFirst(EventExecutorGroup group, String name, ChannelHandler handler);

    ChannelPipeline addLast(String name, ChannelHandler handler);

    ChannelPipeline addLast(EventExecutorGroup group, String name, ChannelHandler handler);

    ChannelPipeline addBefore(String baseName, String name, ChannelHandler handler);

 ...
```

Pipeline 中的 Handler 被封装进了上下文中。通过上下文可以轻松获取当前节点所属的 Channel 及其线程执行器。

```
// todo AttributeMap：让 ChannelHandlerContext 可以存储自定义的属性
// ChannelInboundInvoker：让 ChannelHandlerContext 可以进行 InBound 事件的传播，读事件，read，注册事件或者 active 事件
//ChannelOutboundInvoker：让 ChannelHandlerContext 可以传播写事件
public interface ChannelHandlerContext extends AttributeMap, ChannelInboundInvoker,
ChannelOutboundInvoker {

    //获取 ChannelHandlerContext 所对应的 Channel 对象
    Channel channel();

    //获取事件执行器
    EventExecutor executor();
...
```

责任终止机制包括如下内容。

（1）Pipeline 中的任意一个节点，只要不手动往下传播，这个事件就会在当前节点终止传播。

（2）对于入栈数据，默认会传递到尾节点进行回收，如果不进行下一步传播，事件就会终止在当前节点。

（3）对于出栈数据，头节点使用 Unsafe 对象，把数据写回客户端也意味着事件的终止。

底层事件的传播使用的就是针对链表的操作，如 AbstractChannelHandlerContext 类的 findContextInbound()方法。

```
abstract class AbstractChannelHandlerContext extends DefaultAttributeMap implements
ChannelHandlerContext, ResourceLeakHint {
    private AbstractChannelHandlerContext findContextInbound() {
        AbstractChannelHandlerContext ctx = this;

        do {
            ctx = ctx.next;
        } while(!ctx.inbound);

        return ctx;
}
}
```

16.7 工厂模式源码举例

工厂模式的要点就是将创建对象的逻辑封装起来。

我们最先接触 ReflectiveChannelFactory 就是专门用来创建 Channel 的工厂，它接收一个 Class

对象，然后调用 Class 的 newInstance()方法，将创建好的对象返回去。

```
public class ReflectiveChannelFactory<T extends Channel> implements ChannelFactory<T> {
    private final Class<? extends T> clazz;

    public ReflectiveChannelFactory(Class<? extends T> clazz) {
        if(clazz == null) {
            throw new NullPointerException("clazz");
        } else {
            this.clazz = clazz;
        }
    }

    public T newChannel() {
        try {
            return (Channel)this.clazz.newInstance();
        } catch (Throwable var2) {
            throw new ChannelException("Unable to create Channel from class " + this.clazz,
var2);
        }
    }
}
```

Netty 中的设计模式应用非常多，这里不再一一列举，希望大家花时间去探索，这样才能够有更多的收获。本章旨在为大家打开深入探索 Netty 的大门，有不妥之处请纠正交流。

17

第 17 章 Netty 经典面试题集锦

17.1 基础知识部分

17.1.1 TCP 和 UDP 的根本区别

- TCP 面向连接，如打电话要先拨号建立连接；UDP 是无连接的，即发送数据之前不需要建立连接。
- TCP 提供可靠的服务。也就是说，通过 TCP 连接传送的数据，无差错，不丢失，不重复，且按序到达；UDP 尽最大努力交付，即不保证可靠交付。
- TCP 通过校验和、重传控制、序号标识、滑动窗口、确认应答实现可靠传输。如丢包时的重发控制，还可以对次序乱掉的分包进行顺序控制。
- UDP 具有较好的实时性，工作效率比 TCP 高，适用于对高速传输和实时性有较高要求的通信或广播通信。
- 每一条 TCP 连接只能是点到点的；UDP 支持一对一、一对多、多对一和多对多的交互通信。
- TCP 对系统资源要求较多；UDP 对系统资源要求较少。

17.1.2 TCP 如何保证可靠传输

TCP 对发送的每一个包都进行编号，接收方对数据包进行排序，把有序数据传给应用层。

- 校验和：TCP 将保持它首部和数据的校验和。这是一个端到端的校验和，目的是检测数据在传输过程中的任何变化。如果收到段的校验和有差错，TCP 将丢弃这个报文段和不确认收到此报文段。

TCP 的接收端会丢弃重复的数据。

- 流量控制：TCP 连接的每一方都有固定大小的缓冲空间，TCP 的接收端只允许发送端发送接收端缓冲区能接纳的数据。当接收方来不及处理发送方的数据，能提示发送方降低发送的速率，防止包丢失。TCP 使用的流量控制协议是可变大小的滑动窗口协议。（TCP 利用滑动窗口实现流量控制。）
- 拥塞控制：当网络拥塞时，减少数据的发送。
- 停止等待协议：也是为了实现可靠传输，它的基本原理就是每发完一个分组就停止发送，等待对方确认。收到确认后再发下一个分组。
- 超时重传：当 TCP 发出一个段后，它启动一个定时器，等待目的端确认收到这个报文段。如果不能及时收到一个确认，将重发这个报文段。

17.1.3 Netty 能解决什么问题

Netty 是一款基于 NIO 开发的网络通信框架，相比于 BIO，它的并发性能得到了很大提高。难能可贵的是，在保证快速和易用性的同时，并没有丧失可维护性和性能等优势。

17.1.4 选用 Netty 作为通信组件框架的举例

典型应用：阿里分布式服务框架 Dubbo，默认使用 Netty 作为基础通信组件；还有 RocketMQ，也是使用 Netty 作为通信的基础。

17.1.5 Netty 有哪些主要组件，它们之间有什么关联

- Channel：作为 Netty 中进行网络操作的抽象类，包括基本的 I/O 操作，如 bind()、connect()、read()、write()等。
- EventLoop：主要是配合 Channel 处理 I/O 操作，用来处理连接的生命周期中所发生的事情。
- ChannelFuture：Netty 框架中所有的 I/O 操作都是异步的，因此需要 ChannelFuture 的

addListener()注册一个 ChannelFutureListener 监听事件，当操作执行成功或者失败时，监听就会自动触发返回结果。

- ChannelHandler：充当所有处理入栈和出栈数据的逻辑容器。ChannelHandler 主要用来处理各种事件，这里的事件很广泛，比如连接、数据接收、异常、数据转换等。
- ChannelPipeline：为 ChannelHandler 链提供了容器，当 Channel 创建时，就会被自动分配到它专属的 ChannelPipeline，这个关联是永久性的。

17.2 高级特性部分

17.2.1 相较同类框架，Netty 有哪些优势

- 使用简单：封装了 NIO 的很多细节，使用更简单。
- 功能强大：预置了多种编解码功能，支持多种主流协议。
- 定制能力强：可以通过 ChannelHandler 对通信框架进行灵活的扩展。
- 性能高：通过与其他业界主流的 NIO 框架对比，Netty 的综合性能最优。
- 稳定：Netty 修复了已经发现的所有 NIO 的 Bug，让开发者可以专注于业务本身。
- 社区活跃：Netty 是活跃的开源项目，版本迭代周期短，Bug 修复速度快。

17.2.2 Netty 的高性能体现在哪些方面

- I/O 线程模型：同步非阻塞，用最少的资源做更多的事。
- 内存零拷贝：尽量减少不必要的内存拷贝，实现更高效率的传输。
- 内存池设计：申请的内存可以重用，主要指直接内存。内部实现用一棵二叉查找树管理内存分配情况。
- 串行化处理读写：避免使用锁带来的性能开销。
- 高性能序列化协议：支持 Protobuf 等高性能序列化协议。

17.2.3 默认情况下 Netty 起多少线程，何时启动

Netty 的线程数默认是 CPU 核数的两倍，Bind 完之后启动。

17.2.4 Netty 有几种发送消息的方式

Netty 有两种发送消息的方式。

- 直接写入 Channel，消息从 ChannelPipeline 中尾部开始传播。
- 写入与 ChannelHandler 绑定的 ChannelHandlerContext，消息从 ChannelPipeline 的下一个 ChannelHandler 中传播。

17.2.5　Netty 支持哪些心跳类型设置

- readerIdleTime：为读超时时间（即测试端一定时间内未接收到被测试端消息）。
- writerIdleTime：为写超时时间（即测试端一定时间内向被测试端发送消息）。
- allIdleTime：所有类型的超时时间。

17.2.6　Netty 和 Tomcat 的区别

- 作用不同：Tomcat 是 Servlet 容器，可以视为 Web 服务器，而 Netty 是异步事件驱动的网络应用程序框架和工具，用于简化网络编程，例如 TCP 和 UDP 套接字服务器。
- 协议不同：Tomcat 是基于 HTTP 的 Web 服务器，而 Netty 能通过编程自定义各种协议，因为 Netty 本身自己能编码/解码字节流，所以 Netty 可以实现 HTTP 服务器、FTP 服务器、UDP 服务器、RPC 服务器、WebSocket 服务器、Redis 的 Proxy 服务器、MySQL 的 Proxy 服务器等。

17.2.7　在实际应用中，如何确定要使用哪些编解码器

通过查看 Netty 的 API，不难发现 Netty 自带的编解码器可以解决 99%的业务需求，只有 1%的需求需要自己编解码。

反侵权盗版声明

举报电话：（010）88254396；（010）88258888
传　　真：（010）88254397
E-mail：　dbqq@phei.com.cn
通信地址：北京市万寿路 173 信箱
　　　　　电子工业出版社总编办公室
邮　　编：100036